Glycerine

A Key Cosmetic Ingredient

COSMETIC SCIENCE AND TECHNOLOGY SERIES

Series Editor
ERIC JUNGERMANN
Jungermann Associates, Inc.
Phoenix, Arizona

Other Volumes in Preparation

Glycerine

A Key Cosmetic Ingredient

edited by

Eric Jungermann
Jungermann Associates, Inc.
Phoenix, Arizona

Norman O. V. Sonntag
Consultant
Ovilla (Red Oak), Texas

CRC Press
Taylor & Francis Group
Boca Raton London New York

CRC Press is an imprint of the
Taylor & Francis Group, an **informa** business

First published in 1991 by Marcel Dekker

Published 2019 by CRC Press
Taylor & Francis Group
6000 Broken Sound Parkway NW, Suite 300
Boca Raton, FL 33487-2742

First issued in paperback 2019

No claim to original U.S. Government works

ISBN 13: 978-0-367-45056-4 (pbk)
ISBN 13: 978-0-8247-8465-2 (hbk)

Visit the Taylor & Francis Web site at
http://www.taylorandfrancis.com

and the CRC Press Web site at
http://www.crcpress.com

About the Series

The Cosmetic Science and Technology series was conceived to permit discussion of a broad spectrum of current knowledge and theories of cosmetic science and technology. The series is made up of books by a single author or edited volumes with a number of contributors. Authorities from industry, academia, and the government participated in writing these books.

The aim of this series is to cover the many facets of cosmetic science and technology. Topics are drawn from a wide range of disciplines ranging from chemical, physical, analytical, and consumer evaluations to safety, efficacy, toxicity, and regulatory questions. Organic, inorganic, physical, and polymer chemistry, emulsion technology, microbiology, dermatology, toxicology, and related fields all play a role in cosmetic science. There is little commonality in the scientific methods, processes, or formulations required for the wide variety of cosmetics and toiletries manufactured. Products range from hair care, oral care, and skin care preparations to lipsticks, nail polishes and extenders, deodorants, body powders, and aerosols to over-the-counter products, such as antiperspirants, dandruff and acne treatments, antimicrobial soaps, and suntan products.

Cosmetics and toiletries represent a highly diversified field with many subsections of science and "art." Indeed, even in these days of high technology, "art" and instinct continue to play an important part in the development of formulation, evaluation, and selection of raw materials, although there is a strong move toward the "scientific method," particularly in such areas as efficacy evaluation, claim substantiation, safety testing, and product evaluation and analysis.

Emphasis is placed on reporting the current status of cosmetic technology and science in addition to historical reviews. The series includes books on oral hygiene products, cosmetic product safety, efficacy testing, sunscreen technology, deodorants and antiperspirants, clinical testing procedures in the area of skin care and hair care, preservatives, and cosmetic raw materials. Contributions range from highly sophisticated and scientific treatises, to primers, practical applications, and pragmatic presentations. Authors are encouraged to present their own concepts as well as established theories. They have been asked not to shy away from fields that are still in a state of development or transition, nor to hesitate to present detailed discussions of their own work. Altogether, we intend to develop in this series a collection of critical surveys covering most phases of the cosmetic industry.

This eleventh book in the Cosmetic Science and Technology series, *Glycerine: A Key Cosmetic Ingredient*, comprises seventeen chapters. The first nine chapters cover information on the chemical and physical characteristics, general properties, manufacturing and handling procedures, and background information on glycerine itself. The second half of the book is concerned with the uses and functions of glycerine in cosmetic formulations, ranging from creams, lotions, cleansers, and soaps to hair and oral care products. The book concludes with a review of some alternatives to glycerine that have been used in cosmetic formulations and a discussion of some of the other industrial areas where glycerine and its derivatives play important roles.

I want to thank the contributors and my coeditor, Dr. Norman O. V. Sonntag, for participating in this book. Special recognition is due to the editorial staff at Marcel Dekker, Inc., Ted Brenner and Anita Kuemmel of The Soap and Detergent Association for encouraging this project, and, above all, my wife, Eva, without whose help and constant support I would never have undertaken this project.

ERIC JUNGERMANN, Ph.D.

Foreword

The Cosmetic Science and Technology series provides an important new source of current information for educators, chemists, consultants, and researchers on the art and science of cosmetic chemistry. *Glycerine: A Key Cosmetic Ingredient* is a significant addition to this collection and to the scientific and technical literature in the field.

The progress of the soap and detergent industry over the last 150 years has been integrally related to the development of glycerine as an important commercial chemical. In fact, The Soap and Detergent Association was founded in 1926 as The Association of American Soap and Glycerine Producers. In the early days of the association, glycerine was a surplus commodity resulting from the manufacture of soap. One of the first missions of the association was to develop new applications and markets for the material. Even today, with the cleaning product business dominated by petrochemically derived detergents, glycerine remains an integral part of the industry's structural matrix as well as a prominent industrial material in its own right. Glycerine's continuing relevance to cleaning products is underscored by the fact that one of the four key operating divisions of The Soap and Detergent Association is dedicated to glycerine and other oleochemicals.

Although the fundamentals of glycerine chemistry were described by Scheele, its discoverer, and Chevreul in the late eighteenth and early nineteenth centuries, the full economic and technical development of this material did not occur until after Alfred Nobel's invention of dynamite in 1866. As noted elsewhere in this volume, glycerine now has more than 1500 end uses in addition to its functions in nature where it is a component of all living cells. Major applications fall into some dozen categories, an important one of which is cosmetics. Cosmetic applications represent an estimated 5 to 10 percent of annual glycerine consumption in the United States.

In recent years, U.S. glycerine production has fluctuated around 300 million pounds. Worldwide consumption is in the neighborhood of 1.2 billion pounds per year and growing modestly, particularly in underdeveloped nations where soap production is on the rise. While these figures may not seem impressive next to the other workhorse chemicals of industry, it should be remembered that glycerine is a fine or pharmacopeial material produced in accordance with high standards of purity. On that basis, its volume of consumption is very large indeed, probably the largest of that class.

A unique combination of physical and chemical characteristics has made glycerine a building block for many products over the years. In addition, glycerine is virtually nontoxic to human health and the environment and is free from unpleasant odors and tastes, qualities that are of paramount importance in cosmetic and toiletry applications. In such uses, glycerine may be the material of choice for its humectancy, low vapor pressure, and viscosity. But equally important is glycerine's compatibility with many other substances. This characteristic gives the cosmetic chemist greater freedom in selecting the other ingredients in a formulation and the assurance that glycerine will not clash or interfere with the other product components.

The scholarly efforts of the authors and editors of this work have resulted in a notable contribution to the scientific literature by bringing together in one convenient volume virtually all that is known about this ubiquitous material. They are to be commended for preparing this important new compendium for use by a broad technical readership for many years to come.

THEODORE E. BRENNER
President
The Soap and Detergent Association
New York, New York

Preface

Glycerine: A Key Cosmetic Ingredient is a comprehensive book covering the chemical and physical properties of glycerine and the use of this material in cosmetic and personal care products. Glycerine is like old wine in a new bottle: ever present, ever getting better, and more varied in its applications. Chemists studying the compositions of a wide range of cosmetics and toiletries ranging from creams and lotions to toothpastes, cleansers, haircare products, and many others too numerous to name, will become aware of the ubiquitous nature of glycerine in these product categories.

Glycerine's role in cosmetics is diverse, ranging from moisturizer, emollient, viscosity modifier, solubilizer, humectant, foam booster, and occasionally just as a "magic ingredient." The economics of glycerine tend to fluctuate, depending on the vagaries of market shifts of natural and petrochemical raw materials.

This book is an authoritative resource for the many scientists working in the field of cosmetic and personal care formulations, dermatologists, testing laboratories and manufacturers and marketers of glycerine. In one volume, the book brings together all aspects of this versatile substance. It is of use to both expert and novice. Illustrated with drawings, tables and charts, and an up-to-date bibliography, it provides the reader with immediate insight.

Writing, editing, and producing this text is a labor of love. We would like to thank the contributors who willingly gave many hours of their time: we would like to express our appreciation to Marcel Dekker, Inc., and in particular to Sandra Beberman, for their help and support. Last, but not least, we would like to thank our respective wives, Eva S. Jungermann and Margaret Sonntag, for their encouragement, suggestions, and patience, which contributed heavily to getting this project finished.

ERIC JUNGERMANN, Ph.D.

NORMAN O. V. SONNTAG, Ph.D.

Contents

Contributors

David J. Anneken   Henkel Corporation, Emery Group, Cincinnati, Ohio

Yohini Appapillai   Neutrogena Corporation, Los Angeles, California

S. R. Gregory   Humko Chemical Division, Witco Corporation, Memphis, Tennessee

Dale H. Johnson   Helene Curtis, Inc., Chicago, Illinois

Eric Jungermann   Jungermann Associates, Inc., Phoenix, Arizona

Jon J. Kabara   Lauricidin, Inc., Galena, Illinois

Beth Lynch   Neutrogena Corporation, Los Angeles, California

Rolf Mast   Neutrogena Corporation, Los Angeles, California

Walter H. C. Neumann   Henkel Corporation, Gulph Mills, Pennsylvania

Morton Pader   Consumer Products Development Resources, Inc., Teaneck, New Jersey

Richard A. Reck   Consultant, Hinsdale, Illinois

Norman O. V. Sonntag   Consultant, Ovilla (Red Oak), Texas

David C. Underwood   Procter & Gamble, Cincinnati, Ohio

Glycerine

A Key Cosmetic Ingredient

1

Introduction

Eric Jungermann

Jungermann Associates, Inc., Phoenix, Arizona

Glycerine was discovered more than two centuries ago by the Swedish chemist Scheele (1742–1786) when he heated a mixture of litharge (lead oxide) and olive oil. He extracted and isolated a sweet tasting liquid which he named "sweet oil" (Oelsuess). The French chemist Michel Eugene Chevreul (1786–1889), the father of the chemistry of fats and oil, established the structure of fats as triesters made up of three moles of mixed fatty acids and one mole of "sweet oil," which he renamed "glycerine" after the Greek word for "sweet."

Glycerine (or glycerin) is also referred to in many texts as "glycerol." Chemically it is a tribasic alcohol and more correctly named 1,2,3-propanetriol. Its chemical structure is shown below:

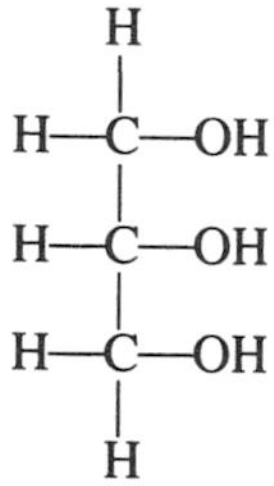

$$
\begin{array}{c}
\text{H} \\
| \\
\text{H--C--OH} \\
| \\
\text{H--C--OH} \\
| \\
\text{H--C--OH} \\
| \\
\text{H}
\end{array}
$$

A ubiquitous substance, glycerine is found widely in nature as a component of thousands of natural substances, which have a broad range of uses in their own

Table 1.1 Sources and Uses of Glycerine and Glycerine Derivatives

Sources	Glycerine	Derivatives
I	II	III
Plant fats and oils	H₂C—OH	Fatty monoglycerides
Animal fats and oils	H—C—OH	Fatty diglycerides
Phospholipids	H₂C—OH	Nitroglycerine Polygylcerols Alkyd resins

Uses and Fields of Application

I	II	III
Food products	Cosmetics	Emulsifiers
Drying oils	Toiletries	Printing inks
Soap manufacture	Tobacco	Explosives
Cosmetics	Tooth paste	Drugs
Plasticizers	Medicinals	Paints
Animal feed	Metal finishing	Cosmetics
Margarine	Food products	Food products
Bakery products	Plasticizer	Detergents
	Ceramics	

right; when isolated in its free form, glycerine itself has found use in a myriad of applications. As a chemical building block, it has been converted into many derivatives with additional applications. Table 1.1 describes sources and uses of glycerine and its derivatives. Table 1 also shows some of the uses and fields of application of three categories of glycerine. The listings are not intended to be all inclusive; the table is primarily designed to illustrate the broad spectrum of products and industries that use glycerine or glycerine-related materials.

Glycerine is a versatile chemical. It is found in baby care products and in embalming fluids used by morticians, in glues that hold things together and in explosives to blow them apart; in throat lozenges and in suppositories. A colorless, viscous liquid, and stable under most conditions, glycerine is nontoxic,

Table 1.2 U.S. Production of Glycerine

Year	lbs. (MM)
1987	307.2
1988	296.8
1989	293.7

Source: From Ref. 1.

easily digested, and is environmentally safe. It has a pleasant taste and odor, which makes it an ideal ingredient in food and cosmetic applications.

Production of glycerine has been relatively stable in recent years, as shown in Table 1.2.

The largest quantity of naturally occurring glycerine is obtained from (a) the manufacture of soap and (b) the production of fatty acids by fat splitting (hydrolysis).

Soap is manufactured by reacting fats and oils with caustic soda [2]

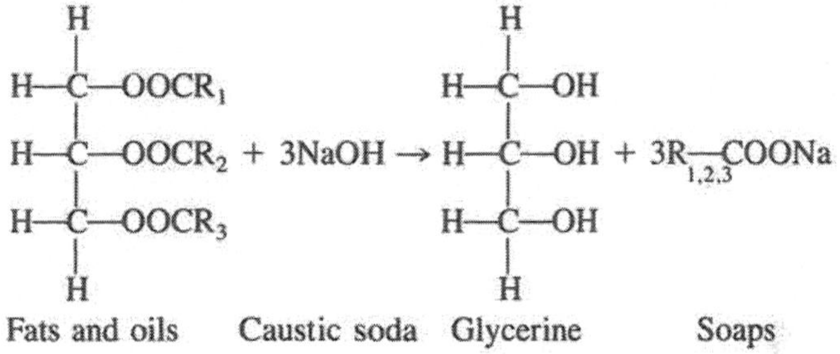

$$
\begin{array}{c}
H \\
| \\
H-C-OOCR_1 \\
| \\
H-C-OOCR_2 \\
| \\
H-C-OOCR_3 \\
| \\
H
\end{array}
+ 3NaOH \rightarrow
\begin{array}{c}
H \\
| \\
H-C-OH \\
| \\
H-C-OH \\
| \\
H-C-OH \\
| \\
H
\end{array}
+ 3R_{1,2,3}COONa
$$

Fats and oils Caustic soda Glycerine   Soaps

Production of fatty acids by fat splitting (hydrolysis) is shown below:

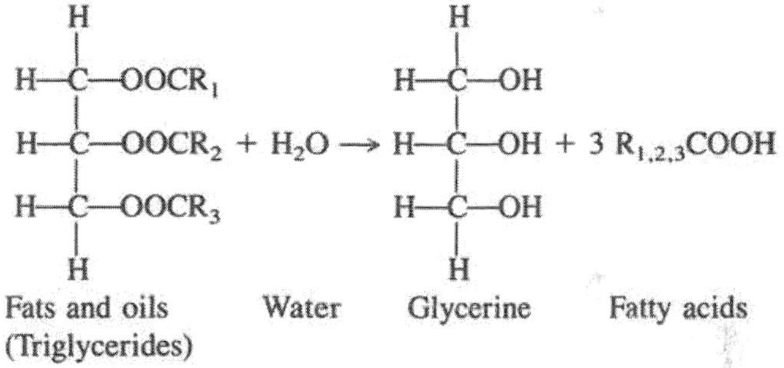

$$
\begin{array}{c}
H \\
| \\
H-C-OOCR_1 \\
| \\
H-C-OOCR_2 \\
| \\
H-C-OOCR_3 \\
| \\
H
\end{array}
+ H_2O \rightarrow
\begin{array}{c}
H \\
| \\
H-C-OH \\
| \\
H-C-OH \\
| \\
H-C-OH \\
| \\
H
\end{array}
+ 3 R_{1,2,3}COOH
$$

Fats and oils   Water   Glycerine   Fatty acids
(Triglycerides)

Some fatty acids obtained by fat splitting are neutralized to make soaps; they also have many applications per se, or are used in the manufacture of derivatives. Another process that results in the formation of glycerine from fats and oils occurs in the manufacture of methyl esters of fatty acids. In this process fats are interesterified with methyl alcohol:

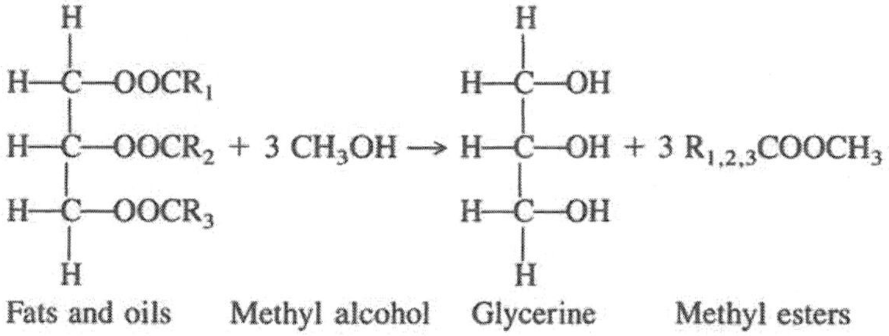

$$
\begin{array}{c}
H \\
| \\
H-C-OOCR_1 \\
| \\
H-C-OOCR_2 \\
| \\
H-C-OOCR_3 \\
| \\
H
\end{array}
+ 3 CH_3OH \rightarrow
\begin{array}{c}
H \\
| \\
H-C-OH \\
| \\
H-C-OH \\
| \\
H-C-OH \\
| \\
H
\end{array}
+ 3 R_{1,2,3}COOCH_3
$$

Fats and oils   Methyl alcohol   Glycerine   Methyl esters

Methyl esters are useful in the manufacture of many fatty derivatives, such as fatty alcohols and amides.

The first part of this volume provides a broad background on the properties of glycerine, methods of manufacture, and some of the historical and economic aspects of this important chemical. Chapter 3 discusses the various processes for producing glycerine from natural fats and oils. Methods of separation from the resultant fatty residues and subsequent purification techniques are reviewed. Chapter 4 is concerned with a review of various industrial methods for manufacturing glycerine from nonfat-based starting materials. Processes utilizing petrochemicals constitute the most important examples, although the chapter will also cover natural carbohydrates as raw materials. The production of synthetic glycerine from petrochemicals had an important impact on that industry in the preoil crisis years of the 1960s. Dow, FMC, Olin, and Shell operated plants in the United States engaged in producing synthetic glycerine. Shifting economic climates have shut down most of these plants. In 1990, only Dow still produces synthetic glycerine in the United States.

Chapters 5, 6, and 7 deal with the chemical and physical properties of glycerine. The chemical reactions of glycerine as used in the production of industrially important derivatives are relatively simple and are discussed in some detail in Chapter 5, as well as in Chapter 12, which is specifically concerned with monoglycerides. Many reactions of historical and scientific interest are also reviewed in Chapter 5. Chapter 6 discusses the physical properties of glycerine. The multiplicity of uses of glycerine are not due to a single property, but rather to a unique combination of properties, such as broad compatibility, blandness, nontoxicity, stability, and hydroscopicity. These are the key physical properties that make glycerine an almost essential ingredient in cosmetic formulations.

The various methods for analyzing glycerine are covered in Chapter 7. These include the traditional "wet" methods, as well as modern instrumental techniques. The analyses used for the detection of various impurities that can be found in glycerine are also discussed. A review on handling, safety, and environmental factors of glycerine are described in Chapter 8. A chapter on economic implications concludes the first part of the volume.

The second half of the text deals with the broad role and the many applications of glycerine in the cosmetics and toiletries industry. Chapters 10 and 11 discuss the functions glycerine fulfills in various formulations, and evaluation techniques used to demonstrate efficacy. Chapters 13, 14, and 15 cover specific product areas, such as skin, hair, and oral care products, and soaps. The role of glycerine in the formulations of these products is discussed in detail. Chapter 16 considers some alternatives to glycerine use in cosmetic formulations. The volume concludes with a broad-based review of some of the other fields and areas where glycerine is widely used.

REFERENCES

1. Bureau of the Census, U.S. Department of Commerce, Current Industrial Report Series M20K, Fats and Oils.
2. Jungermann, E., Soap, in Bailey's Industrial Oil and Fat Products, 4th ed. (D. Swern, ed.), John Wiley & Sons, New York, 1979, Chap. 8.

2

Glycerin(e) and Its History

Walter H. C. Neumann

Henkel Corporation, Gulph Mills, Pennsylvania

As early as 600 A.D., the Phoenicians were reported to have an "alchemist knowledge" of soap making, which in later centuries found its way north via Marsilia into Gallic and Germanic customs [1]. No references are known, however, to a "by-product of soap making" named "glycerine or glycerol," whose existence twentieth-century man has taken for granted. From pre-Renaissance Italy, references to soap making are found in Central European and German regions in later centuries. Under Charles I (14th century), the English crown monopolized the commercial trade with soap and even extracted an excise tax from people with a preference for cleanliness.

Inquisitive minds on the Continent and in England paved the way from alchemy to chemistry. A historical example is C. G. Geoffrey [2] (1741), whose intensive studies of the nature of fatty substances unlatched the door for the discovery of glycerine.

Less than 40 years later, Karl Wilhelm Scheele (1742–1786), a scientifically curious young pharmacist in Sweden, achieved the historic distinction of discovering that natural oils and fats contain what we refer to as "glycerine," a slightly sweet substance which he called "the sweet principle of fats."

In 1783, Scheele published a very short description of one of his many scientific endeavors under the title "Experiment on a special sugar material in pressed oils and fats." He wrote: "Presumably, it is not yet known that all solid and pressed oils have a natural sweetness which differs in its behavior and properties from the sugar-type substances normally found in materials of vegetable origin. The sweetness arises when such oils are boiled with lead calx and water." While Scheele was pursuing a soap plaster, at the time, he did not stop with the extraction of his "sweet principle" from olive oil [3]. Scheele also used such other substances as almond oil, lard, and butter, to confirm its existence in vegetable oils *and* animal fats.

He concluded from his experimental work that [4]:

All fatty oils contain a sweetness which differs from that of sugar and honey in—
1. that it cannot be made to crystallize,
2. that this sweetness withstands more heat and passes into the receiver in unchanged form,
3. that it cannot be fermented, and,
4. that it mixes with alcoholic alkali.

In a sense, Scheele's [3] chemical pursuit duplicated a process of nature, namely the breakdown of oils and fats—known to modern man to exist of triglycerides—into glycerine and its various fatty radicals.

Much later, on May 20, 1933, the *Journal of the American Medical Association* makes reference to that process, pointing out that glycerine is a "normal digestion product," that is a byproduct of natural decomposition (fermentation) in the human body.

Glycerine, or glycerol, as it is known in its pure chemical form, occurs in combined form in all animal and vegetable fats and oils. It is rarely found in the free state in these fats, but is usually present as a triester combined with such fatty acids as stearic, oleic, palmitic, and lauric acids. Generally, these are mixtures or combinations of glycerol esters of several fatty acids. Such oils as coconut, palm kernel, cottonseed, soybean, and olive oil yield larger amounts of glycerol than do such animal fats as lard and tallow. Glycerol also occurs naturally as triglycerides in all animal and vegetable cells in the form of lipoids, such as lecithin and cephalins. These complex fats differ from simple fats in that they invariably contain a phosphoric acid residue in place of one fatty acid residue [5].

One assumes that glycerol and/or triglycerides are formed as part of the photosynthesis in plants with carbon dioxide and water as building blocks.

Glycerol plays a considerable physiological role in animals. For instance, acting as a converter of toxic substances, generated metabolically, in bears during hibernation [6]. That other animals can survive arctic cold is related to the presence of glycerol in their bloodstreams [7] (e.g. "Kaefer").

We owe it to the great French scientist, Michel Eugene Chevreul (1786–1889) that, in 1811, the colloquial term of "Scheele's sweet" was substituted by the name, derived from the Greek language (glycos = sweet, i.e., "glycerine''). Chevreul, in 1823, obtained the first patent dealing with the manufacture of glycerine [8] and went on to become one of the founders of scientific research on fats and soaps.

In 1836, the French chemist Pelouze established the empirical formula of glycerol, or propane -1, 2, 3-triol [I]; and, in 1883, Berthelot and Luce published its structural [II] formula [8].

$$
\begin{array}{ccc}
\text{CH}_2\text{OH} & & \overset{\displaystyle \text{H}}{\underset{\displaystyle |}{\text{H---C---OH}}} \\[2mm]
| & & | \\
\text{CH·OH} & & \text{H---C---OH} \\[2mm]
| & & | \\
\text{CH}_2\text{OH} & & \text{H---C---OH} \\[2mm]
& & | \\
& & \text{H} \\[4mm]
[\text{I}] & & [\text{II}]
\end{array}
$$

Louis Pasteur, the great pioneer in microbiology, found in 1858 [3] that sugar, when present in wine and grape juice, through "fermentation" yields small amounts (1/30) of glycerine.

However, it was not until the end of the nineteenth century that glycerine entered the Technical Age!

It was another Swedish chemist who made history (1867) by successfully transforming A. Sobrero's (Italian chemist, who in 1846 discovered what is known as "nitroglycerine") unstable trinitroglycerol into the safely transportable explosive, called "dynamite." His name was Alfred Nobel [8]. The impact of this development went beyond military interests.

The importance of Nobel's discovery can hardly be exaggerated. Without Nobel, the Industrial Revolution could not have been accomplished. From here on, glycerine found its way into many, vital industrial applications.

Apart from explosives, one of the earlier large single uses was in "alkyd" resins, a market which, however, has shown a declining trend since. Although a modern type "alkyd" resin was produced as far back as in 1901, the full commercial potential was not realized until 1927, when Kienle of General Electric Company registered his first patent covering modifications of the basic resin [8]. Today, the pharmaceutical and personal care industry dominate in the end use of pharmacopoeia-grade glycerine. One of the largest end uses being the manufacture of toothpaste.

USP-grade glycerol, for example, is approved by the FDA for use in many topical and internal preparations. Since 1959, glycerine is generally recognized as safe (GRAS) as a miscellaneous or general food additive under CFR (18), and also is permitted to be used in certain food packaging materials [5].

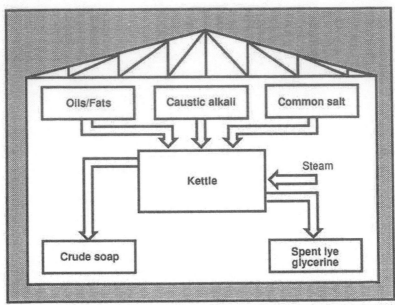

Figure 2.1 Production diagram showing the formation of spent lye glycerine in the manufacture of soap. (From Ref. 4.)

Another major use is in tobacco processing either as such or as triacetin. Many other derivatives of glycerine, especially esters and polyesters, have found a wide ranging field of applications. Back in 1872, the world consumption of glycerine had risen to approximately 10,000 tons per year. Today, it is estimated to have surpassed the 500,000 tons per year mark.

In the early 1870s the first U.S. patent for "recovery of glycerine from soap lyes by distillation" was registered. Around the turn of the century, in Europe, as well as in North America, commercial soap making lead to large-scale glycerine production. Some well-known names in today's soap and detergent industry go back to those times: Procter & Gamble, Unilever, Colgate, and Henkel.

As the soap industry developed, two separate sources of glycerine and its derivatives became industry standards.

One method was to isolate glycerine, common salt, and other impurities from "spent lyes," obtained in the manufacture of soap. Depending on the type of fat used, 5-8% glycerine was contained in the soap lyes [4] (Fig. 2.1).

This process was perfected in England in 1883, by Runcorn, who achieved the final goal after earlier experiments had failed [4].

However, at an earlier date, crude glycerine had been made from so-called "sweet waters," arising from the manufacture of stearine, used in the manufacture of candles. Procter & Gamble entered the history of glycerine via this route.

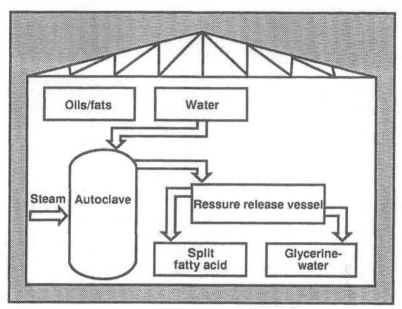

Figure 2.2 Production diagram of autoclave splitting. (From Ref. 4.)

In the search for better production processes, a breakthrough occurred in 1900. The fat-splitting process by Twitchell, Emery's first chemist, was introduced. Although commercially utilized for many decades, this process, using dilute sulfuric acid and naphthalene sulfonic acid, was not optimal as to the desirable, light color in the generated fatty acids, and subsequent color of the finished soap. However, the glycerine yielded was of fairly good quality and usable for dynamite, as well as for some pharmaceutical preparations. However, improvements were sought, in the years before the World War I, leading to the "batch autoclave process." Stearine manufacturers (candles) had for some time utilized the relatively more expensive autoclave route. Technically speaking, "autoclave splitting" demonstrated great superiority over the Twitchell method [4] (Fig. 2.2).

Finally, in the 1930s, with the development of stainless steel and its use as a corrosion-resistant material for autoclaves, it became possible to progress to high-pressure splitting with water, without the addition of chemical accelerators as in the Twitchell process. A new era began [4]!

By using distilled water, a "low ash" glycerine resulted. Waste water problems were minimized! Fatty acid quality improved, also. Heat generation inherent in the splitting process led to the evaporation of the glycerine water. By combining splitting and fatty acid distillation with repeated cycles, the quality of glycerine was steadily improved, meeting the Pharmacopoeia requirements in Europe and North America.

However, in spite of the progress in process technology, the industry had to revert to older, less efficient processes during times of war and economic hardships, when fats and oils became rare commodities. So, during both world wars, some European manufacturers resorted to processing of sugars via the fermentation process, the so-called "Protol process" (see L. Pasteur) [4], to produce glycerine.

As alternate routes, creative minds developed or selected, depending on requirements, various kinds of substitutes for *natural* glycerine. The literature shows ample examples between 1910 and 1939 varying from akin molecules, such as glycols, sorbitols, manitols, and erythritols, to mere, none affiliated replacements of organic or inorganic matter. The especially high demand for glycerine during war times, stimulated creative chemical minds to develop *synthetic* routes for the manufacture of glycerine. A great number of British, German, and U.S. patents were recorded between 1937 and 1945 and extended into the 1960s.

As early as during the first world war, BASF, a part of I. G. Farben, had developed a synthetic glycerine process based on the use of "allyl alcohol" [8]. Similar efforts were repeated during the World War II, documented by a variety of patents granted.

Another route recommended the use of "anthrachinone" in lieu of hydrogen peroxide [8]. The ample and low cost availability of "propylene," a byproduct of petroleum refining or a coproduct of ethylene production, opened the door for the manufacture of synthetic glycerine! In the late 1930s, Shell Development Company developed an "allylchloride–epichlorohydrin process" leading to the first full commercial synthetic glycerin plant in the United States (1948). In the 1960s, Shell commercialized the "acrolein-allyl alcohol process." None of these processes are practiced any longer. Shell exited the U.S. market with both operations in mid-1980.

Dow followed with an "epichlorohydrin process," which is still used in the United States and Germany. Shell (Holland) and Solvay (France) pursued the same route in Europe, but only the latter company is still in operation. Shell's operation in England, utilizing the "allyl chloride" process, closed down in the early 1980s.

In the United States, FMC started another process with "allylalcohol and peracetic acid," but exited the market in mid-1982. Olin, formerly Olin-Mathieson Chemical Corporation, developed a process involving the "isomerization of propylene" in the United States in the 1960s. But this technology was practiced only for a very short time [5].

Kashima became the first Japanese manufacturer of synthetic glycerine. Daicel, also in Japan, was reported to have used a similar process like FMC but, today, practices Dow technology [9].

While today, Dow is the only U.S. producer of "synthetic glycerine," the number of natural crude glycerine producers amounted to 16 as of 1986, nine of these produce refined glycerine up to a purity of 99.8%. Expressed in production volume, the ratio of natural to synthetic glycerine is approximately 2:1, with natural output growing at a faster rate than synthetic.

In general, the economic trend favors the production of natural glycerine, the more so as many countries rich in natural (especially vegetable oil) resources have started to expand their economies into the oleochemical market. It is estimated that by the early 1990s, the existing and planned natural glycerine capacities could fully satisfy the world demand.

When "synthetic glycerin capacities" were created, the traditional manufacturers of *natural* glycerine thought of ways to improve their process technologies, to improve the quality of glycerine to the highest standards, and to optimize the manufacturing costs, at the same time. Glycerine, although a byproduct of soap manufacture, or fat splitting, became a vital part of oleochemistry.

Continuous fat-splitting processes were introduced, so were transesterification processes, the latter using the same fats and oils as raw materials to form methylesters, an intermediate step in the manufacture of fatty alcohols. These processes constitute today's main sources for natural glycerine, whereas the soap production process plays only a minor role.

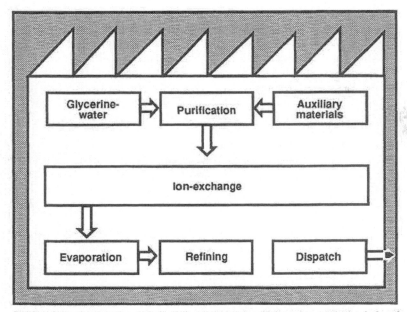

Figure 2.3 Production chart of the processing of glycerine with the help of ion-exchangers. (From Ref. 4.)

Further progress was industrially utilized in the mid-1960s by introduction of ion exchange technology [4], which dominates modern refining of *naturally* produced glycerine (Fig. 2.3).

The ion-exchange process made bleaching and distillation widely superfluous. Its principle had been used for some time in water softening. Separating inorganic salts, fatty acid, soap residues, and other contaminants out, yields an extremely clean and colorless glycerine.

Strangely, in the 1980s the "low-energy" Twitchell process regained in status in view of worldwide concern about energy consumption.

Like all major raw materials, cost is a most important factor. Glycerine pricing is influenced significantly by market forces, yet each manufacturer of natural or synthetic glycerine needs a satisfactory return on his investment on plant equipment and need cover his raw material and other operating costs.

Natural glycerine, as a byproduct of soap, fatty acids or fatty alcohol manufacture, can be valued in many different ways by its manufacturer, whereas synthetic glycerine is more dependent on propylene and equipment costs.

With the rising costs for petrochemical raw materials and their diminishing availability over the next century, glycerine derived from abundantly available natural raw materials seems to be the choice of the future.

REFERENCES

1. Thanner, E., Kleine Kulturgeschichte der Seife, Kunst und Aufbau, 3: 16–17, 1946.
2. Riggs, A. S., Soap, to you . . . , *Catholic World* 149: 184–1885, 1939.
3. Leffingwell, G., and Lesser, M., Glycerin, Its Industrial and Commercial Applications, Chemical Publishing Company, Brooklyn, NY, 1945, Chap. 1.
4. Bohmert, F., and Steinberner, U., *75 Years of Henkel Glycerine*, Company Archives of Henkel KGaA, Duesseldorf, West Germany, 1986.
5. Kern, J. C., Glycerine Producers Association, *Glycerol*, Volume 11, p. 921–932, Encyclopedia of Chemical Technology.
6. *Compar. Physiol.*, 155: 175.
7. Miller, L. K., Umschau—Heft 3, 1971.
8. Newman, A. A., *Glycerol, The History of* . . . , CRC Press, Cleveland, 1968, Chap. 1.
9. Dosch, F. O., Marketing View on the International Glycerine Situation, Bombay (India), Henkel KGaA, Duesseldorf, West Germany, 1986.
10. Heinz, H. J., and Henkel, KGaA, 75 Years of Henkel Glycerin, Anniversary address, 1986.

3

Manufacture of Glycerine from Natural Fats and Oils

David J. Anneken

Quantum Chemical, Emery Division,* Cincinnati, Ohio

I. INTRODUCTION

The raw materials for production of ''natural'' glycerol are the wide variety of animal fats and vegetable oils harvested throughout the world (Table 3.1). Depending upon the animal or vegetable feedstock, the ''fatty'' moiety of natural fats and oils varies widely from 6 to 22 carbon atom compounds [1], but their alcohol portion† is essentially pure 1,2,3-propanetriol, which simplifies the glycerol refining process from a chemical perspective. Properly recovered and refined, glycerine is one of the highest purity chemicals obtained from the processing of fats and oils. Modern glycerine refineries are capable of producing 99.7% glycerol purity by specification. As with most chemicals manufactured in large volume, commercial glycerol products are available in a wide range of grades and purities. Glycerol is and has been produced from other natural [2] and synthetic raw materials, using several different manufacturing processes; however, for this chapter, only the manufacture of natural glycerol from fats and oils is discussed.

Any discussion of the manufacturing processes of glycerol requires definition of many terms that are commonly encountered. The common name

---

*In March 1989, the Emery Division of Quantum was purchased by Henkel Corporation.
†Notable exceptions include lanolin, jojoba, sperm whale and some other marine oils.

Table 3.1 World Production of Selected
Fats and Oils, 1985/1986 (10^9 lbs.)

Soybean	30.0
Palm	18.2
Tallow/grease	14.1
Sunflower	14.0
Rapeseed	13.8
Lard	12.1
Butter fat	11.9
Cottonseed	7.6
Coconut	7.3
Groundnut	6.9
Olive	3.3
Fish	2.9
Palm kernel	2.5
Corn	2.4
Linseed	1.5
Totals	148.4

Source: J. Am. Oil Chem. Soc., 64: 706
1987.

"glycerol" refers to the pure chemical compound 1,2,3-propanetriol. When Scheele, a Swedish chemist, discovered glycerol in 1779 [3], he identified it as "the sweet principle of fat." Later, in 1811, it became known as glycerol from the Greek word "glukeros" which means "sweet tasting." In the trade, glycerol is commonly referred to as "glycerine" or "glycerin." Depending on the method for manufacturing glycerol, any number of intermediates are formed. These are discussed in Section II and include such names as: sweetwater, crude, semi- or half-crude, soap-lye crude, saponification crude, and alcoholysis crude. Ion exchange and distillation are the two final refining processes used by the industry. The ion-exchange process removes contaminants by passing the glycerine through a series of resin beds; in the distillation process the impurities are removed by vaporizing and condensing the glycerine.

To meet the requirements for the various uses of glycerol, a number of grades are commercially available which differ from one another in purity, odor, color, and taste. These properties are directly related to the glycerol source and the method of manufacture. Table 3.2 summarizes some terms commonly encountered among product grades in the industry.

II. CRUDE PRETREATMENT

Glycerol produced or "recovered" from the processing of natural fats and oils usually is first obtained as an aqueous solution containing both organic and

Table 3.2 Common Terms of the Glycerol Industry

Source	Intermediates	Refining	Grades
Natural:	Sweetwater Semicrude Crude —Soap lye —Alcoholysis —Saponification	Ion-exchange Distillation	Chemically pure U.S. Pharmacopoeia British Pharmacopoeia ACS reagent High gravity
Synthetic:	Allyl chloride Epichlorohydrin Allyl alcohol Acetone	Distillation	(same)

inorganic contaminants. Purification techniques employed depend on three very important factors: (1) the process used to separate or "split" the fat into glycerol and oleo products, (2) the type of fat or oil processed, and (3) the type of final refining operation to which the glycerol will be subjected.

There are three primary fats and oils processes from which glycerol is recovered: (1) hydrolysis, to produce fatty acids; (2) saponification, to produce soap; and (3) alcoholysis, to produce fatty esters. The relative importance of each of these processes to the glycerol industry is shown in Table 3.3. Prior to the 1950s, the hydrolysis reaction was carried out by boiling fat with water in an open kettle. The process employed a sulfonate catalyst and sulfuric acid [3], and the glycerol/water seat recovered contained varying amounts of these catalysts in addition to emulsified fat and fatty acids. By the 1950s, both batch and continuous high-pressure hydrolysis techniques had been developed for production of fatty acids [4,5]. These have increasingly displaced the older

Table 3.3 Glycerine Manufacture, Capacity, and Production 1985 (10^6 lbs.)

Type	Capacity		Production	
	U.S.	Other[a]	U.S.	Other[a]
Hydrolysis	105	—	—	—
Soap lye	100	—	—	—
Alcoholysis	45	—	—	—
Total, natural	250	523	203	392
Total, synthetic	110	224	108	149

[a] Western Europe, Japan, and Canada.

Source: Estimated from production figures in various government reports.

Table 3.4 Typical Characteristics of Recovered Sweetwaters and Finished Crude
Glycerol from Fats/Oils Processing

Process	Glycerol (%)	pH	Salts (%)	Organics (%)	Ionic load (mEq/L)
Sweetwaters					
Hydrolysis					
Twitchell	8–12	<2	0–0.2	0–2	50–200
High pressure	10–20	4–5	0–0.2	0–2	20–100
Saponification	4–20	>12	7–14	0.5–2	500–2000
Alcoholysis	40–60	a	1–2	2–10	100–500
Crude					
Hydrolysis	85–95	—	0.2–1	0.5–1	30–100
Sap/Alcoholysis	80–85	—	4–10	0.5–2	500–1500

a Catalyst-dependent.

open-kettle or "Twitchell" processes. The glycerol solutions obtained from high-pressure processes usually contain a higher concentration of glycerol than the older hydrolysis products, and thus are more economical to process. If the high-pressure process employs a catalyst, varying amounts of catalyst residue, in addition to emulsified fat and fatty acid, will be found in the recovered glycerol solution.

Direct alkaline saponification of fats and oils, generally achieved by boiling in kettles with caustic soda, is used to produce fatty acid soaps and glycerol. Glycerol is recovered from this saponification process by a variety of extraction and precipitation techniques [3,6,7], and contains varying amounts of both soap and free alkali or other salts.

Transesterification of fats or oils to produce fatty esters also produces a glycerine "seat." The glycerol recovered from transesterification processes usually contains catalyst residues and varying levels of the alcohol used, plus fat and ester.

Typical properties of the glycerol recovered from the various fats and oils processing techniques are summarized in Table 3.4.

Since the glycerol recovered from any fat or oil process always contains some entrained or emulsified fatty material, the types of contaminants to be removed in the refining process depend, to some extent, on the fat or oil feedstock. Both the chemical composition of the fatty contaminants and the nature of any other organic matter are feed dependent. Processing of refined, food-grade oils usually introduces very little organic contamination into the glycerol products other than the oil itself or normal fatty acids. Processing of crude, inedible, soapstock, or rendered fats and oils introduces a variety of contaminants which require special processing techniques for purification of the glycerol products. Some of the contaminants which are introduced from inferior grades of fats and oils include phosphatides, sulfur compounds, proteins and other nitrogenous compounds, aldehydes and ketones, oxidized fatty matter, and fermentation or bacterial products such as trimethylene glycol and formic acid. Extraneous matter such as dirt, minerals, bone, or fibers can also cause special handling problems.

A number of conventional unit operations, both chemical and physical, are employed in the treatment of crude glycerine prior to the refining operation. Chemical operations include coagulation or precipitation reactions, bleaching, and saponification/acidulation. Physical operations include decantation, filtration, centrifugation, and evaporation/distillation. Relatively newer unit operations introduced in recent years include ultrafiltration and reverse osmosis [8]. Before the individual operations are discussed, it is essential to note that the type and sequence of treatments used depend on the type of final refining process employed. Ion-exchange refining methods, for example, treat glycerol/ water solutions and, thus, an evaporation step is not required in the crude process except for very dilute solutions. For distillation refining methods the water must be removed so as not to overload the vacuum systems: fat or fatty acids which might codistill must also be removed by pretreatment. Crude requirements for the different refining methods are discussed in later sections, but the objectives of any crude glycerol pretreatment operation are always to maximize yield and quality and to minimize processing costs for the final refining process employed.

A. Chemical Treatments

Chemical treatment techniques for crude glycerols or sweetwaters were primarily developed for removal of the salts and catalysts employed in the fat-processing step. Classical inorganic salt reactions to form insoluble precipitates, or flocculation techniques to remove organics are the most common treatments. Mineral acidity in glycerine products must be neutralized prior to subsequent processing since acid catalyzes the dehydration of glycerol to acrolein. Twitchell sweetwaters, which contain free sulfuric acid, are treated with lime to precipitate calcium sulfate: the treatment usually includes some alum or ferric chloride to precipitate fatty acids prior to a filtration step. The Twitchell treatment has, to some extent, continued to be used on high-pressure hydrolysis sweetwaters. Since calcium sulfate is soluble to an appreciable extent in sweetwater solutions, lime has no advantage over neutralizing with sodium or potassium hydroxide at low levels of acidity. Table 3.5 shows solubility data for calcium sulfate and sodium sulfate in glycerol solutions. These data show that at 60°C, for example, there is no advantage for lime if free acidity of the glycerine is below 17.5 mEq/L (850 ppm as H_2SO_4) since all the salt formed is

Table 3.5 Solubility of Sulfates in Aqueous Solutions of 40% Glycerol

	—CaSO$_4$—		—Na$_2$SO$_4$—	
	(wt %)	mEq/L	(wt %)	mEq/L
20°C	0.22	35.0	14.4	2197
60°C	0.11	17.5	12.9	1967

Source: Quantum Chemical Corp., Emery Division, Technical Center.

solubilized and sodium hydroxide neutralization results in the same net ash or salt content. At free acidity levels above 17.5 mEq/L, lime is the preferred additive, followed by removal of insoluble calcium sulfate.

Use of flocculating agents for Twitchell or high-pressure sweetwaters is also a questionable practice, offering advantages only in certain applications. Solubility data for normal fatty acids and their metallic soaps (Table 3.6) are similar, and so their presence in solution is virtually the same in either form. There are some conditions, however, in which conversion to the soaps is preferred. One such situation is for distillation refining, because the soap volatility (and hence, carryover to distillate) is much lower than that of the corresponding fatty acid; simple neutralization to the sodium soap is satisfactory for this purpose. More on this subject is presented in Section III. Another situation in which soap formation is desirable is for the direct sale of crude glycerol (nonrefined form). While fatty acids are as insoluble as their salts, they are liquids at the lowest temperatures generally employed, and are difficult to totally remove from glycerine solutions: any residual fatty acids cause a haze in the final glycerol product. The salts of these fatty acids, on the other hand, are solids (Table 3.7) and can be removed by filtration techniques to yield a clear product. If the crude glycerol is to be subsequently refined by deionization, it is apparent that the net ionic load is the same whether fatty acids or fatty salts are present: the fatty acids, though, present no cationic load (as discussed in Sect. IV).

Crude glycerine recovered from the saponfication or alcoholysis of fats and oils usually contains large quantities of alkali soaps, which must be removed chemically. The chemical treatments most commonly used involve coagulation and precipitation of the fatty materials with aluminum sulfate or ferric chlo-

Table 3.6 Aqueous Solubility of Carboxylic Acids and Calcium Soaps (g/100 g H$_2$O)

Chain Length:	C$_3$	C$_5$	C$_{12}$	C$_{16}$	C$_{18}$
Acid (60°C)[a]	Total	>3.7	.0087	.0012	.0005
Calcium salt (hot)[b]	56	7.4	.05	.003	—

From [a] Ref. 9; [b] Ref. 10.

Table 3.7 Melting Point (MP) (°C) of Carboxylic Acids and Calcium Soaps

Chain length	C_9	C_{12}	C_{16}	C_{18}
Acid MP[a]	12.5	44.2	63.1	69.6
Calcium salt MP[b]	216	182–3	153–6	150–4
Aluminum salt MP[c]			200	103

Source: [a] From Ref. 9, Part 1.
      [b] From Ref. 9, Part 2.
      [c] From Ref. 10.

ride. A number of procedures are described in the literature [3,6]. Proper control of pH is essential to minimize residual salt levels in the finished crude, and to minimize glycerol degradation due to dehydration (at low pH) or polymerization (at high pH). Modern soap-making plants use centrifugal separators to extract soaps from the glycerine. These separators can lower the net soap content to a level where simple acidulation is sufficient to separate fatty acids from the crude glycerol, thus foregoing the complicated precipitation/filtration procedures. Simple acidulation may also be sufficient for glycerine solutions obtained from transesterification processes. Although a much simpler process, acidulation results in higher residual salt content (NaCl, Na_2SO_4) in the finished glycerine product than the classical treatments described above.

An attractive alternative to simple acidulation involves batch treatment of soap–lye or transesterification glycerine solutions with a cation-exchange resin. The resin acidulates the soaps, allowing them to separate as fatty acids just as in normal acidulation, but also removes the sodium or potassium ions by cation exchange to yield a very low salt level in the glycerine product. The resin is recovered and regenerated with mineral acid for reuse. The advantage of this operation over normal ion-exchange treatment of acidulated crudes is that the demineralization is accomplished without use of an anion resin. Process equipment for this type of treatment is described in the literature for ion-exchange installations [11,12].

Other processes employed for purification of glycerine solutions containing high levels of soap or salt are ion exclusion and ion retardation [13–15]. These are continuous operations employing an ion-exchange resin, and are similar to chromatographic separations; no net ion exchange occurs, so resin regeneration is not necessary. All of the glycerine treatments employing resins are very effective techniques for crude pretreatment. They are all capable of removing ash or soap content to levels below 1%, allowing for economical refining of glycerine solutions which would have little value to the glycerine refiner without the benefit of these processes.

Crude glycerine solutions obtained from any of the above processes are generally bleached to remove additional impurities. The most frequently used bleaching agent is activated carbon, but certain clays have also been employed

Table 3.8   Viscosity and Boiling Point Properties of Glycerol/Water Solutions

Glycerol (wt %)	Viscosity (CP)			B.P. (°C)		
	20°C	40°C	80°C	760 mm	355 mm	92 mm
90	219	36	11	140	116	80
60	11	3.8	1.8	110	89	58
40	3.7	1.6	0.9	105	84	53
10	1.3	0.7	—	101	80	50

Source: Glycerine Producers' Association (New York) Handbook, *Physical Properties of Glycerine and Its Solutions*.

with success. Color bodies, primarily pigments and residues from the fat or oil originally processed, are effectively removed by these bleaching treatments. Some residual fatty material, especially the odiferous low-molecular-weight acids, may also be removed by absorption onto bleaching agents. Bleaching operations are normally performed at 165 to 185°F to facilitate filtration while minimizing any thermal degradation. Other chemical bleaching treatments are generally not useful on glycerine due to the relative ease with which the glycerol itself can be degraded or reacted, but such methods have been described in the literature [6].

B.   Physical Treatments

Physical treatments used in the processing of crude glycerine include decantation, filtration, centrifuging, and evaporation/distillation. These physical operations are used to (a) remove fatty, insoluble, or precipitated solids, and (b) remove water. An important parameter for all physical operations is temperature: at low temperatures, glycerine is very viscous and difficult to handle or transfer (Table 3.8); at high temperatures, glycerine is unstable and subject to dehydration or polymerization. For these reasons, most glycerine processing is done at temperatures of 150–200°F. In this temperature range the water evaporation processes must be conducted under vacuum (Table 3.8).

Any glycerine solutions recovered from fats or oils will contain entrained, soluble, and/or emulsified fat and oil. The entrained or emulsified materials can usually be separated readily by settling or centrifuging if the fat or oil processed is of a highly refined grade; traces of soap, phosphatides, monoglycerides, or other natural emulsifiers make the separation more difficult. Sometimes a pH adjustment can facilitate the separation. In most operations, some type of separation to remove fats and oil is the first step in the crude glycerine process: if the glycerine product is recovered at temperatures above the boiling point or pressures above atmospheric, it will first be "flashed" to atmospheric pressure and then pass to the oil/water separation process. The separation can be accomplished by coalescers, centrifuges, or simple gravity settling and de-cantations. Additional oil is generally "kicked-out" of solution during subsequent concentration operations, and so multiple oil/water separation steps are commonly used throughout the glycerine processing plant.

If the glycerine is subjected to chemical treatment for precipitation of salts or bleaching, then a filtration step is necessary. Any of a number of industrial filtration devices are satisfactory including pressure-leaf, plate and frame, or rotary equipment; however, a discussion of the various types of filtration equipment is beyond the scope of this chapter. It is sufficient here to point out that solids removal is the object of the filtration, and equipment must be specified to perform the task satisfactorily. A major consideration in filtering glycerine solutions recovered from natural fats and oils is that fatty contaminants or soaps can blind the filter media and reduce filtration rates. To overcome this problem, large quantities of filter aids are sometimes used. Increased efficiency in the oil/water separation techniques discussed above can minimize filtration problems.

The most common unit operation in the crude glycerine process is evaporation of water. The recovered glycerine solutions, after being subjected to one of the oil/water separation processes, can be concentrated by evaporation of water either before or after chemical treatment and is sometimes concentrated in separate steps, both before *and* after these treatments. Concentration is usually done early in the crude process because it allows for smaller tanks and vessels downstream; however, most of the chemical treatments used are ionic equilibrium reactions and require that some water be present. The glycerine becomes quite viscous as concentration increases, and a compromise must be reached as to extent of water removal in the process. The type of final refining operation which is used is also quite important in deciding on an optimum concentration process: for distillation refining the final water content must be low (below 15%) to avoid vacuum problems in continuous stills, but for deionization refining an aqueous solution ($\geq$ 60% water) is required to obtain optimum flow and ion exchange. All of these factors enter into the design and specifications for the concentration process.

A number of different concentrator designs are suitable for removing water from glycerine, but energy efficiency is an important factor in modern plant design. The units generally operate under a vacuum to remove water without thermally degrading the glycerol product. Table 3.8 shows the effect of vacuum on boiling temperature of glycerol/water mixtures. Both batch and continuous units are in commercial use, employing single or multieffects or stages. Because of the large water to glycerol ratios that are usually encountered, most modern plants use continuous multieffect evaporators for economy of operation. A relatively recent energy-saving development is the use of "vapor-phase recompression" in multieffect evaporator designs [16]. If the glycerine contains high levels of salts, then salt boxes must be included in the evaporator design. Figures 3.1 and 3.2 show representative examples of evaporators used for glycerine concentrations. Water recovered from the evaporator is usually clean and soft, and is utilized in another process operation. This water may contain small quantities of glycerol or fatty acid, particularly if concentration is done at an acidic pH, but it is generally of sufficient quality to be recycled to the original fat and oil process.

Figures 3.3 and 3.4 are representative depictions of modern glycerine pretreatment plants for the processing of sweetwater from high-pressure hydrolysis. Treatment of soap-lye crudes can be quite different, as seen in Figure 3.5. The unit operations involved in these processes include decantation, centrifuging, chemical treatment, filtration, and concentration. Numerous references are found in the literature [2,3,6,8] for the use of these types of unit operations for the recovery of glycerol from a variety of natural sources.

Modern developments in reverse osmosis, ultrafiltration, or biotechnology will have important impacts on the glycerine process industry. Reverse osmosis has potential for concentration or desalination of glycerol solutions, while ultrafiltration has the ability to separate fats or other high-molecular-weight contaminants. The field of biotechnology may find ways to alter the nature of the fats and oils which provide glycerol, or find new ways to manufacture glycerol from natural materials. Supercritical fluid (SCF) or other extraction methods can be utilized to remove contaminants which are resistant to normal refining techniques [17,18]. Only time will show the impact these emerging technologies will have on the unit operations now employed in glycerine treatment or recovery.

III.   REFINING BY DISTILLATION

Simple distillation is a very effective method of purification for most crude glycerine, but several factors influence both the practicality and the economics of the distillation process. The high boiling point (refer to Table 3.9) of glycerol, along with its relatively low decomposition temperature (ca. 200°C), necessitate the use of vacuum distillation systems to lower the temperature requirements. Steam is usually injected into the distillation column to lower the distillation temperatures and to minimize dehydration and polymerization of the glycerol. Water content of the crude glycerine feed must be kept low so as not to overload the vacuum system. Since some water (usually 5–10%) and stripping steam are desirable to avoid product degradation, condensers and vacuum systems must be designed to remove all water from the distilled product if a high-density (high-purity) glycerol is to be produced.

Any of the standard industrial distillation designs (bubble-cap tray, sieve tray, packed column, etc.) are satisfactory for distilling glycerol. Modern

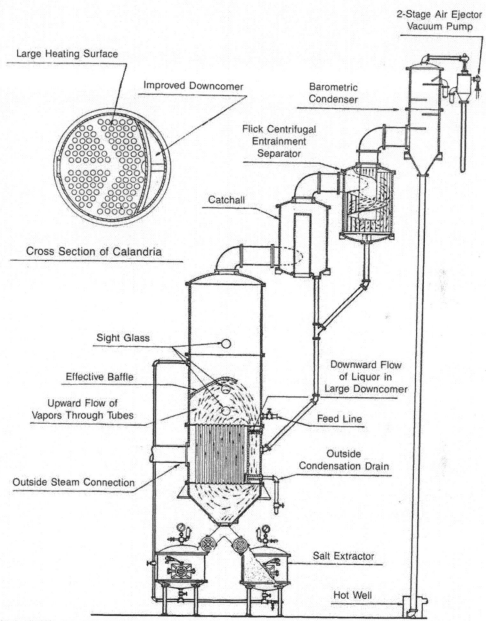

Figure 3.1 Wurster and Sanger single-effect glycerine evaporator. (Reprinted with permission of *Journal of the American Oil Chemist's Society.*)

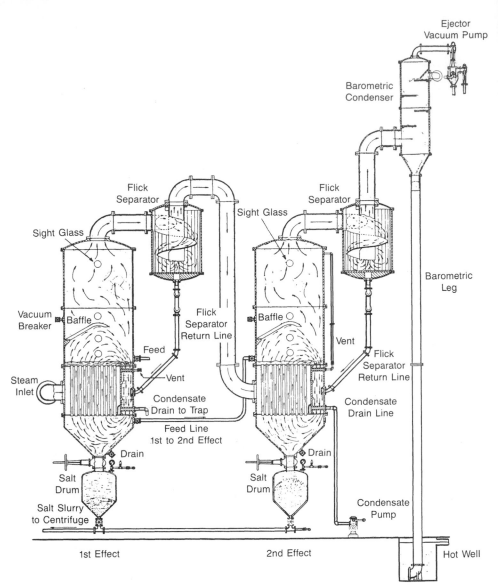

Figure 3.2 Wurster and Sanger double-effect glycerine evaporator. (Reprinted with permission of *Journal of the American Oil Chemist's Society.*)

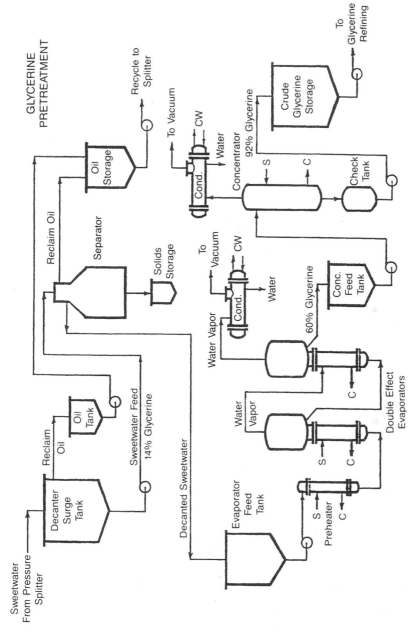

Figure 3.3 Flow diagram of a plant for the treatment and concentration of sweetwaters. S = steam, C = condensate, CW = cooling water. (Courtesy of Quantum Chemical Corporation, Emery Division.)

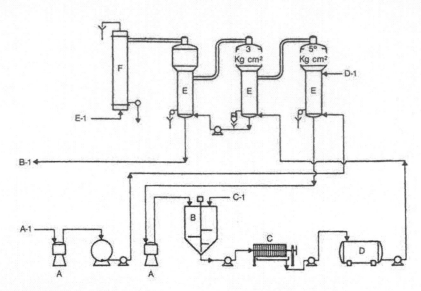

A. CENTRIFUGAL SEPARATOR
B. PURIFIER
C. FILTER PRESS
D. CONCENTRATED WATER TANK
E. TRIPLE-EFFECT EVAPORATOR
F. CONDENSER

A-1 SWEETWATER INLET
B-1 CONCENTRATED GLYCERINE OUTLET
C-1 LIME INLET
D-1 INLET FOR 7 kg/cm² STEAM
E-1 WATER INLET

Figure 3.4 Flow diagram of a plant for the treatment and concentration of sweetwaters. (Reprinted with permission of *Journal of the American Oil Chemist's Society*.)

continuous distillation units operating under high vacuum can be designed for high throughput with minimal loss or degradation of product. Although capital costs for distillation units are generally high compared with other refining processes (ion-exchange units), distillation remains the refining method employed by most major glycerine producers due to the economies of scale achieved in these modern continuous units.

Distillation is a very effective refining technique for glycerine derived from natural fats and oils because the contaminants normally present are salts, glycerides, and soaps of the fatty acids. These are much less volatile than glycerol and are removed as residue from the distillation separation. However, the lower molecular weight fatty acids are considerably more volatile, and can codistill with the glycerol. For this reason it is common practice to add a caustic solution to the crude glycerine feedstock sufficient to convert all fatty acids to soap. It has been found that simple conversion to soap is not sufficient to totally prevent contamination of the distilled glycerol with fatty acid, since an equilibrium is established at distillation conditions to form volatile glyceride. This equilibrium is dependent upon both pH and total soap content (see

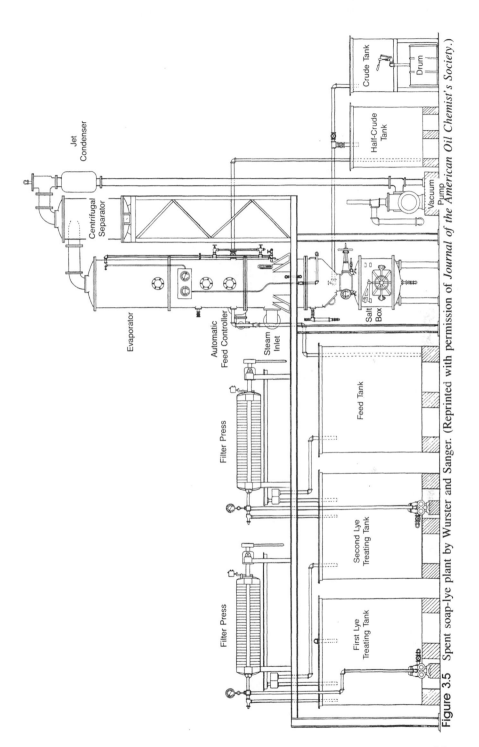

Figure 3.5 Spent soap-lye plant by Wurster and Sanger. (Reprinted with permission of *Journal of the American Oil Chemist's Society*.)

Table 3.9 Boiling Point of Pure Glycerol
at Various Pressures

mmHg	°C
760	290
500	274
100	222
50	204
20	181
10	166
6	156
4	148

Source: Glycerine Producers' Association
(New York) Handbook, *Physical Properties
of Glycerine and Its Solutions*.

Fig. 3.6). While higher pH levels reduce the fatty carryover, the free caustic catalyzes polymerization of the glycerol and results in substantial loss of yield. A similar problem exists with ash or salt content of the crude glycerol: while salts are readily removed in the distillation residue, each pound or unit of salt results in a loss of several pounds of glycerol because the residue cannot be distilled completely dry. From the above discussion it is apparent that both the fat and salt content of crude glycerines must be minimized in the crude processing steps, discussed in Section II, if satisfactory yields are to be obtained by the distillation process.

Volatile contaminants such as water, glycols, and lower fatty acids must also be removed by the distillation process in order to achieve the high purity required for most of today's glycerol markets. These voltaile contaminants are generally removed using a multiple condenser design, a multiple distillation unit, or a separate stripping/deodorization operation. The multiple condensers or still columns are operated at different temperatures to collect glycerol of varying purity. Figure 3.7 shows an example of this type of design for glycerine distillation. The lower purity cuts may be reprocessed to increase the yield of high purity glycerol.

While the distillation unit is designed to produce glycerol of very high purity from virtually any crude feedstock, the distilled product usually contains some color or odor which renders it unacceptable for most applications. These remaining color and odor contaminants are removed by steam deodorization and/or carbon bleaching techniques which are usually an integral part of the glycerine distillation refinery. Figures 3.8 and 3.9 depict typical distillation refineries which produce CP/USP-grade glycerine using the techniques discussed above. Many other types of distillation units are described elsewhere

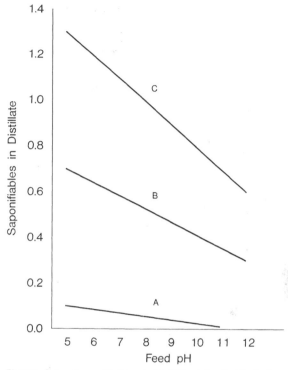

Figure 3.6 Saponifiables (mEq/100 g) in distilled glycerine versus pH. (A) refined glycerine control; (B) refined glycerine plus 0.2% sodium acetate; (C) refined
glycerine plus 0.5% sodium acetate. (Courtesy of Quantum Chemical Corporation, Emery Division, Technical
Center.)

[2,3,6,8,19]. The distillation towers generally operate at pressures of 10 to 30
mmHg absolute, with bottom temperatures of 160–200°C. Construction materials must be able to withstand the corrosive soap-lye crudes, as well as any
fatty acids which might be present from the fats and oils feedstock.

To summarize, distillation of glycerine refining offers the advantages of
continuous, high-capacity plants with the capability of producing very high-
purity glycerol products through a variety of design options. The disadvantages
of the distillation process are that high capital investment is required for maintaining modern plants, several byproducts (residues and toppings) of low purity
are produced, yields are greatly diminished by salt deposits or alkaline residues resulting from neutralization of fatty acids, and multiple processing steps
are required to meet the color and odor requirements of today's marketplace.

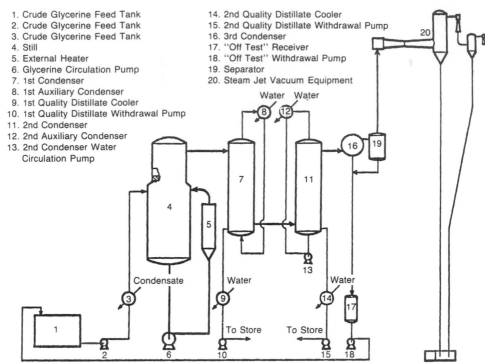

1. Crude Glycerine Feed Tank
2. Crude Glycerine Feed Tank
3. Crude Glycerine Feed Tank
4. Still
5. External Heater
6. Glycerine Circulation Pump
7. 1st Condenser
8. 1st Auxiliary Condenser
9. 1st Quality Distillate Cooler
10. 1st Quality Distillate Withdrawal Pump
11. 2nd Condenser
12. 2nd Auxiliary Condenser
13. 2nd Condenser Water
    Circulation Pump

14. 2nd Quality Distillate Cooler
15. 2nd Quality Distillate Withdrawal Pump
16. 3rd Condenser
17. "Off Test" Receiver
18. "Off Test" Withdrawal Pump
19. Separator
20. Steam Jet Vacuum Equipment

Figure 3.7 Scott single still, with multiple condensers. (Reprinted with permission of CRC Press.)

The distillation refining process does offer the versatility required to manufacture refined grades of glycerol from virtually all types of natural fats or oils process sources.

IV. REFINING BY ION EXCHANGE

Since most contaminants found in crude glycerine derived from natural fats and oils are ionic in nature, it was reported in the 1940s that such glycerol solutions could be economically purified by passing them through ion-exchange resins [20]. The first commercial plant for ion-exchange refining of glycerol was reportedly built in 1951 [21], and resin improvements over the years have made ion exchange an efficient technique for producing refined glycerine. The process has several unique advantages and disadvantages, which are discussed below.

The ion-exchange process consists of passing the glycerol solution through alternating pairs of cationic and anionic resins to effect total demineralization,

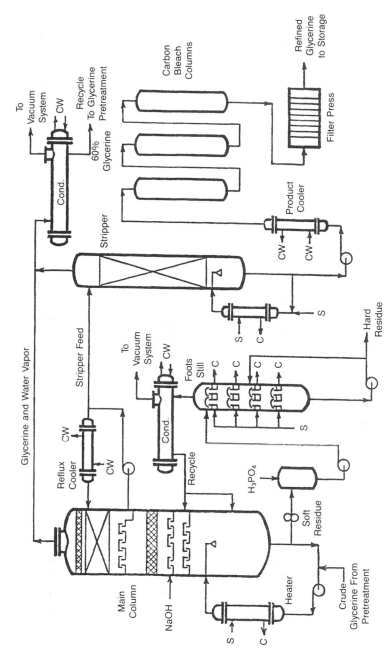

Figure 3.8 Flow diagram of a plant for the distillation refining of natural glycerine. S = steam; C = condensate; CW = cooling water. (Courtesy of Quantum Chemical Corporation, Emery Division.)

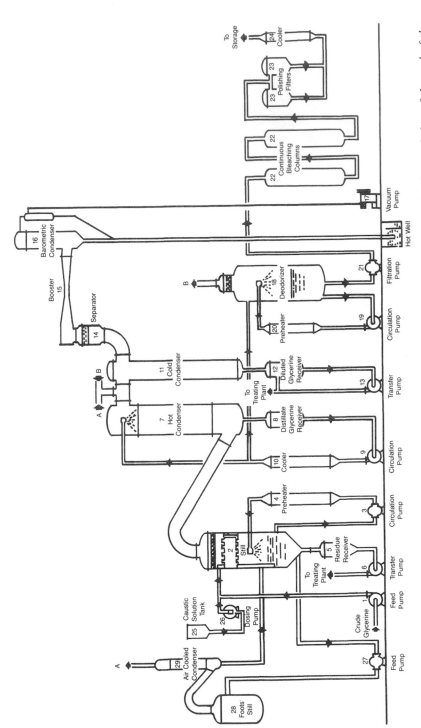

Figure 3.9 Continuous glycerine distillation, deodorizing, and bleaching plant "DG-A." (Reprinted with permission of *Journal of the American Oil Chemist's Society.*)

and the operation is more correctly identified as "deionization" (or DI) to distinguish it from more common exchange processes such as water-softening. The term "deionization" (DI) will thus be used in this discussion to describe the process of refining glycerol by use of ion-exchange resin treatment. The process is a batch or semicontinuous operation because the resin has a finite capacity: once this capacity is exhausted the flow of glycerine is stopped and a regeneration sequence is performed before the purification process can be repeated. Because the process involves exchange or reaction of ions, a dilute solution of glycerine is used as feedstock since the water of dilution increases the extent of ionization: the viscosity of glycerine (Table 3.8) is also a concern, and high pressure drops through the packed resin beds are encountered with very high concentrations of glycerol. Solutions containing 15–35% glycerol are generally employed in the DI process. This crude glycerine solution is passed through a series of one, two, or more pairs of cationic/anionic resin beds. One or more mixed resin beds, containing both anionic and cationic exchangers, are used to obtain very high degrees of purity [22,23]. The overall plant design and type of resins used depend on the purity requirements of the final product and the needed capacity. Much of the theory and practice of glycerine refining by DI is similar to water purification techniques, and information is available in the literature of the resin producers: problems unique to glycerine purification are addressed below. After purification through the "resin-train," the glycerine solution is concentrated to the final purity requirements by removal of water.

The DI plant usually consists of only a few resin beds and evaporation equipment, therefore capital costs are significantly lower than for a distillation refinery unless the bed volume requirement is extremely high. The regeneration of resin each time its capacity is exhausted produces large volumes of wastewater which add to operating costs. The capital and operating costs must be factored into all plant design decisions, but the DI refining process is usually most attractive for refining smaller volumes of glycerol (< 25 tons/day) where the economies of scale achieved in modern distillation units cannot be realized. Because the DI separation is based on polarity or ionic properties rather than volatility, it may also be the preferred refining method for removal of certain contaminants which cannot be readily separated by distillation refining. This same factor becomes a disadvantage for DI if the glycerine to be refined contains glycols or other nonionic contaminants. Properly designed and operated DI plants are capable of manufacturing very high-purity glycerol and, except for the wastewater from resin regeneration, have the unique advantage of producing no byproducts.

The DI plant consists of a feed tank, a series of ion-exchange resin vessels, product-holding tanks, and final concentration/evaporation equipment. Figure 3.10 depicts a typical plant schematic. Except for water extraction, which

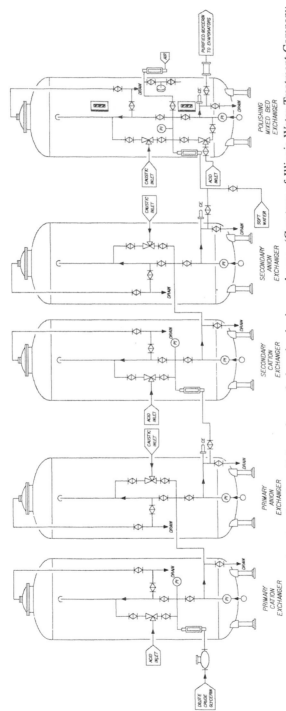

Figure 3.10 Flow diagram of a plant for the refining of natural glycerine by ion exchange. (Courtesy of Illinois Water Treatment Company, Rockford, IL.)

occurs in the final concentrator, all purification takes place in the resin vessels. These resin vessels are generally cylindrical and are equipped with distributors at the top and bottom to achieve proper fluid-flow properties. In addition to piping for product flow, lines are provided for rinse water and regenerant solutions to each bed. The resin vessels are lined with rubber or other suitable coatings to withstand the acidic and basic environments produced during operation and regeneration. The vessels are usually sized larger than the resin volume to allow for bed expansion during backwash of the resin. The resins may shrink or swell in size as they change ionic form, and these volume changes must be considered in the overall vessel design volume. The deionization process is usually designed for downflow operation. A typical vessel is shown in Figure 3.11. Other designs are available such as upflow or continuous with resin replacement [11,12] which can be used to advantage with certain types of resins or processes. Vessels for mixed resin bed operation are more complex in design due to the necessity of separating the resins, regenerating both resin layers, and remixing prior to use [22,23]. The final product concentrator in the DI refinery can be any of the common types discussed in Section II.

Three types of resin are commonly used in the glycerine DI process: strong acid, strong base, and weak base. Many variations of these types exist and may be utilized, but a detailed explanation of ion-exchange theory and mechanisms will not be attempted here. To effect deionization, the cationic resins are used in the acid form and the anionic resins in the base form. As the crude glycerine solution is passed through an acidic cationic resin, any cations (Na^+, Ca^{2+}, etc.) are replaced with H^+ to generate acid. The acidified stream is then passed through a basic anionic resin. If a strong base resin is used, any anions (Cl^-, SO_4^-, etc.) are replaced with OH^- to form water (from acids) or hydroxides (from salts). When a weak base resin is used, the acid reacts with the amine groups of the resin but any salts will pass through. Either a strong or weak base resin can be used in combination with an acidic resin: generally a weak base resin will be used to react with the free acid formed in the *first* cation bed because the weak base resins have higher capacities and are lower in cost. Strong base resins have more utility in removing trace anion contaminants, and are effective for removal of weakly acidic contaminants such as silica and fatty acids: the strong base resins are usually employed in secondary or tertiary bed pairs. A mixed bed of anionic and cationic resins is used as the last bed in a DI process to achieve very high levels of purity. A small-volume plant processing low-ionic-content feed could limit itself to mixed beds only, but much of the design and resin flexibility found in paired beds would be lost. Mixed beds are also more difficult to regenerate, usually resulting in significant capacity losses [22,23]. Color bodies and other polar organic contaminants can be absorbed by the resins, especially by the

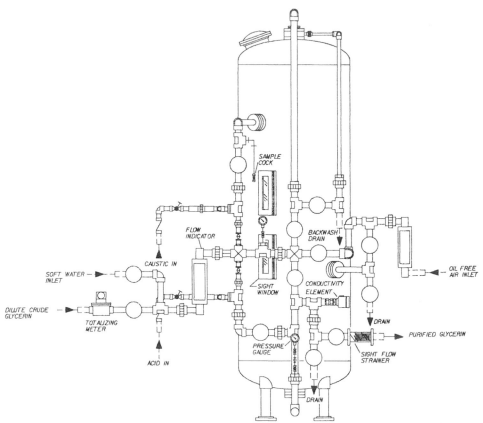

Figure 3.11 Typical monobed construction for downflow ion-exchange operations. (Courtesy of Illinois Water Treatment Company, Rockford, IL.)

anionic resins, and special color-absorbing resins are available and sometimes used in a separate bed to decolorize very dark feed streams. Proper choice of resins and resin pairs allows for manufacture of a glycerine product meeting the highest quality requirements of today's marketplace.

Because the resins have a finite capacity for removing impurities and must be regenerated at regular intervals, the DI glycerine refining process is usually discussed in terms of a "cycle." A cycle consists of one complete exhaustion/regeneration sequence for the exchange resins. The cycle begins with introduction of the crude glycerine solution into the active resin bed train. The crude feed solution may have been subjected to any of the processing treatments discussed in Section II to minimize impurities and/or ionic load to the resins. It is very important that insoluble fatty acids or oils be removed from the feed solution, as they readily foul the resin and form a coating which prevents the

Table 3.10 Specific Conductance (μmho) and Specific Resistance (Ω cm) of Sodium Hydroxide and Sodium Chloride Solutions at 25°C

Dissolved solids	Sodium hydroxide		Sodium chloride	
(ppm)	Conductance	Resistance	Conductance	Resistance
0.1	0.6	1,666,670	0.2	5,000,000
2	12	80,650	4	250,000
10	62	16,130	22	45,450
50	310	3,230	108	9,260

Source: Interpolation of data from *International Critical Tables*, Vol. VI, McGraw Hill, New York and *CRC Handbook of Chemistry and Physics*, 42nd ed., The Chemical Rubber Company, Cleveland (Ref. 10).

regenerant solutions from contacting the resin surface. Flow rates through the DI beds are generally established by design at 0.5 to 2.0 gpm/ft^2 of bed surface area to achieve acceptable exchange equilibria. At start-up, the resin beds contain water from the previous regeneration, because they are allowed to stand covered with water between cycles to minimize the risk of fluid channeling upon start-up and to eliminate hydraulic stresses on the resin from swelling/shrinking due to dehydration. The presence of water in the beds dilutes the glycerine as it is introduced. Frequently the diluted glycerol effluent is collected separately and used to dilute subsequent feed batches. Once the effluent reaches a predetermined glycerol concentration (10–15%), flow is switched to a product collection tank and/or to the product concentrator; this start-up procedure is referred to as "sweetening on" in glycerine DI.

Flow is continued until effluent product quality indicates the resins have been exhausted, thereby allowing contaminants to pass through. Product quality is continuously monitored by measuring conductivity or resistivity of the effluent. As indicated above, the final quality checks on the finished glycerine include other tests for purity to insure the absence of other nonionic materials. Conductivity (or resistivity), however, is a very effective way to monitor performance of the resin beds which function by removing ionic (conductive) species. Table 3.10 shows the relationship between conductivity and salt content. Effluent from the last resin bed is monitored until a pre-established conductivity limit is reached. The flow to the product tank or concentrator is then stopped and the regeneration sequence begins. While conductivity need only be monitored off the final resin bed, common practice is to install meters after each resin pair in the plant. This allows for effective troubleshooting in the event of poor performance, by providing information as to which resin bed or beds are exhausting prematurely. The additional meters can also be used during regeneration to determine when all mineral acidity or alkalinity has been sufficiently rinsed from the resins before "sweetening on."

Before starting the regeneration process, the glycerine remaining in the resin vessels is removed by feeding water through the beds. The glycerine rinsed from the beds is returned to the feed tank: this step is known as "sweetening off." When all the glycerine has been removed from the resins, each bed is backwashed to remove accumulated dirt or other solid particulates. The backwash is generally done with a water flow sufficient to cause a 50–100% expansion of the resin bed volume, and is maintained for 15–30 minutes. Care must be taken in design and operation to prevent loss of resin during the backwash step. The resin bed is then allowed to settle, and in the case of a mixed resin bed, classification and separation of the different resins occurs. Water is drained just to the top of the resin beds to minimize dilution of the regenerants, then regenerant solution is introduced. The regenerant solution is usually added at the same velocity and in the same upflow or downflow direction as the glycerine process stream. Resin manufacturers have also reported benefits from regenerating counterflow to the process [12]. The cationic resins are returned to the acidic form by treating with acid, normally either sulfuric or hydrochloric acid. The anionic resins are returned to the basic form with alkali, generally sodium or potassium hydroxide. The regenerants are added as an aqueous solution; solution strengths and volumes are usually specified by the resin manufacturer, but typical dosage is on the order of 12 to 15 gallons of regenerant solution per cubic foot of resin for 4–5% solutions of acid or base. Some special care or precautions are required during regeneration to avoid fouling of the resins. If the crude glycerine feed contains calcium ions, insoluble sulfates are formed when regenerating the cation bed with sulfuric acid. Hydrochloric acid is the preferred regenerant to prevent this from occurring. If economics dictate the use of sulfuric acid, then a multistep regeneration is necessary using dilute acid in the first steps to maintain $CaSO_4$ levels below the solubility limit. The hydroxide regenerants used for the anionic resins can also cause calcium or magnesium to precipitate from hard water, and it is recommended that softened water be used for regenerating and rinsing of anion beds.

The excess regenerant is rinsed from the resin bed prior to beginning the next cycle. The rinse requirement is usually set by the resin manufacturer, and is measured by titrating the acid or base remaining in the bed effluent or by monitoring conductivity. Rinse water quality must be compensated for in setting rinse specifications for strong acid and strong base resins, since these resins will exchange water hardness to form acid or base, respectively. A series rinse or total water recirculation step is usually the final operation to condition the DI unit for the next batch of crude glycerine feed. During the recirculation step, water of very high purity is produced and can be used elsewhere in the plant or for resin rinses in subsequent cycles. The "sweeten-on" step marks the beginning of the next cycle. Depending on the capacity requirements of the plant, a dual DI train may be installed so that one is operating while the other is regenerating.

The purified glycerine solution from the DI process is fed to an evaporator to remove water. The evaporator may be of batch or continuous design, but a holding tank for the feed is required for a continuous unit due to the intermittent process/regeneration cycles of the DI operation. The evaporators must operate under high vaccum to achieve water removal at temperatures low enough to prevent degradation of the purified glycerol product (Tables 3.8, 3.9). This evaporation step also functions as an efficient deodorizing process, but improper operation of the evaporator can cause formation of color, taste, or odor bodies through thermal degradation of the glycerol. Some special handling precautions to prevent deterioration in the evaporator include short heating cycles, rapid product cooling, and use of inert atmospheres on product tanks. All of these capabilities are designed into a plant for manufacturing the highest purity (99.5% and higher) glycerol products.

While the operating procedures and most of the theory of glycerine DI are identical to water purification technology, several unique distinctions apply to the purification of glycerine solutions derived from natural fats and oils. Two important considerations are resin fouling due to fatty acids, oils, or soaps in the sweetwater and capacity limitations due to high ionic loads. Both of these factors must be carefully controlled using proper pretreatment techniques (as discussed in Section II) for economical efficient operation of the DI refinery. Dirt and inert materials can usually be removed during the resin backwash procedures, but insoluble fatty acids can be absorbed by the resin to impart a hydrophobic character which prevents the regenerant solution from contacting the resin. Soaps in the sweetwater feed are acidulated by the strong acid resin to form fatty acids, and thus have the same effect. Soluble fatty acids, silicas, or other weakly acidic contaminants are removed by the strong base resins, but these same contaminants can irreversibly react with both strong and weak base resins and foul the bed. Modern resin technology has made available many types of open-pore or ''macroreticular'' resins that enable the regenerants to more efficiently remove the bound foulants; however, some fouling is inevitable due to the nature of the resins and the types of contaminants found in natural glycerine products [24–26]. Certain clean-up techniques can be employed to recover lost capacity, but the overall resin activity must be closely monitored in order to implement these techniques at the first sign of fouling; otherwise, the contaminants will accumulate to a level that the clean-up solutions cannot penetrate. Caustic or brine solutions are the most frequently used clean-up solutions, but new techniques employing surfactants are very interesting [27,28]. Caustic is used on cationic resin to saponify and wash out any fatty acids, as well as to convert the resin to the sodium form: the resin's affinity for the sodium ion can be strong enough to effect a displacement of other cationic foulants. This same theory applies to the use of brine to convert strong base anionic resin to the chloride form. Hydrochloric acid is sometimes used on anionic resins to remove iron or other minerals.

The resin activity or exchange capacity must be monitored to detect when a fouling problem occurs, and a number of simple techniques are available. Since the ionic content of the feed stream is converted to acidity in the cation bed, a simple acid/base titration of the cation bed effluent from the beginning part of the cycle (before resin exhaustion becomes significant) is a good approximation of the feed ionic load. This value, multiplied by the feed volume processed in a cycle, gives a total activity or capacity value for the DI train, which can be expressed in any suitable units. Monitoring of pH and conductivity after each resin pair at breakthrough can help to pinpoint which beds are exhausting prematurely. The resins themselves can be analyzed in a laboratory for activity, capacity, or foulants. Strong base resins require a more complex analysis for salt-splitting (strong) capacity versus acid-neutralizing (weak) capacity, since the strong base resins frequently decompose to amines which retain only weak base activity. Standard test procedures for these measurements are found in the resin literature [29,30]. By monitoring resin activity and using clean-up techniques effectively, the life of the resins can be appreciably extended; eventually, however, the resins will foul or deteriorate to the point that replacement becomes necessary. Feed quality and operating practices have great impact on resin life, and both must be optimized to avoid excessive resin replacement costs. Resin life can range from 50 to more than 500 cycles, and must be maximized by study of each individual plant installation.

In summary, ion-exchange resins have been used commercially for almost 40 years to refine the glycerine obtained from fats and oils. The process can obtain a product which meets the highest purity requirements for most applications, and has several unique advantages and problems. Advantages offered by DI refining include lower capital investments for plant construction, high product yield since no byproducts are produced, and excellent odor quality of finished product due to the nature of the process. Problems which must be considered in overall plant design are the generation of large volumes of wastewater during resin regeneration, the "batch" nature of the process cycle, inability of DI to remove nonionic contaminants, and possible color degradation during the final concentration operation. Strong mineral acids and alkali must be handled during regeneration, and because the process uses large volumes of water it is important to recover and recycle this wherever possible.

V. STORAGE AND OTHER PROCESS CONSIDERATIONS

Dilute solutions of glycerol are subject to bacterial fermentation, forming glycols, acids, and/or gasses [31]. In addition to reducing yield, these fermentation products can be very difficult to remove by distillation and the glycols are impossible to remove by DI. To prevent bacterial degradation, dilute solutions (<50%) of glycerol are generally held above 150°F, requiring that some method be devised of heating the storage tanks. At higher concentrations the glycerol becomes quite viscous and, although bacterial growth is not a problem at the higher glycerol levels [32], the storage tank might still be heated to keep the product reasonably fluid. Typical handling or pumping temperatures for concentrated glycerol are in the range of 100–140°F.

Since dilute solutions of glycerol are such excellent growth media for bacteria, carryover of small amounts of glycerol to cooling towers or other process water supplies must be monitored closely to prevent bacterial contamination and fouling of heat exchangers or other surfaces. On the other hand, process water-treatment programs and chemicals must be chosen with caution to avoid potential contamination of the glycerol, especially in the manufacture of pharmaceutical or food-grade products.

Although glycerol and glycerol solutions are inert to most types of metals used for tanks or process equipment [33], many of the refining operations require that special consideration be given to the materials of construction. Acids and/or alkalies are frequently used in the crude pretreatment process; soaplye crudes usually contain residual alkalinity from the soap process and high levels of sodium chloride; sweetwater from Twitchell or high-pressure hydrolysis of fats and oils contains residual fatty acids which can be corrosive to medium or light weight steel. Any of these environments obviously require special materials. Distillation and deionization equipment is usually of special construction, as it is subject to the harsh conditions noted above. Still condensers, final concentrators of a DI plant, and finishing equipment can be constructed of light weight steel or aluminum since they are exposed only to the purified glycerol. Even here, however, trace fatty acid contaminants or moisture could cause corrosion problems. Glycerol can react with iron or iron compounds to form highly colored complexes [34], and proper construction materials are imperative to produce high-purity grades and prevent degradation.

The hygroscopic nature of high-purity glycerol is another concern in the design of product storage tanks. Glycerol readily absorbs moisture from the atmosphere (Table 3.11), therefore the storage tank must be closed and vented through a dessicant or other checking device; alternatively, the tank may be purged and kept under a positive pressure with dry inert gas. Where refined glycerol is transported commercially in drums, tank trucks, or rail cars, similar design and handling practices must be exercised during transfer and shipment of the high-purity grades to prevent contamination by the environment.

VI. BYPRODUCTS

Both the distillation and the deionization processes for refining have proved capable of manufacturing glycerine of 99.5% or higher purity from natural fats

Table 3.11   Equilibrium Relative Humidity Over Glycerol-Water Mixtures at 25°C

Wt. % Glycerol	100	95	92	84	72	51
Relative humidity	0	10	20	40	60	80

Source: *Physical Properties of Glycerine and Its Solutions*, published by the Glycerine Producers' Assoc., New York.

and oils. The two processes do, however, present different byproduct and waste-stream problems during their operations. Both plants may generate spent bleaching carbon or other types of filter cake and sludge, particularly during the crude pretreatment operations. In addition to carbon and filter aids, these solid wastes may contain fats and oils, soaps, protein matter, phosphatides, gums, and other contaminants indigenous to the starting raw material. While the deionization plant in theory produces no byproducts, large volumes of acidic and alkaline wastewater streams are produced during the resin regeneration cycles. The resins employed have a finite life and, ultimately, must be disposed of and replaced. Refining by distillation produces a residue product containing large amounts of ash, soaps, and/or polymerized glycerine. A dilute glycerine topcut or condensate is also generated by most distillation units.

To some extent, the byproduct glycerol streams can be reprocessed to produce lower grades of saleable product, but some portion will have to be discarded. Landfill, incineration, or biodegradation of liquid effluent are the usual disposal techniques for these byproducts. With the environmental concerns associated with disposal of waste materials and attendant costs, it is very important to look for alternative methods for disposing of these materials. The solid wastes can be extracted to remove any residual glycerol [17,18]; sludges remaining in filter cakes may have value as a source of gum or phosphatides in the case of vegetable feedstocks, or as a feed meal or fertilizer due to bone or protein matter from animal feedstocks. Glycerol-containing residues and topcuts from distillation may be utilized in asphalt/tar/pitch markets where low-quality grades of glycerine have been adequate. The ash and soap content of these glycerine residues can be reduced by application of ion-exclusion or ion-retardation separation techniques [13–15].

Minimizing costs and risks associated with glycerine byproducts is an ever-increasing challenge to maintaining a profitable glycerine refinery. This same challenge will play a major role in selecting the method of refining for new plant construction.

REFERENCES

1.   Bailey's Industrial Oil and Fat Products, 3rd Ed., (D. Swern, ed.), John Wiley & Sons, New York, 1964.

2. Newman, A. A., *Glycerol*, CRC Press, Cleveland, 1968.
3. Miner, C. S., and Dalton, N. N., Glycerol, Reinhold Publishing Corp., New York, 1953.
4. Lawrence, E. A., Hydrolysis methods, J. Am. Oil Chem. Soc., 31: 542, 1954.
5. Sturzenegger, A., and Sturm, H., Hydrolysis of fats at high temperatures, Ind. Eng. Chem., 43: 510, 1951.
6. Ziels, N. W., Recovery and purification of glycerol, J. Am. Oil Chem. Soc., 33: 556, 1956.
7. Lamborn, L., Modern Soaps, Candles, and Glycerin, 3rd Ed., D. Van Nostrand Company, New York, 1920.
8. D'Souza, G. B., The importance of glycerol in the fatty acid industry, J. Am. Oil Chem. Soc., 56: 812A, 1979.
9. Markley, K. S., Fatty Acids, Parts 1 and 2, Interscience Publishers, New York, 1960.
10. Handbook of Chemistry and Physics, 45th Ed., The Chemical Rubber Co., Cleveland, 1964.
11. Feser, R. et al., Closed ion exchange system, Chem. Proc. (Chicago), 40(1): 36, 1977.
12. Marquardt, K., Back-rinsable solid-bed upflow ion exchange, Ultrapure Water, 3(5): 28, 1986.
13. Wheaton, R. M., and Bauman, W. C., Ion exclusion, Ind. Eng. Chem., 45: 228, 1953.
14. Prielipp, G. E., and Keller, H. W., Purification of crude glycerin by ion exclusion, J. Am. Oil Chem. Soc., 33: 103, 1956.
15. Simpson, D. W., and Bauman, W. C., Concentration effects of recycling in ion exclusion, Ind. Eng. Chem., 46: 1958, 1954.
16. Tucker, J. F., et al., Evaporative recovery system slashes fuel costs, Chem. Proc. (Chicago), 46(7): 94, 1983.
17. Diaz, Z., and Miller, J. H. (Shell Oil Co.), Glycerol Purification Process, U.S. Patent 4,683,347, 1987.
18. Blytas, G. C. (Shell Oil Co.), Recovery of glycerin from saline waters, U.S. Patent 4,560,812, 1985.
19. Brockmann, R., et al., (Henkel), Glycerol distillation process, U.S. Patent 4,655,879, 1987.
20. Stromquist, D. M., and Reents, A. C., C. P. glycerol by ion exchange, Ind. Eng. Chem., 43: 1065, 1951.
21. Busby, G. W., and Grosvenor, D. E., The purification of glycerin by ion exchange, J. Am. Oil Chem. Soc., 29: 318, 1952.
22. Reents, A. C., and Kohler, F. H., Mixed bed deionization, Ind. Eng. Chem., 43: 730, 1951.
23. Kunin, R., and McGarvey, F. X., Monobed deionization with ion exchange resins, Ind. Eng. Chem., 43: 734, 1951.
24. Frederick, K. H., Inorganic fouling of high purity mixed beds, Ultrapure Water, 5(2): 44, 1988.
25. Lefevre, L. J., Ion exchange: problems and troubleshooting, Chem. Eng., 93(13): 73, 1986.

26. Frisch, N. W., and Kunin, R., Long-term operating characteristics of anion exchange resins, Ind. Eng. Chem., 49: 1365, 1957.
27. Staff, Ion exchange washing additive reduces bead fouling effects, Chem. Proc. (Chicago), 46(10): 50, 1983.
28. Bornak, W. E., and Griffin, J. W., Resin fouling and cleaning—selecting the best economic alternative, Ultrapure Water, 4(6): 39, 1987.
29. Dow Chemical Company, A Laboratory Manual on Ion Exchange, Midland MI, 1971.
30. Resinous Products Division of Diamond Shamrock Chemical Company, Duolite Ion Exchange Manual, Chemical Process Co., Redwood City, CA, 1969.
31. Rayner, A., The occurrence, properties, and uses of trimethylene glycol, and the fermentation of glycerin lyes, J. Soc. Chem. Ind., 45: 265T, 1926.
32. Mariani, E. J., Jr., Libbey, C. J., Litsky, W., Antimicrobial activity of commercial grade glycerin, Dev. Ind. Microbiol., 14: 356, 1973.
33. Glycerine Producers' Association Handbook, Glycerine: Terms, Tests, and Technical Data, New York, New York.
34. Cotton, F. A., and Wilkinson, G., Advanced Inorganic Chemistry, 3rd Ed., Interscience Publishers, New York, 1972, p. 864.

4

Manufacture of Glycerine from Petrochemical and Carbohydrate Raw Materials

Norman O. V. Sonntag

Ovilla (Red Oak), Texas

I. MANUFACTURE FROM PETROCHEMICAL RAW MATERIALS

Synthetic glycerol, for the purposes of Section I of this chapter, is defined as the product manufactured from petrochemical raw materials such as propylene, allyl chloride, epichlorhydrin, glycerol chlorohydrins, acrolein, or allyl alcohol. These syntheses have provided a classic example of the interplay of the American petrochemical and natural oleochemical industries in providing, subject to the economics prevailing, the required glycerol for both domestic and international market demand.

A. Historical Development

Prior to 1948, glycerol was available only as a byproduct from soap manufacture and from fatty acid-splitting operations, of which, at that time, the volume from Twitchell fat-splitting processing was decreasing, and the annual volume from the continuous Colgate-Emery fat-splitting process was increasing. Essentially, very little natural glycerol was obtained from fat methanolysis processing, and practically none from soapstocks (acidulated soapstocks from animal and vegetable fat and oil refining, containing about 25% triglycerides).

U.S. sales of soap in the decade 1940–1950 had decreased from 3,200 to 2,486 $\overline{M}$ lbs./year as synthetic detergent sales increased from about 25 to 1,093

Table 4.1 U.S. & International Soap and Synthetic Detergent Sales For Selected Years in 1940–65 Period ($\overline{M}$ lbs.); 1988–90 ($\overline{B}$ lbs.).

	1940	1950	1951	1952	1953	1954	1955	1956	1957	1958	1959	1960	1961	1962	1963	1964	1965	1988[a]	1989[b]	1990[c]
U.S.																				
Synthetic Detergents	25	1093	1250	1532	1867	2070	2312	2689	2911	2453	3213	3310	3467	3747	3868	4033	4161	5.27	6.31	6.47
Soap	3200	2486	2056	1876	1644	1441	1349	1285	1188	1137	1065	1054	1014	1042	1025	1100	968	1.23	1.29	1.32
Europe																				
Synthetic Detergents						1276	1606	1830	2068	2288	2413	2700[d]	2900[d]	3130	3337	3549	3145	5.47	6.50	6.89
Soap						3491	3463	3590	3425	3122	3287	3207	3132	3132	3056	3087	2851	1.58	1.56	1.60
Japan																				
Synthetic Detergents				16.6	23.9	26.9	34.0	49.4	54.0	67.4	106.2	189.6	331.0	416.0	567.1	609.1	787.0	2.38	2.41	2.60
Soap				330.5	422.0	521.9	614.2	652.6	703.4	762.4	836.5	764.3	656.6	569.4	494.2	442.5	378.0	0.429	0.432	0.430

a. Actives in Synthetic Detergents from "Surfactant" sales less soap (no builders, enzymes, additives, etc.) *C&E News*, Jan. 23, 1989.
b. Actives in Synthetic Detergents from "Surfactant" sales less soap *C&E News*, Jan. 29, 1990.
c. Estimated N.O.V. Sonntag, Jan 30, 1990.
d. Approximate only.

Table 4.2 Selected Average Annual Glycerol Prices For USP Grade (¢/lb.)

Year	Price/lb.
1948	38
1949	38
1950	24
1951	53
1952	54
1953	44
1954	33
1955	28
1965	23
1969	23.5
1984	68.1
1986	87.75
1987	87.75
1988	89.5
1989	80.25

$\overline{M}$ lbs./year. Over the next ten years, soap sales had declined precipitously to 1,054 $\overline{M}$ lbs./year, while synthetic detergent sales had shown a phenomenal growth to 3,310 $\overline{M}$ lbs (Table 4.1). Faced with erratic glycerol prices, (Table 4.2), and a distinct shortage during the 1940–1950 decade, primarily due to the loss of natural glycerol as a byproduct from decreasing soap production, the American petrochemical industry responded vigorously. In the majority of the programs that were developed and translated into commercial production, synthetic glycerol was an offshoot of a complex, integrated chemical program in which the volumes provided of other products substantially exceeded that of the glycerol produced.

Shell Chemical pioneered in 1948 with construction of a 54.5 $\overline{M}$ lb/year plant at Deer Park, Texas (near Houston) employing propylene as raw material, and a developed technology initiated about 1937 based upon the route: propylene → allyl chloride → chlorohydrin → epichlorohydrin → glycerol. Within eleven years, a second Shell plant costing $8.5 $\overline{M}$ with a capacity of 25 $\overline{M}$ lbs./year was onstream at Norco, Louisiana, based upon an entirely different synthetic route: propylene → acrolein → allyl alcohol → glycerol. The earlier plant was part of an integrated complex in which dichloropropane and dichloropropylene (as soil fumigants), and epichlorohydrin (for epoxy resins) were also manufactured; the latter plant was part of a second complex in which acrolein, acetone, isopropanol, hydrogen peroxide, and allyl alcohol were co-produced. Availability of increasing annual quantities of essentially dry byproduct natural glycerol from methyl ester processing and other economic disadvantages caused Shell to shut down its Deer Park synthetic glycerol operation in 1976; the Norco, LA synthetic glycerol plant survived until 1980. Today, a Shell plant in Rotterdam, The Netherlands produces synthetic glycerol with technology similar to the Deer Park operation. Its estimated capacity is 55 $\overline{M}$ lbs./year.

Dow Chemical joined the ranks of synthetic glycerol producers in 1959 with a large plant at Freeport, Texas, initially of 36 $\overline{M}$ lbs./year capacity, and destined to be enlarged several times. Essentially, its technology was similar to Shell's first process, but with subtle differences in later processing steps and in the glycerol/polyglycerol ratios in the recycle streams. Dow's U.S. plant still operates with a stated 1988 capacity of 110 $\overline{M}$ lbs./year. Dow's West German plant has a 44 $\overline{M}$ lbs./year estimated capacity. According to Thwaites [1], a French Solvay plant capacity 33 $\overline{M}$ lbs./year, operates on technology similar to the Dow Chemical process.

Olin-Mathieson's synthetic glycerol venture, at the Doe Run plant in Brandenburg, Kentucky, started in late 1961, and was sized at 40 $\overline{M}$ lbs./year capacity. The technology route was allyl alcohol → chlorhydrins → glycerol, with a unique first step, and with the second step similar to Dow's and Shell's. The Olin-Mathieson program was shortlived; by 1964 it had been shut down.

FMC Corporation initiated synthetic glycerol production in 1969 with a 40 $\overline{M}$ lbs./year plant at Bayport, Texas (also near Houston). At the time, U.S. synthetic glycerol capacity was increased to 290 $\overline{M}$ lbs./year. FMC's technology was entirely free of chlorine dependency: propylene oxide → allyl alcohol → glycidol → glycerol. Although high expectations were held for this approach, the plant was compelled to cease production in 1982.

Two Japanese plants are listed in 1984 as current synthetic glycerol producers: Daicel, with an estimated capacity of 33 $\overline{M}$ lbs./year, presumably uses technology involving the oxidation of allyl alcohol with peracetic acid in ethyl acetate to glycidol followed by hydration with efficient recovery of allyl alcohol and water. Kashima, at a estimated 26 $\overline{M}$ lbs./year capacity, apparently operates with licensed technology.

It is interesting to trace the changes in the proportion of natural and synthetic glycerol produced in the United States, Europe, and worldwide after 1948, when synthetic glycerol first appeared. Table 4.3 illustrates some of the U.S. trends in selected years of the period 1945–1988. The period 1980–1988 appears to be one in which natural glycerol dominated (greater than half); the years 1963–1969 (54–58%) typify those in which synthetic glycerol production was prominent. Today, the trend is undoubtedly strongly for natural glycerol production. In the United States this is due primarily to the increasing volumes of dry, byproduct crude available from fat methanolysis, while on a worldwide scale the trend is rooted in the substantial increase in Southeast Asian natural fatty acid and methyl ester production (Philippines, Malaysia,

Table 4.3 U.S. Production of Natural and Synthetic Glycerol ($\overline{M}$/lbs.)

Year	Natural	Synthetic
1945	172.4	0
1950	190	36
1951	171	40
1952	152	36
1953	172	43
1954	147	60
1955	148	80
1956	143	101
1957	135	105
1958	132	81
1959	147	118
1960	151	150
1961	138	141
1962	131	118
1963	141	151
1964	150	178
1965	145	201
1969[a]	153.2	195.0
1984[b]	223.9	57.7
1985	246.3	64.2
1986[b]	263.1	57.7
1987[b]	248.8	58.4
1988[b]	249.3	47.5
1989	256.7	48.9
1990	259.3[b]	47.5

Source: period until 1965, Glycerine Outlook, H. W. Zabel (Roger Williams T & E Serv.), at SDA 40th Annual Conv., New York, January 25, 1967.
period 1969–1988.
[a] J. D. Thwaites (Shell, Rotterdam) Chem Ind (London), 48(8–9): 508–509, 1969.
[b] N.O.V. Sonntag estimate, January 1, 1989.

and Indonesia). Table 4.4 illustrates worldwide synthetic glycerol trends as estimated by three authors during the 1969–1988 period.

B. Synthetic Glycerol Technology

Figure 4.1 outlines the various possible chemical routes that have been developed for synthetic glycerol manufacture. It is obvious that a considerable variety of raw materials and intermediates, as well as chemical reactions plus a

Table 4.4 Estimates of Worldwide Trends in Natural and Synthetic Glycerol Production ($\overline{M}$ lbs.) 1969–1988

	Worldwide	U.S.		Europe	Japan	Other areas
1969[b]						
Natural Oils and Fats	54%	37%	128.9	74%		
Carbohydrates	4%	7%	24.3	—		
Synthetic	42%	56%	195.0	26%		
Total			348.2[a]			
1984[c]						
Natural	75%	825				
Synthetic	25%	275				
Total[a]		1.1$\overline{B}$				
Consumption ($\overline{M}$ lbs.)			280	325	<100	400
1988[d]						
Natural	76%	874	a			
Synthetic	24%	276				
Total		1.15$\overline{B}$				
1995[e]						
Natural	79%	956				
Synthetic	21%	254				
Total		1.21$\overline{B}$				

Key:
[a] Source: Glycerine & Oleochemicals Division, Soap & Detergent Association, New York.
[b] J. D. Thwaites (Shell, Rotterdam), Chem. & Ind. (London), 48(8–9), 508–509, 1969.
[c] C. Gentry (Procter & Gamble), AOCS Short Course, Paper 21, King's Island, Cincinnati, Ohio, Sept. 26, 1984.
[d] N.O.V. Sonntag, estimate based upon new, on-stream (1984–1988) international production at 65% of plant capacity overall, Jan. 1, 1989.
[e] N.O.V. Sonntag, future estimate based on projected and announced glycerol byproduct from international oleochemical programs plus 5 $\overline{M}$ lbs/yr from P & G's "Olestra" program.

plethora of reagents and reaction conditions have been investigated to initially achieve acceptable product quality, yield, and economic soundness in synthetic glycerol manufacture.

Table 4.5 summarizes the patented, international technology that has developed beginning in 1930 and continuing through 1988 for manufacture of syn-

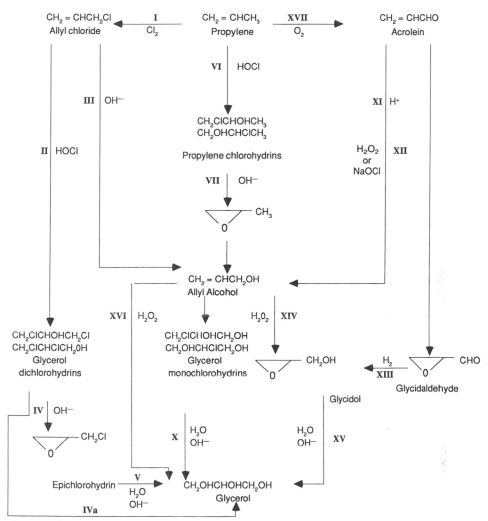

Figure 4.1 Chemical routes for manufacture of synthetic glycerol.

thetic glycerol by the so-called "epichlorohydrin route" and its variations and innovations that involve a chlorination step. Shell originated it; Dow currently operates a variation of it.

Table 4.6 is a summary of the patented, international technology developed during the same time period for the nonchlorine route through acrolein, allyl alcohol, or other intermediates. Shell's Norco plant produced synthetic glycerol by this kind of processing.

Table 4.5 International, Patented Technology for Synthetic Glycerol Manufacture by Epichlorohydrin Route or by Innovations and Variations Thereof

Processing steps[a]	Raw material	Reagent	Conditions	Patent no. Date	Assignee	Reference
IV–V, X	5 Chlorinated precursors	H_2O	Removes HCl in absence of hydrogen halide binders	British 536,428 (5/14/41)	H. Dreyfus	12
X	Glycerol mono-chlorohydrin	H_2O	(Same as above)	U.S. 2,311,741 (2/23/43)	Celanese Corp. America	13
X	Glycerol chlorohydrin	$NaOH/Na_2CO_3(H_2O)$	Continuous; pH control, alkalinity between pH of 0.018 N $NaHCO_3$ and 0.042 N Na_2CO_3	U.S. 2,318,032 (5/4/43)	Shell Dev. Corp.	14
I–(II)–IV–V	Propylene	I: SO_2+Cl_2 (II)–IV–V: $NaHCO_3(H_2O)$	I: 90% H_3PO_4; irrad. 20°, 5.5 h (II)–IV–V: Hydration and hydrolysis [88 cc propylene → 18 cc glycerol]	U.S. 2,378,104 (6/12/43)	C. L. Horn	15
I–(II–IV–V)	Propylene	I: Cl_2	I: SO_2, TeO_2 or SeO_2 used as cat., 0–100°, 1.5–16.5 h into H_2SO_4 or H_3PO_4	U.S. 2,407,344 (9/10/46)	C. L. Horn	16
IV–V	Glycerol dichlorohydrin	$NaOH/Na_2CO_3$ (H_2O)	Uses either alkali or admixture in continuous addition	Swedish 119,077 (6/25/47)	Svenska Cellulose Aktiebolaget	17

Route	Starting material	Reagent	Details	Reference	Company	No.
IV–V, IVa	Glycerol dichlorohydrin & HOCH—$(CH_2Cl)_2$	IVa: NaOH IV–V: H_2O (NaOH)	Combined routes IVa & IV–V w. salt removal. 4% dichlorohydrin soln. IVa $\longrightarrow$ 3% glycerol and 7% salt; w. 50% IV diversion, 5.3% glycerol and 5.3% salt; 3.5% dichlorohydrin soln IVa $\longrightarrow$ 2.6% glycerol, 6.1% salt; w. 50% IV diversion, 4.6% glycerol, 7.2% salt; w. 65% IV diversion, 6% glycerol, 8.8% salt.	U.S. 2,665,293 (7/29/52)	Shell Dev. Corp.	18
V–V, IVa or combined	Allyl chloride	H_2O (NaOH)		Dutch 83,156 (11/15/56)	Dutch Shell	19
IX–X; IV–V	Monochlorohydrins (I) & dichlorohydrin (II)	NaOH or Na_2CO_3	100 Parts allyl alcohol $\xrightarrow{HOCl}$ 2000 parts aqueous mixture containing 7.8% mixture of I & II yields 34 parts epichlorhydrin & 105 parts glycerol	French 1,328,311 (5/31/63)	Scientific Design, Inc.	20
IX–X	Chlorohydrins	Na_2CO_3	200 and 220°C, 1.5 min gives 88% glycerol 1% polyglycerol and 90% polyglycerol 5% polyglycerol, resp	British 940,284 (10/30/63)	Olin-Mathieson	21

Table 4.5 (Continued)

Processing steps[a]	Raw material	Reagent	Conditions	Patent no. Date	Assignee	Reference
IX–X; IV–V	Dichlorohydrins and monochlorohydrins	NaOH	Several combined step conversions	German 1,156,774 (11/7/63)	Halcon, Intl.	22
V	Epichlorohydrin	Two steps 1. *hydration:* HCl (H_2O), 0.2% H_2SO_4 or 0.25% H_3PO_4. 2. *hydrolysis:* 30% Na_2CO_3	1. 15–45 min addition, 80–105°, then 3.5 h, reaction time. 2. 1 h Glycidol formation minimized; product glycerol has <1% polyglycerols	Czechoslovakia 104,347 (1/28/62)	I. Andrus & O. Lustik	23
V	Epibromohydrin	Two steps 1. anionic exchange resin 2. cationic exchange resin	1. 20–60°C, Product— 68% glycidol and 19.5% glycerol. 2. 89.2% theo. yield of glycerol	British 971,633 (9/30/64)	Standard Oil Co.	24
V	Epichlorohydrin	$Na_2CO_3(H_2O)$	Vapor phase into liquid aq. alkali through 40–220 μm diam. glass pores at 1 mol/L soln/h 50 pts. epichlorohydrin → 48 parts glycerol	Polish 50,594 (2/25/66)	Politechnika Szczechinska	25

V	Epichloro-hydrin	$Na_2CO_3(H_2O)$	Two steps 1. 5–10% soln hydrolyzed at atm pressure 2. treat with epichlorohydrin at elev. temp. and pressure.	U.S.S.R. 129,197 (11/17/66)	V.P. Choporov, et al.	26
V	Epichloro-hydrin	$Na_2CO_3(H_2O)$	pH<9; flow-type autoclave; affords low ether content during hydration and hydrolysis	French 1,537,588 (2/14/69)	T. Reis	27
IX,X	Chlorohydrins	3:1 $NaOH/Na_2CO_3$	3:1 $NaOH/Na_2CO_3$—96.8% glycerol; 9:1 $NaOH/Na$ Na_2CO_3—76.8% glycerol	Japan Kokai 73, 96,509 (12/10/73)	Nippon Soda Co., Ltd.	28
V	Epichloro-hydrins	aq. alkalis	Alkali hydrolysis w. 0.1—0.2% ''Pluronic''-type surfactants	Czechoslovakia 153,376 (5/15/74)	Peterka	29
IV, V	Dichlorohydrins and/or epichlorohydrin	Na_2CO_3	Solid slurry reaction in glycerol-immiscible solvent like heptane, xylene, cyclohexane.	Japan Kokai 76, 125,309 (11/1/76)	Osaka Soda Co., Ltd.	30
V, X	Epichlorohy-drin, mixed chlorohydrins	Na_2CO_3	50–70°C; selectivity for glycerol >90%	U.S. 4,053,525 (10/11/77)	Shell Oil Co.	31
Chlorination of allyl chloride, IV	Allyl alcohol	I. Cl_2 IV: $Ca(OH)_2$	1. Chlorination in acidic (HCl) medium 2. Dehalogenation w. aq. alkaline suspension of $Ca(OH)_2$	French Demande Fr 2,565,229 (12/6/85)	N. Negato, et al. Showa Denko K.K.	32

[a]For delineation of process routes, see Figure 1.

Table 4.6 Patented International Technology for Synthetic Glycerol Manufacture by the Nonchlorine Route through Acrolein, Allyl Alcohol, or Other Intermediates

Processing steps[a]	Raw material	Reagent	Catalyst conditions	Patent no.; date	Assignee	Reference
XVI	Allyl alcohol	H_2O_2	OsO_4, V_2O_5, Fe or Cu salts	British 540,370 (10/15/41)	Dutch Shell	33
XVI	Allyl alcohol	H_2O_2	Tungstic catalysts/7 days → 78.9% yield	U.S. 2,373,942 (4/17/45)	Shell Dev. Co.	34
Fringe technology for integrated program	Hydrocarbons → peroxides	O_2	Hydrocarbons → H_2O_2 440–500°C; organic peroxides at 325–400°C	U.S. 2,376,257 (5/15/45)	Shell Dev. Co.	35
Fringe technology	Propylene → acrolein	O_2	0.4% Cu_2O on SiC/4–8% by vol. diluent gases, 1–10 atm, 368°C, 0.1–2 s contact time: 65% yield acrolein	U.S. 2,451,485 (10/19/48) See also U.S. 2,486,842 (11/1/49) U.S. 2,606,932 (8/12/52)	Shell Dev. Co.	36 37 38
Fringe technology	CH_2=CHCHO → CH_2=CHCH_2OH	C_2H_5OH	MgO, ZnO (4:1)/vapor phase molar ratio 2.3:1; rate 0.099 mol/100 ml cat/min, 311–401°C; 77% yield	British 619,014 (3/2/49)	Dutch Shell	39
Fringe technology	Olefins	Peroxides	Organic sulfonic acids catalyze polyhydroxy compound formation	U.S. 2,731,502 (1/17/56)	Shell Dev. Co.	40

	Substrate	Oxidant	Conditions/Catalyst	Patent	Company	Ref.
XVI	Allyl alcohol	H_2O_2 (anhy.)	Oxides of Os, Ti, Zr, V, Nb, Cr, Mo, W, U, Ta, Ru; pref. OsO_4, RuO_4, V_2O_5, Mo oxide, or CrO_3/unreactive organic solvents	U.S. 2,414,385 (1/14/47)	Research Corp., NY	41
XVI	Allyl alcohol	H_2O_2	HCOOH	U.S. 2,500,599 (3/14/50)	Shell Dev. Corp.	42
XVI	Allyl alcohol	H_2O_2	MoO_3 or $WO_2(H_2SO_4)$ 70–80°C	British 654,764 (6/27/51)	Distillers Co.	43
XVI	Allyl alcohol	H_2O_2	Tungstic acid	British 725,375 (3/2/55)	Dutch Shell	44
XVI	Allyl alcohol	H_2O_2	Unstable peroxy acid-forming metal oxide catalysts; i.e., oxides of W, Os, Mo or V. 40–60°, then 60–100°C	British 730,431 (5/15/55)	Dutch Shell	45
XVI	Allyl alcohol	H_2O_2	OsO_4, +Mn or Ce cpd; 30°C, 5 h, distillation: 74% (based on allyl alc.) glycerol yield	German 907,774 (3/24/54)	Badische Anilin & Soda Fabrik Akt-Ges	46
(XVI)	Allyl alcohol	HCOOOH	Three steps (thru formate) 1. +$HCOOH(H_2O_2)$ 2. Dehydrating 3. Further hydroxylation, based on allyl alcohol is 94%	U.S. 2,739,173 (3/20/56)	Allied Chem & Dye Corp.	47

Table 4.6 (*Continued*)

Processing steps[a]	Raw material	Reagent	Catalyst conditions	Patent no.; Date	Assignee	Reference
XVI	Allyl alcohol	aq. H_2O_2	Heteropolyacid cat. ctg Co, Mo, W, and S, Se, Te/w. or without solvent (selenotungstic acid) 30°C +34% aq. H_2O_2 2 h, −50°C 1 h−70°C → 89.9% yield	U.S. 2,754,325 (3/20/56)	Shell Dev. Corp.	48
XVI	Allyl alcohol	H_2O_2	HCOOH+H_2O_2 75–80% yield by distillation	Dutch 83,175 (11/15/56)	Dutch Shell	49
XII	Acrolein (A)	H_2O_2(H_2O)	OsO_4, 5 mol/L. H_2O, add cat as 2% soln +0.88 m. aq. H_2O_2/mol A dropwise, 25–30, → 77.5% glyceraldehyde	U.S. 2,718,529 (9/20/55)	Shell Dev. Co.	50
XVI	Allyl alcohol	H_2O_2	H_2WO_4 cat.	U.S.S.R. 107,762 (10/25/57)	Sargeav, etal.	51
Several steps for recycle study	Allyl alcohol (B)	several	84% yield (USP) based on (B).	British 889,613 (2/21/62)	Olin Mathieson	52
α-hydroxyalde-hyde to polyol	Glyceraldehyde isopropyl hemiacetal	H_2	Raney nickel hydrogenation of hemiacetal, 20–300°, 50–5000 psig to glycerol.	U.S. 3,168,579 (2/2/65)	Shell Oil Co.	53
XVI	Allyl alcohol	H_2O_2	NaHWO$_4$; 2 stages	Polish 48,251 (5/11/64)	Instytut Ciezkiej Organicnej	54

XVI	Allyl alcohol	H_2O_2	Tungstic acid; 2 stages 1. $=>$ 40°C 2. $=>$ 70°C	U.S.S.R. 166,009 (11/10/64)	W.G. Markina, et al.	55
XIV–XV	Allyl alcohol	H_2O_2	Tungstic acid 45°C, 7 h; Two steps 27% glycidol is hydrolyzed in 87.3% yd. (based on H_2O_2)	Netherlands Appl. 6,408,089 (1/18/65)	Chemische Werke Huels A.-G.	56
(VI–VII)–VIII– XVI.	Propylene		1. Epoxidized w. org. peroxide using Ti, V, Mb, Se, Mo, W, or Re cats. 2. Isomerization to allyl alcohol 3. Dihydroxylate to glycerol	Netherlands Appl. 6,513,621 (5/10/66)	Halcon Intl Inc.	57
IX–X	Allyl alcohol	1. HOCl 2. H_2O, NaOH/Na_2CO_3	Simult. prodn of chlorohydrin & glycerol IX:25% aq. allyl alc. 15°C, X: alk. hydrolysis NaOH/Na_2CO_3(H_2O) of mixed chloro cpds 10:1 NaOH/Na_2CO_3, 100°C	French 1,484,819 (6/16/67)	Romano Minstry of Chem. Ind.	58
XIV–XV	Allyl alcohol	XIV: CH_3COOOH XV: (H_2O)	XIV: 0.7 m allyl alcohol +5 m HOOAc at 25–100°C in Me_2CO, 2 h at 69°C → 94.5% glycidol (from allyl alcohol 92.7%) XV: 72 pts. glycidol, 4 h at 60°C w. 1 pt. HCOOH & 432 parts H_2O → 96% glycerol	French 1,509,277 (1/12/68)	W.C. Fisher, FMC Corp.	59

Table 4.6 *(Continued)*

Processing steps[a]	Raw material	Reagent	Catalyst conditions	Patent no.; Date	Assignee	Reference
XVI	Allyl alcohol	CH_3COOOH	Peracid in org. solv. [Me_2CO], 50–70°C, ion exchange & C cleanup	French 1,509,278 (1/12/68)	Liao, et al. 70, FMC Corp.	60
XIV–XV	Allyl alcohol	cumene hydroperoxide	Pr vanadate; yield=99.9%	French 1,548,678 (12/6/68)	J. Barthoux, et al., PROGIL.S.A.	61
XVI	Allyl alcohol	$H_2O_2(H_2O)$	Tungstic acid 3 h at 40°C w. 30% H_2O_2	Polish 57,650 (7/30/69)	T. Beres, et al Inst. Ciezkiej Syntezy Organicznej	62
XVI	Allyl alcohol	$AcOH + O_2$	Pd cpds; product is glycerol acetate	Japan 71, 42,884 (12/18/71)	Kuraray	63
XV	Glycidol	H_2O	Hydrolyze 40–150°C w. alkali metal bicarbonates	U.S.S.R. 236,454 (4/11/72)	Mendeleev Chem-Technol. Inst. Moscow	64
XV	Glycidol	H_2O	Basic catalyst; 200°C, 6–12 atm CO_2	U.S.S.R. 322,973 (8/15/72)	N.N. Lebedev, et al Mendeleev, Chem-Technol. Inst. Moscow	65
IV–XV	Allyl alcohol	1. Ethylbenzene-hydroperoxide 2. basic alkalies	1. 0.4% V napthenate 2. Hydrolysis	Netherlands Appl. 75 (7/31/75)	Halcon Intl Inc.	66
IV–XV	Allyl alcohol	1. AcOOH, 5–40% 2. Aq. alkalis	1. Epoxide in diisobutyl ketone with HOAc removal. 2. Hydrolysis	U.S. 3,954,815 (5/4/76)	W.C. Fisher FMC Corp.	67

IV–XV	Allyl alcohol	1. Organic hydroperoxide 2. Styrene divinyl-benzene resin	1. Oxidation: Gp V–VIII metal cats.; 2 stages 2. Hydrolize with macro-porous vinylbenzene resin.	U.S.S.R. 360,336 (11/18/72)	M.I. Farberov, et al. Yaroslavl Technol. Inst.	68
IX	Allyl alcohol	NaOCl	OsO_4 oxidization hypochlorite in Me_3COH: 97.8 yield	U.S. 3,846,478 (11/5/74)	R.W. Cummins, FMC Corp.	69
XIV–XV	Allyl alcohol		Continuous manufacture of glycerol 1. AcOOH in EtOAc solv. 2. hydration recovers allyl alcohol and H_2O	Japan Kokai 77 33,705 (8/6/77)	Y. Kinoto et al., Daicel Ltd.	70
XV	Glycidol		W. sulfopolystyrene cat. ionic exchange, 90–95% conv. at 80°C	No patent	—	71
XVI	Allyl alcohol	OsO_4	OsO_4 hydroxylation with Me_3COOH at 25–55°C, pH>8	U.S. 4,049,724 (9/20/77)	Atlantic Richfield Co.	72
XVI	Allyl alcohol	organic peroxide	Vanadyl sulfate; org. hydroperoxide (et benzene hydroperoxide). 2 h, 90°C, yield=89.3%	Japan KTK JP 82 77,636 (5/15/82)	Sumitomo Chem. Co., Ltd.	73
Epoxide	Glycidol	H_2O	basic N-containing catalyst (anthranilic acid) 100°C/1 h	Japan KTK JP 86, 271,230	Matsui Toatsu Chemicals Inc.	74

Table 4.7 includes a technology different from that contained in Tables 4.5 and 4.6. Finally, Table 4.8 summarizes miscellaneous, patented, international technology used to upgrade or purify various glycerol mixtures and prepare or regenerate catalysts and other incidental technology.

1. The "Epichlorohydin Route" Process Description

Early reports in 1948–1949 described the original Shell Deer Park, Texas plant and process [2–5]. Faith, Keyes, and Clark [6] also detailed the original Shell process in 1964. Yerman (Dow) briefly described the Dow epichlorohydrin process in 1974 in Considine's *Chemical and Process Technology Encyclopedia* [7]. Other reports [8–10] illustrate the variability in some of the processing steps.

Chemical Reactions:

$$CH_2 = CHCH_3 + Cl_2 \rightarrow CH_2 = CHCH_2Cl + HCl$$
$$\text{Allyl chloride}$$

$$CH_2 = CHCH_2Cl + HOCl \rightarrow CH_2OH \cdot CHCl \cdot CH_2Cl$$ 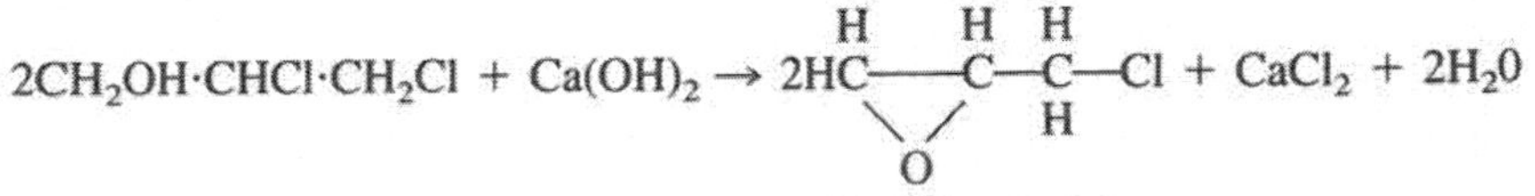
$$\text{Dichlorohydrin}$$

$$2CH_2OH \cdot CHCl \cdot CH_2Cl + Ca(OH)_2 \rightarrow 2HC\underset{O}{\overset{H}{\diagdown}}\overset{H}{\underset{/}{C}}\overset{H}{\underset{H}{C}}Cl + CaCl_2 + 2H_2O$$
$$\text{Epichlorohydrin}$$

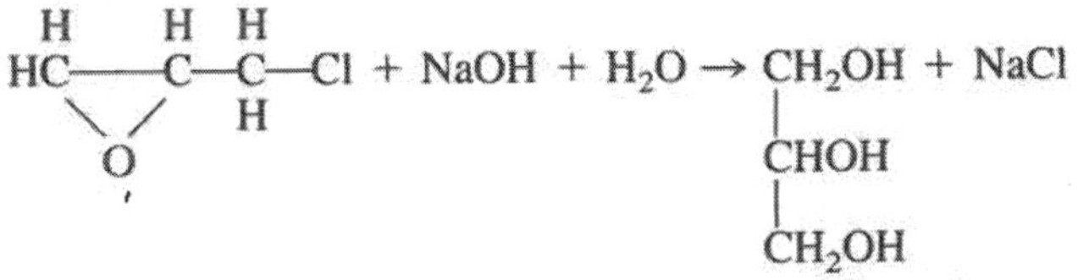

$$HC\underset{O}{\overset{H}{\diagdown}}\overset{H}{\underset{/}{C}}\overset{H}{\underset{H}{C}}Cl + NaOH + H_2O \rightarrow \begin{array}{c} CH_2OH \\ | \\ CHOH \\ | \\ CH_2OH \end{array} + NaCl$$
$$\text{Glycerol}$$
$$\text{75–80\% yield}$$

Material Requirements:

Basis—1 ton glycerol (99%)

Propylene	1,250 lb	Sodium Hydroxide	900 lb
Chlorine	4,000 lb	Hydrated lime	900 lb

Process

High-temperature chlorination of propylene gas is accomplished by preheating to about 400°C and passing previously vaporized liquid chlorine through a water heater, at a ratio of 4:1 propylene:chlorine. The inlet chlorine jets

Table 4.7 Patented, International Technology for Synthetic Glycerol Manufacture by Miscellaneous Routes

Raw material	Reagent	Catalyst/conditions	Patent no. (date)	Assignee	Reference
Carbon monoxide	H_2	Cobalt acetate or 7 others; "oxo" synthesis >150°C, 2M–5M atm	British 655,237 (7/11/51)	DuPont de Nemours	75
Formaldehyde and CO	H_2	Co cpds; 500–700 atm, yields ethylene glycol and glycerol, low yield	U.S. 2,605,333 (10/12/48)	DuPont de Nemours	76
Glyceraldehyde	H_2	$[HOCH_2CHOHCHO \rightarrow HOCH_2CHOHCH_2OH]$ Raney nickel; 2 reactors in series, 90°C, efficient catalyst usage.	French 1,335,313 (8/10/63)	Dutch Shell	77
Selected organic compounds	—	Electrolytic cell for oxid'n or redn of org. cpds	U.S. 3,119,760 (1/28/64)	Standard Oil Co. (Ohio)	78
Hydroxylation of hydrocarbons	alkalies	15–50% Ni, & 55–50% r-Al_2O_3 180–220°C, 100–50 atm	U.S.S.R. 436,811 (2/25/74)	T.I. Poletaeva, et al.	79
Propylene oxide[a]		Cats ctg. NH_2 and COOH groups; anthranilic acid in H_2O, 160 l h, 82.8:15.6:1.6 of propylene, dipropylene, tripropylene glycols, resp. 99.8% conv.	Japan KTK JP 86, 271,229 (12/1/86)	Matsui Toatsu Chemicals Inc.	80
DL-glyceraldehyde	—	Electroreduction	No patent		81

[a]Potentially applicable to glycidol conversion to glycerol, but no data.

Table 4.8 Separating Intermediates, Catalyst Preparation and Regeneration in Synthetic Glycerol Manufacture

Subject	Conditions	Patent no. (date)	Assignee	Reference
Upgrading technical process glycerol	Heated w. ammonium molybdate and vanadate to 65°C, 3 h at 90°, rectified 70–90 mm, 180°C, treated with C	German 1,156,773 (11/7/63)	Halcon Intl Inc.	82
Recovery and regeneration of OsO_4 catalyst	Removal of trace volatile organics, H_2O_2 treatment to convert OsO_2 to OsO_4, distillation of OsO_4.	U.S. 2,773,101 (12/4/56)	Shell Dev. Co.	83
Epoxyallyl cpds	Allyl cpds + aldehyde, nitrile, and O-ctg. gas heated for acylperoxy or peracid → epoxyallyl cpds	Canada 940,935 (1/29/74)	Union Carbide Corp.	84
Per-p-toluic acid catalyst preparation	p-MeC$_6$H$_4$CHO + air in anhyd. acetone 28–30°C/30 kg/cm^2 with 81 mol% peracid, used for propene to propylene oxide or allyl alcohol to glycidol (→ glycerol)	German 2,515,033 (10/1/75)	Mitsubishi	85
Removing epichlorohydrin from gases and wastewaters	4–6 h extract w. Na$_2$CO$_3$ soln → glycerol	Czechoslovakia CS 220,530 (11/15/85)	J. Hires, et al.	86
Separation of epichlorohydrin and chloropropanes	Separation of five byproducts from manufacture of glycerol from allyl chloride and apparatus therein described	Romania 64,299 (12/15/78)	"Petrochim"	87

receive the halogen at 60 psi and 20°C. The propylene/chlorine mixture reacts in a steel tube type reactor. The high exothermic nature of the chlorination requires insulation of the reactor for temperature control; ordinarily the reactants are between 400–500°C for an estimated time of only 2–3 seconds at 15 psig. Figure 4.2 shows a schematic outline of the chlorination and the succeeding process steps. Chlorine utilization is over 99%. Two reactors are run alternately to afford an opportunity for cleanouts, which are required about twice a month. Although the chemical reaction shows only allyl chloride as the product, this amounts to only about 80% of the chlorinated products. There are also chloropropenes, mixed dichlorides, trichlorides, and heavier residues. The largest of the organic byproducts consists of about 15% *cis-* and *trans-*1,3-dichloropropenes and 1,2-dichloropropane. Hydrogen chloride is the largest of the inorganic byproducts.

Distillation is used to separate the series of byproducts and allyl chloride. Unreacted propylene and hydrogen chloride are readily volatile; the former is returned to the propylene supply after passing through the HCl absorber and scrubber and being compressed, the latter is absorbed in water and recovered eventually as commercial 32% muriatic (hydrochloric) acid.

Crude allyl chloride is separated from the other organic materials by successive fractionation with two fractionating columns hooked up in series. The vapor feed is supplied to the middle plate of the first column, its high boiler cut feeds the center plate of the second column. Chloropropenes are removed as a fraction from the top of the first column, dichloropropanes as bottoms off the second column. Based on propylene used, the yield of allyl chloride approximates 85%.

Allyl chloride is fed from the top of the second column to the hypochlorination reactor as a condensed liquid. It is then reacted with freshly prepared hypochlorous acid (from chlorine and water) in a slurry tank reactor with stirring at 85–100°F. The product is a mixture mostly of dichlorohydrins. Reaction occurs largely in the aqueous phase with allyl chloride concentration in the tank kept to a minimum to avoid side reactions. The aqueous reaction phase is recycled back to the hypochlorous acid reactor. Milk of lime, reacted at below 140°F, converts the organic phase over to epichlorohydrin. Steam distillation removes epichlorohydrin and a small amount of byproduct dichlorohydrins as water azeotropes and the epichlorohydrin is upgraded to 98% pure product by distillation.

Finally, 10% aqueous caustic soda converts epichlorohydrin at 150°C in 30 minutes in almost theoretical yield. The product, aqueous glycerol, contains salt and is concentrated by evaporation and then conventional distillation.

The foregoing description applied to the early synthetic glycerol processing as originally practiced at Shell's Deer Park plant. In the course of the

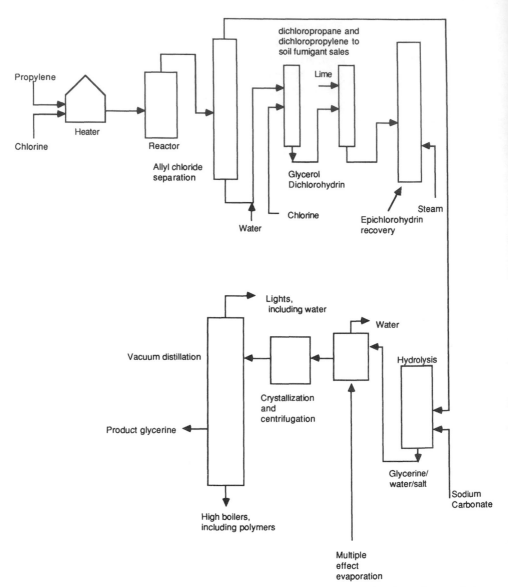

Figure 4.2 Process schematic for epichlorohydrin process for synthetic glycerol manufacture. (Shell's Deer Park, TX and Dow's Freeport, TX processes.)

operation some improvements resulted in changes in recycle procedures. Shell's Rotterdam plant also operates an innovation of the described process. Dow Chemical's plant process operating Freeport, Texas differs only slightly from Shell's; nor can the European Dow Chemical plant in West Germany be assumed to operate identical to the American plant.

2. The Nonhalogen Route from Propylene via Acrolein

Chemical Reactions:

$$CH_2 = CHCH_3 + O_2 \rightarrow CH_2 = CH - CHO + H_2O$$
$$\text{Acrolein}$$

$$CH_2 = CH - CHO + CH_3CH(OH)CH_3 \rightarrow CH_2 = CHCH_2OH + CH_3COCH_3$$
$$\text{Isopropanol} \qquad \text{Allyl alcohol} \qquad \text{Acetone}$$

$$CH_2 = CHCH_2OH + H_2O_2 \rightarrow CH_2OH - CHOH - CH_2OH$$
$$\text{Glycerol}$$
$$50\% \text{ yield}$$

Material Requirements:

Basis—1 ton glycerine (99%)
(plus 1,980 lbs. acetone)

Propylene	1,850 lbs.	Isopropanol	2,200 lbs.
Oxygen	460 lbs.	Hydrogen peroxide (100%)	970 lbs.

Process

Propylene and steam, with the latter in slight excess, are mixed with 25% oxygen based on weight of propylene and pumped to a reactor. Twenty percent of the propylene is oxidized to acrolein over a supported copper oxide catalyst. At a residence time of only 0.8 seconds at 350°C/2 atm pressure, the yield is 85%. The reaction mixture is cooled and distilled to separate by distillation.

The purified and distilled acrolein at 400°C in the vapor phase is isomerized by reaction with isopropanol (2 to 6 mol) over a catalyst of uncalcined MgO and ZnO. Product allyl alcohol and acetone are separable by distillation; yield of the former is 77% based on acrolein charged.

The distilled allyl alcohol is oxidized with a 2 M aqueous solution of H_2O_2 containing 0.2% tungstic oxide. The glycerol water mixture, which is generated within 2 hr reaction time at 60–70°C is distilled to afford high-purity glycerol. The filtered catalyst is recycled. Yield of glycerol (based on allyl alcohol) is 80–90%; the overall yield of glycerol based on propylene is about 50%. Isopropanol and hydrogen peroxide auxiliary raw materials can be produced from propylene; acetone is highly marketable.

There are a number of uncertainties as to the precise steps which were involved in Shell's integrated Norco, Louisiana plant process. Considerable speculation in the chemical press always attends those instances in which large chemical corporations do not disclose their technology. One evaluation published in 1956 [11] postulates that Shell's U.S. patent 2,731,502 (1/17/56) affords it the necessary final step in an integrated overall program that included improved synthetic glycerol manufacture among other products manufactured in large volume (Fig. 4.3). This patent describes a possible processing sequence that could involve the following:

Propylene to isopropanol—by conventional addition of sulfuric acid to form the sulfate, followed by hydrolysis

Propylene to acrolein—direct air oxidation at 200–400°C yielding 86% acrolein; U.S. patents 2,451,485 (10/19/48); 2,486,842 (11/1/49); 2,606,932 (8/11/52)

Isopropanol to acetone and hydrogen peroxide—liquid phase oxidation of isopropanol with air to give acetone and hydrogen peroxide; U.S. patent 2,376,251 (5/15/45)

Acrolein isomerized to allyl alcohol—yields acetone and allyl alcohol; British patent 619,014 (3/2/49)

Allyl alcohol to glycerol—organic sulfonic acids, particularly alkylaryl sulfonates like *p*-toluenesulfonic acid, promote hydroxylation; U.S. patent 2,731,501 (1/17/56) [These are readily available, economical, simple to use and presumably, eliminate the need to recover catalyst.

3. The Propylene Oxide Route To Glycerol (U.S. Defunct Chlorination of Allyl Alcohol Route)

Olin Mathieson Chemical Corporation combined isomerization of propylene oxide to allyl alcohol with a hypochlorination step in its overall route to synthetic glycerol at its Brandenburg, Kentucky Doe Run plant from late 1961 to 1964; British patent 940,284 (10/30/63).

Chemical Reactions:

$$CH_3CH\overset{O}{\overbrace{}}CH_2 \xrightarrow{\text{isom}} CH_2 = CHCH_2OH$$

$$CH_2 = CHCH_2OH + HOCl \rightarrow \begin{array}{c} HOCH_2CHOHCH_2Cl \\ \text{or} \\ HOCH_2CHClCH_2OH \end{array}$$

$$\begin{array}{c} HOCH_2CHOHCH_2Cl \\ \text{or} \\ HOCH_2CHClCH_2OH \end{array} + H_2O + \tfrac{1}{2}Na_2CO_3 \rightarrow \begin{array}{c} HOCH_2 - CHOH - CH_2OH \\ + NaCl + \tfrac{1}{2}CO_2 + \tfrac{1}{2}H_2O \end{array}$$

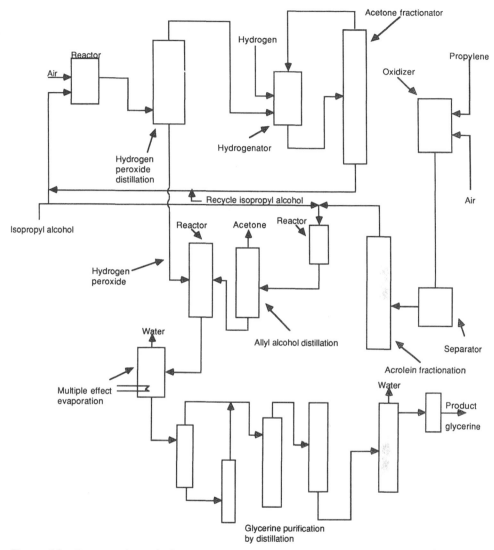

Figure 4.3 Process schematic for synthetic glycerol based on allyl alcohol. (Shell's Norco, LA process).

Process

Figure 4.4 is a schematic diagram of this process route. The isomerization of propylene oxide to allyl alcohol was a catalytic method using lithium phosphate on an inert support. The allyl alcohol was hypochlorinated with hypo-

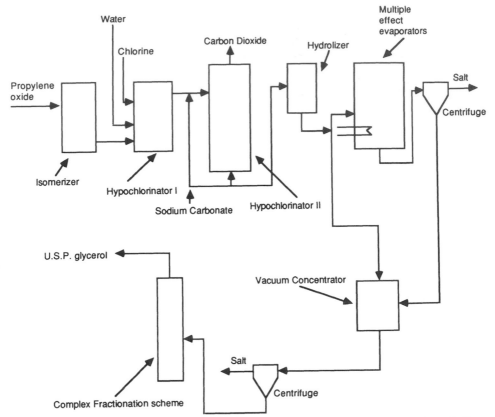

Figure 4.4 Chlorination of allyl chloride route for synthetic glycerol. (Olin-Mathieson's Bran-denburg, KY process.)

chlorous acid to mixed monochlorohydrins, and these were hydrolyzed in aqueous sodium carbonate to glycerol.

4. Allied Chemical and Dye Corporation's Route from Allyl Alcohol by Epoxidation, Dehydration, and Further Hydroxylation

Allied developed technology in the 1950s, not commercialized in the United States, which was essentially a two-step route from allyl alcohol, [U.S. patent 2,739,173 (3/20/56)].

Chemical Reactions:

1A $CH_2 = CHCH_2OH + HCOOH + H_2O_2 \rightarrow HOCH_2 - CHOH - CH_2OCOH + H_2O$

1B $CH_2 = CHCH_2OH + 2HCOOH + H_2O_2 \rightarrow HCOOCH_2 - CHOH - CH_2OCOH + 2H_2O$

2A $HOCH_2CHOHCH_2OCOH + CH_3OH \rightarrow HOCH_2 - CHOH - CH_2OH + HCOOCH_3$

2B $HCOOCH_2CHOHCH_2OCOH + 2CH_3OH \rightarrow HOCH_2 - CHOH - CH_2OH + 2HCOOCH_3$

Process:

Figure 4.5 is a schematic of the process for synthetic glycerol from allyl alcohol by epoxidation, dehydration, and further hydroxylation. Glycerol is synthesized from allyl alcohol (I) by admixing it with HCO_2H (II) and H_2O_2, thereby effecting hydroxylation of I to produce glycerol formate (III), distill-

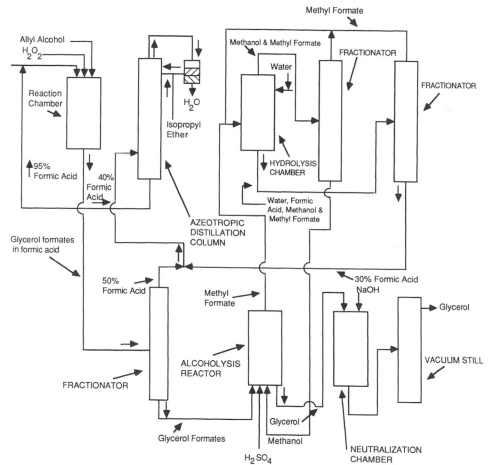

Figure 4.5 Process for synthetic glycerol from allyl alcohol by epoxidation, dehydration and further hydroxylation (Allied Chem. & Dye Corp.)

ing an aqueous solution of II from the mixture, dehydrating the dilute II solution by azetropic distillation, recycling the concentrated II produced to effect further hydroxylation of I, adding MeOH and a small amount of acid (IV) with the reaction product containing III to effect alcoholysis of the III, neutralizing the glycerol thus produced, vacuum distilling the neutralized glycerol, hydrolyzing the HCO$_2$Me (V) to MeOH and II, and returning the MeOH to the alcoholysis stage and the II to the azeotropic distillation stage. Thus, a charge of 46 parts by weight I and 555 parts 98% II in a vessel equipped with a stirrer is treated intermittently, over a period of 15 minutes, with 89 parts by weight of 30% H$_2$O$_2$, and the mixture agitated; the exothermic reaction raises the temperature to about 46°C, where the temperature is maintained by passing water through a cooling coil immersed in the liquid mixture. After 70 minutes the mixture is distilled at no higher than 150°C to vaporize an aqueous II solution of about 50% concentration. The bottoms containing III are then admixed with 120 parts by weight MeOH and 2 parts by weight of 66 Be' H$_2$SO$_4$ (IV) and the mixture, in a distillation column with reflux, is heated to distill the V formed, NaOH is added to the distillation bottoms to neutralize the small amount of acid present, and neutralized bottoms are heated to distill off the excess MeOH. The separated V, together with added water and a trace of IV is heated to 105°C to hydrolyze it to MeOH and I, the MeOH separated from II, and the latter dehydrated by azeotropic distillation with (Me$_2$CH)$_2$O to produce concentrated II, thus making the MeOH and concentrated II available for use in further alcoholysis of III and hydroxylation of I, respectively. Dilute II recovered in the forepart of the operation is also dehydrated by azeotropic distillation with (Me$_2$CH)$_2$O and the concentrated II is recycled for further hydroxylation of I. After alcoholysis of III with MeOH and removal of V and excess MeOH from the mixture, the bottoms containing glycerol are distilled in vacuo to produce substantially pure glycerol. The yield of glycerol, based on the I charged, is approximately 94%.

5. The Propylene Oxide Route to Glycerol (U.S. Defunct Allyl Alcohol-Peracetic Acid Route)

This route was used by FMC Corporation at Bayport, Texas between 1969 and 1982 (Fig. 4.6). Obviously, the location was selected mainly because of the indirect availability of low cost Gulf coast propylene for propylene oxide manufacture (available from the adjacent Oxirane Corporation plant at Bayport). FMC Corporation's integrated program involved a plant to produce peracetic acid from acetaldehyde (available from Celanese Corp. of America Chemical Division Bayport facility) and an associated plant to use peracetic acid to produce epoxidized soybean oil. Some new international synthetic glycerol plants employ the basic three-step scheme.

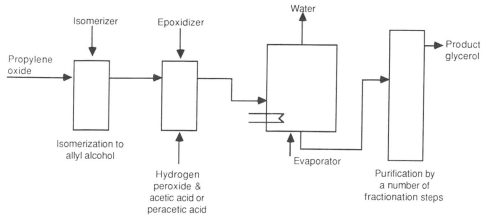

Figure 4.6 Allyl alcohol-peracetic acid route for synthetic glycerol. (FMC Corp. Bayport, TX process.)

Chemical Reactions:

$$CH_3CH\overset{O}{\overbrace{}}CH_2 \xrightarrow{\text{vapor phase}} CH_2 = CHCH_2OH$$

$$CH_2 = CHCH_2OH + CH_3COOOH \rightarrow CH_2\overset{O}{\overbrace{}}CHCH_2OH + CH_3COOH$$

$$CH_2\overset{O}{\overbrace{}}CHCH_2OH + H_2O \rightarrow HOCH_2 - CHOH - CH_2OH$$

Process

Propylene oxide was isomerized over a lithium phosphate catalyst deposited on an inert support to allyl alcohol. The isomerization occurs at 280°C and a space velocity of 1600 reciprocal hours. Yield of allyl alcohol is 94–98% at a conversion of 20–30% per pass. Allyl alcohol, was epoxidized in aqueous media to glycidol with peracetic acid [French patents 1,501,277–8 (1/12/68) (aqueous epoxidations) although FMC Corp. has an alternate solvent epoxidation route U.S. patent 3,954,815 (5/4/76)] and this hydrolyzed to glycerol with aqueous formic acid [U.S. patent 1,509,277 (1/12/68)]. An alternative route involving chlorination of allyl alcohol to glycerol dichlorohydrins by liquid phase chlorination and thence to epichlorohydrin and glycerol, as practiced by Shell and Dow, was apparently considered uneconomical by FMC Corp.

II. MANUFACTURE FROM CARBOHYDRATE RAW MATERIALS

Over the years certain carbohydrate materials such as glucose, sucrose, molasses, starch, or cellulose which contain single or multiple C_3-linked structures, have served as potential sources for manufacture of glycerol. Despite the fact that a surprisingly extensive technical and patent literature on technology for both fermentation and hydrogenolysis processing has been available, no present commercial large-scale production of glycerol now utilizes either method. In the United States hydrogenolysis was the route used by Atlas Chemical Industries, Inc. at its Atlas Point Plant (New Castle, DE) for production of an estimated 25 $\overline{\text{M}}$ lbs./year volume during the period 1962–1969. At that time, this was estimated [1] as 7% of the total U.S. glycerol production for 1969. It is obvious that neither fermentation or hydrogenolysis can compete economically today with either synthetic glycerol production from raw materials such as propylene, acrolein, allyl alcohol, and other petrochemical feedstocks, or natural glycerol from fats and oils.

A. Hydrogenolysis of Various Carbohydrates

The chemical equations for hydrogenolysis of carbohydrates are not thoroughly descriptive of all the complex changes which occur during the reactions. Lenth and Du Puis [88] assume the equations for the principal reactions occurring during hydrogenolysis of dextrose to glycerol to be a two-step sequence

Possibly, glycerol may be reduced to propylene glycol, as follows

Alternatively, the substantial quantities of byproduct propylene glycol produced in most instances could be formed as follows

$$
\begin{array}{c}
\text{H}\ \ \text{H}\ \ \text{H}\ \ \text{H}\ \ \text{H}\ \ \text{H} \\
|\ \ \ |\ \ \ |\ \ \ |\ \ \ |\ \ \ | \\
\text{H}-\text{C}-\text{C}-\text{C}-\text{C}-\text{C}-\text{C}-\text{H} + 2\text{H}_2 \\
|\ \ \ |\ \ \ |\ \ \ |\ \ \ |\ \ \ | \\
\text{O}\ \ \text{O}\ \ \text{O}\ \ \text{O}\ \ \text{O}\ \ \text{O} \\
|\ \ \ |\ \ \ |\ \ \ |\ \ \ |\ \ \ | \\
\text{H}\ \ \text{H}\ \ \text{H}\ \ \text{H}\ \ \text{H}\ \ \text{H}
\end{array}
\rightarrow
\begin{array}{c}
\text{H}\ \ \text{H}\ \ \text{H}\ \ \text{H}\ \ \text{H}\ \ \text{H} \\
|\ \ \ |\ \ \ |\ \ \ |\ \ \ |\ \ \ | \\
\text{HC}-\text{C}-\text{C}-\text{C}-\text{C}-\text{C}-\text{H} + 2\text{H}_2\text{O} \\
|\ \ \ |\ \ \ |\ \ \ |\ \ \ |\ \ \ | \\
\text{O}\ \ \text{O}\ \ \text{H}\ \ \text{H}\ \ \text{O}\ \ \text{O} \\
|\ \ \ |\ \ \ \ \ \ \ \ \ \ \ |\ \ \ | \\
\text{H}\ \ \text{H}\ \ \ \ \ \ \ \ \ \ \ \text{H}\ \ \text{H}
\end{array}
$$

$$\downarrow \text{H}_2$$

$$
2\left[
\begin{array}{c}
\text{H}\ \ \text{H}\ \ \text{H} \\
|\ \ \ |\ \ \ | \\
\text{H}-\text{C}-\text{C}-\text{C}-\text{H} \\
|\ \ \ |\ \ \ | \\
\text{O}\ \ \text{O}\ \ \text{H} \\
|\ \ \ | \\
\text{H}\ \ \text{H}
\end{array}
\right]
$$

The process used by Atlas Chemical Industries, Inc. for the production of sorbitol and/or glycerol was a semisynthetic one, because, while the raw material could vary from a wide choice of suitably priced natural carbohydrates, the operation was degradative; that is it consisted of hydrolysis and hydrogenolysis. Atlas' two Belgian patents [89,90] outline how reducible sugars are hydrogenated in the presence of nickel catalyst and cracked in the presence of CaO, Ca(OH)$_2$, or CaCO$_3$ at 190–230°C in a continuous one-stage process. Products high in glycerol content are obtained. Thus, nickel on diatomaceous earth and a mixture of 0.5% CaO and 0.5% CaCO$_3$, by weight of glucose, were suspended in 50% aqueous glucose. The mixture was introduced into the first of four reactors kept at 200°C under a hydrogen pressure of 140 kg/cm^2 at 6 1./h, and hydrogen was introduced at 38.5 m^3/h to give a product at 77.8% conversion and 51% scission. Alternatively, reducible sugars could be treated with hydrogen in a continuous process in described equipment at 210–230°C under $\geqq$ 35 kg/cm^2 in the presence of a Ni and a Ca salt such as CaO, Ca(OH)$_2$, CaCO$_3$, or Ca(OAc)$_2$, the glycerol and glycol products removed, and the unconverted starting material could be hydrogenated in the presence of a Ni-Ca salt catalyst mixture.

Atlas' U.S. patent [91] describes a glucose to glycerol conversion which essentially was the basis for the now defunct Delaware operation. Glucose solution and hydrogen are fed to the first of a series of four pressure reactors in the presence of catalyst and cracking additive. The catalyst is enhanced nickel containing small amounts of Cu and Fe promoters and is used so as to furnish 0.5–4% Ni based on the sugar weight. The cracking additive is an alkaline earth oxide, hydroxide, or weak acid salt at 0.25–1.0% CaO equivalent based on sugar weight. The reaction is carried out at 190–230°C and at pressures >500 psi (usually 1500–2000 psi). The equipment is described [92]. Previous processes have employed two distinct steps to convert reducing sugars to glycerol. Thus (all percentage volumes based on glucose) 50% aqueous glucose solution was made into a slurry with a reduced Fe-promoted Ni catalyst supported on diatomaceous earth. The amount used gave 1.5% Ni, and 0.5% CaO and 0.05% $CaCO_3$ were added as cracking additives. At a rate of 6 l./h, this slurry was fed to the first of four cylindrical reactors each being 35/16 in inner diameter and 6 ft high. Each reactor temperature was at 200°C, and the hydrogen pressure was 2000 psig. To insure intimate mixing of hydrogen and feed, hydrogen was continuously fed at a rate of 1360 ft^3 (STP) per hour, into the reactor through a distributing nozzle. The product from the third reactor had a conversion of 66.2% and 0.60 product split. Product split is the percentage glycerol in product divided by the percentage of all components which are not higher-boiling than glycerol. The product from the fourth reactor had a conversion of 77.8% and a product split of 0.51. The process technology was very flexible. Atlas made its own hydrogen for the process by conventional reforming of natural gas. Makeup and recycle hydrogen were combined, compressed, and fed to a reactor. Carbohydrates along with a proprietary nickel catalyst were fed to the reactor as a slurry in water. A small amount of heat was evolved during the hydrogenation. Temperature and pressure in the reactor affect the product mix. After unreacted hydrogen was vented, the product mix was cooled and purified by a complex series of steps. Purification depended somewhat on the market for the products, but in all cases, the water content of the glycerol was minimized by distillation and evaporation. If the glycerol required a high purity, a series of distillations was used typical of synthetic or natural glycerol production. Figure 4.7 shows a simplified schematic of the operation.

Atlas [93] describes a tungsten oxide-promoted nickel catalyst for use with polysaccharides such as sugar for the production of polyols. Polysaccharides underwent hydrolysis at 140–80°C and hydrogenolysis at 210–235°C prior to hydrogenation over the same catalyst. It is doubtful if this catalyst was employed during the life of the New Castle, Delaware plant.

Internationally, several other organizations developed processing methods variably suitable for the hydrogenolysis of sugars and related materials. Table 4.9 outlines representative technology.

The limitations of hydrogenolysis of carbohydrates for the commercial production of glycerol have now been appreciated. The variety of reactions which occur, namely, cracking, isomerization, acid formation, hydrogenation, dehydrogenation, and dehydration have been characterized [166] and, it is ap-

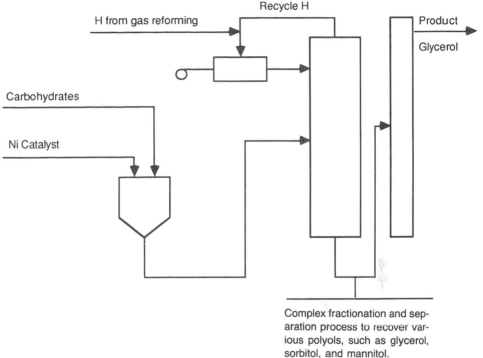

Figure 4.7 Atlas made glycerol by hydrogenating carbohydrates. Atlas process at Atlas Point plant, New Castle, DE.

parent that overall economics for this processing requires the development of markets for several byproducts, for example, propan-1,2-diol, ethylene glycol, tetritols, pentitols, methyl fructiside, and hexitols [167].

B. Fermentation Technology for Glycerol Production

Microbiological processes developed for glycerol production also suffered from a number of disadvantages: nonspecificity of bacteria used for glycerol formation, relatively low yields, multiplicity of byproducts formed in many instances (i.e., propylene glycol, ethylene glycol, acetaldehyde, and others), and difficulty of separating and purifying glycerol from the numerous byproducts and other organic matter contained in the crude reaction medium. Table 4.10 outlines the international technology at various times considered for the manufacture of glycerol from carbohydrate raw materials by fermentation. Few of these processes ever reached the pilot plant stage of development.

One technology approach developed through the continuous pilot plant stage in 1958–1960 by U.S. Forest Products Laboratory, USDA, was the

Table 4.9 International Technology for Glycerol Manufacture by Hydrogenolysis of Carbohydrates

Organization	Raw materials	Catalyst	Conditions	Yields	Reference
Commercial Solvents Corp., Terre Haute, IN	Sucrose, dextrose molasses	CuO/CaF$_2$, or Cu silicate	Sucrose-1800 lb/in^2 to 2000 psi	From sucrose: 40.8% propylene glycol From dextrose 34.3% PG and 24.4% glycerol	94
Hercules Powder Co., Wilmington, DE	"Sugars"	Raney Ni	500–10,000 psi		95
Robert R. Bottoms, Crestwood, KY	Xylitol, sorbitol	Ni with methylamine HCl or halogen acid salt	25–300 atm		96
Assoc. of American Soap & Glycerine Producers, Inc., New York, NY	(a) Sugar or molasses	Cu-Al oxide or Cu Ba chromite	900–400 psi with O$_2$ contact, 150–300°C		97
	(b) Dextrose	Cu aluminate	1700 psi with MeOH with Na$_2$CO$_3$, 250°C		98
Bombrini Parodi-Delfino	Glucose or other sugars	Cu and Ni or Cu and Co catalysts	<240°	Glycerol +propandiol	99
The Miner Laboratories, Chicago, IL	Sucrose or dextrose	Cu/Al oxides	240°C, 1500 psi, in MeOH suspension	From sucrose 25.1% glycerol and congeners	88
Inventa A.-G., Zurich, Switzerland	(a) Saccharose	Ni on kiesel-guhr, Ni/Al$_2$O$_3$, or RaNi	50–250 atm 180–250°C	From saccharose 45% glycols 34% glycerol	100

Source	Substrate	Catalyst	Conditions	Results	Ref.
	(b) Same	same	Improvement on above-continuous feed sucrose, recycle H	30% glycerol	101
Farbwerke Hoechst, Frankfurt-am-Mein, Germany	Sugar	Ni or Cu +catalyst poison on ZnO carrier	200 atm 108°C, 4 h then 130°C, 2 h	Glycols 10.5%; glycerol up to 80%; 68% the rest hexite 21.5% (sorbite, mannite pentites)	102
Hydrocarbon Research, New York, NY	Polysaccharides w. at least one D-glucose unit/corn starch	Solid particulate hydrogenated catalyst in expanded ebullate state	Pass upward thru catalyst bed w. H_2 gas, 200–500°F, 700–3500 psi	From corn starch: (wt. basis) glycerol 50, PG 20, MeOH, EtOH, and i-PrOH 20, others 10%	103–105
N.A. Vishyunina, et al. (Russ.)	Xylitol	Ni on kieselguhr	200 atm	Propyleneglycol 22%, ethylene glycol 26%, glycerol 5%, erythritol 5%, monoatomic alcohols 2.2%. glycerol yields as high as 40%	106
Kyowa Chemicals, Ltd., Japan	(a) Hexahydic alcohols: sorbitol	Ni on diatomaceous earth	H_2 and N_2 mixed gas; 230°C, 159 kg/lcm²	Glycerol yield optimized	107
	(b) Same	Same	Removal of EG, PG, hexitol, 1,2,4-butantriol from glycerol in series of fractionated dist/towers	Glycerol is 99.2% pure dist. yield is 90.8%	108

Table 4.9 (*Continued*)

Organization	Raw materials	Catalyst	Conditions	Yields	Reference
Inst. Org. Chem. (Russian)	Heterogeneous catalyst and alkali	Improvement, preheat soln at 75–175°C			109
N.A. Vasyunina, et al.	Carbohydrates	Catalysts and alkalis	Study of kinetics and mechanism		110
N.E. Vishnevskii, et al.	Sugars	Alkalis and catalysts	Stirring Reynolds crit. at 100,000–200,000; 100–150 atm, preheat to 150°C		111
E.F. Stefoglo, et al.	Glucose	Ni/kieselguhr +Ca(OH)$_2$	230°C, theo. approach to reactor design	Calcd theo. yields of glycerol 33% on wt. glucose	112
N.M. Sokolov, et al.	Sugar	Catalysts and alkalis	Improvement: fractionated dist. of glycerol, propylene glycol, and ethylene glycol		113
All-Union Sci.-Res. Institute of Protein Biosynthesis	Sugars		Improvement: use of high-speed stirring at 60,000 rpm		114
G.L. Grigor'ev, et al.	Glucose	Ni-Al,-Ca catalyst		Optimizes glycerol yields	115

Table 4.10 Glycerol From Carbohydrates By Fermentation Technology

Organization	Raw Material	Catalyst or Medium	Conditions	Yield	Reference
K. Lüdecke	Sugars from light-catalyzed condensation of CH_2O.				116
Norddeutsche Helfeinindustrie A.G., Germany	(a) Sucrose		(a) Neut. mash w. $MgCO_3$		117
	(b) Sucrose	(b) +3% NaCl w. Na_2SO_3, $NaHCO_3$ $(NH_4)_2SO_4$, $MgSO_4$ and yeast	(b) Multinutrient media: pH 7.2–7.5, 37°C		118
Soc. industrielle de nouveaux appareil (S.I.N.A), Paris, France	Fermented "must"	(a)Yeast & nutrients	(a) Lime treatment; recycle a portion		119
		(b) Vinasse treatment	(b) Upgrading to distd glycerol product		120
C.C. Mariller	Ferment vinasses		Uses concd alcohol as additive for inorg and organic sepn		121
Henkel & Cie, GmbH, Dusseldorf-Holthausen, Germany	Molasses and other ferments		Vinasse mixed w. kieselguhr; extr. w. acetone		122
Deutsche Gold-und Silber-Scheideanstalt vormas Roessler (to the Distillers Co. Ltd)	Distillery slop or fermented molasses	(Upgrading process)	Dry slop, add kieselguhr, then extracts G_3 with pyridine, pref. countercurrent		123

83

Table 4.10 (*Continued*)

Organization	Raw Material	Catalyst or Medium	Conditions	Yield	Reference
Hugo Haehn Finckenkrug, Germany	Mold yeast ferment. of sugars	Ox. catalysts Fe or Mn salts	Alkaline media with aeration	after 8 hrs. 21.5% Glyc. (wt/basis from sugar)	124, 125
Speas Development Co., Missouri, U.S.	Blackstrap molasses + yeast	Na_2SO_3	Interrupts fermentation to vac. distill off G_3; recharge and re-ferment		126
United States Ind. Chemicals, Inc., Delaware, U.S.	Low sugar Yeast mash	$(NH_4)_2SO_4$ w. aeration	Add molasses after 12–24 h, $+NH_3$ to pH 5.6–6.6+lime, total ferment time 60–70 h	5–7% EtOH, 2.6–3.4% glycerol	127
Aktieselskabet Dansk Gaerings-Industri	Sugar + yeast (25% wt. ferment sugar)		Weakly alkaline medium, pH 7–8; but 6–7 at end, 32°C	After 48 h, 12.1 wt% glycerol & 44.8% EtOH	128
I.G. Farbenind. A.-G. Frankfurt/Main, Germany	Sugar	Yeast + nutrients	Constant aeration w. sulfides, S, or polysulfides with fresh sugar added intermittently or continuously	Enhances glycerol yield	129
The Distillers Co., Ltd., Edinburgh, Scotland.	(Upgrading process)		Extract yeast ferment. residue with solvents, heat to eliminate H_2O, ppt. salts and filter		130

Organization	Substrate	Additive / Yeast	Conditions	Results	Ref.
Overly Biochemical Research Foundation New York, NY	Starch hydrolysate		Air sparge, continuous ne tr. with bases	Glycerol yield is 22%	131
Iowa State College Research Foundation Ames, IA	(a) Sugar or molasses (b) Sugar	With sli. sol. sulfites, acclimatized yeast +$(NH_4)_2SO_3$	<20% Sugar medium, pH<7, 30°C 30°C, pH 6–7, 20% sugar concn	72 H cycle time-CH_3CHO + glycerol	132 133
Sylvania Industrial Corp., Fredericksburg, VA	Purified molasses ctg 75–80% fermented sugars	Yeast, 15–20% sugar soln NH_4Cl or $(NH_4)_2SO_4$	pH 7.0–8.5 by addn of K_2CO_3+KOAc to neutral AcOH, Membrane purification by dialysis	4–6% Glycerol in broth	134
Neuberg & Roberts New York, NY	Glucose, sugar, or molasses,	Brewers yeast fermentation + 10% bakers yeast + Na_2SO_3 w. continuous CO_2 gas	pH 7.0 to 7.4 at end, 30–2°C, solvent extraction for glycerol recovery	49 h cycle-Wt. yields CH_3CHO 18.3%, EtOH 15.3%, glycerol 37%	135
United States Ind. Alcohol Co.	Sugars	Yeast	pH 6.4 to 7.0, by addn of NH_4 cpds as neutralizers	Optimizes glycerol yields	136
Imperial Chemical Industries, England	(Upgrading process)		Glycerol extracted w. H_2O-immiscible solvent like aniline, then G_3 extract from aniline with H_2O and concd		137

Table 4.10 *(Continued)*

Organization	Raw Material	Catalyst or Medium	Conditions	Yield	Reference
United States Alcohol Co.	Sugar solns	Yeast	Uninterrupted yeast ferment. w. simultaneous addn of NH_3 or NH_4 salts above that reqd for nutrient effects		138
Distiller's Co. Edinburgh, Scotland	Sugar solns	Yeast	Addition of NaOH to fermenting mash	Affords 40.6% EtOH and 8.12% glycerol (sugar bases)	139
Honorary Advisory Council For Scientific and Industrial Research, Ottawa, Canada	Nutrient sugar solutions	*B. subtilus*	5% Sugar soln, with $CaCO_3$, 37°C	Wt. yields: glycerol-29.4% 2,3-butandiol 28.5% lactic acid 11.6% EtOH 2.2%	140
C.C. Cornee	Sugar	Yeast	$+Na_2SO_3$, pH 8.0, 34–5°C	5 days 33–35% glycerol	141
S. Yasuda	Glucose	Yeast operating in high NaCl media	W. orthoclase, NaCl concn decreases with N containing salts + alk. salts		142
American Viscose Corporation, Wilmington, DE	(Upgrading process)		Filtration, acidification, volatilize off EtOH, ion exchange, decolorization with charcoal.		143

G.R. Laborusse	(Upgrading process)		After fermentation, neutralization, and vacuum steam distillation, alcohol extract glycerol		144
Glycerine Corp, of America, Inc. New York, NY	Molasses ctg. 50% fermenting sugars	Yeast fermentation	pH 7.5, 85–95°F, filter, dist. off alcohol, add extra molasses two more times, concn. to 20%, dialyze and distill	4,000 gal molasses → 3,000 lbs distd glycerol	145, 146
Backhefe GmbH. Frankfurt am Mein, Germany	Sugar		20–5°C, temp. increases in the late stages	Glycerol yields increased by 20–25% over "conventional".	147, 148
Miroluboff and J. Colpaert	(a) Ferment. malt		(a) Eliminate AcOH by cooling to ppt. salts, distillate and filtrate rectified together		149
	(b) Cellulose	(b) Nitrates, phosphates, dimethylcyclohexanedione	(b) Minimizes EtOH formation	(b) Glycerol + butanol	150
Sala and Furster	Starch hydrolyzate ctg. 6% reducible sugars	S. cerevisiae S. ellipsoideus	pH 6.5–8.0, supplement with N and salts		151
Imperial Chemical Industries, Ltd. London, England	(Upgrading process)		Remove volatile products after ferment. by sand filtr. and treatment with superheated steam	Glycerol 2.3-butandiol	152

Table 4.10 (*Continued*)

Organization	Raw Material	Catalyst or Medium	Conditions	Yield	Reference
Egerov, et al	Sugar ctg. solutions	"Fixes" alcohol intermediates w. hydroxylamine, hydrazine, or cpds w. a free amino-group.			153
Frankel and Goldheim	Pure sugar	$Na_2SO_3/NaHSO_3$	Evaporate off AcOH and EtOH, residue containing glycerol + Na_2SO_3 ion-exchanged yielding glycerol and H_2SO_3; elim. SO_2 but recover and reuse as $NaHSO_3$		154, 155
National Research Council Canada, Ottawa, Canada	Glucose	*S. rouxii* or *S. mellis* + corn steep liquor, urea or $(NH_4)_2SO_4$	Aerated in baffled fermenter	20% Concn glucose → 8% concn each D-arabitol and glycerol	156, 157
Dyr and Zvacek	Sugars	Yeast + sulfite	Regenerate with fresh sugar and yeast and continue; repeated 4–8 times		158

Organization	Substrate	Organism	Conditions	Products/Yield	Ref.
Commercial Solvents (Great Britain), Ltd. New Ferry, Cheshire, England	Beet molasses	*Saccharomycos rouxii* or *S. mellis*	35–7°C w. urea	8 Days=glycerol yd inc. from, 2.28 to 5.65% w. beet molasses	159
L. Basseguy	Various sugars, molasses, etc.	3% Dry beer yeast, vitamins B_1 & B_6, adenosine triphosphate, glutathione, amino acids, etc.	pH 6–7, fermentation time decreased 50%	Glycerol yields at 40–45% of sugars consumed	160
Noda Industrial Science Research Laboratory, Chiba-ken, Japan	glucose, fructose, mannose	*Pichia miso* & *Debaryomyces mogii*		Glycerol arabitol erythritol	161
A.L. Malchenko, et al.	beet sugar			5–10% Based on reducing sugars	162
Halcon International, Inc., DE	glycerol from molasses fermentation	(Upgrading process)	$+H_2O_2$ w. $(NH_4)_6MO_7O_{24} \cdot 6H_2O$ and NH_4VO_3, also Na tungstate and K_2CrO_4 used as catalysts		163
Hindustan Lever Research Center, India	molasses/corn steep liquor	(a) *A. anomala*			164
		(b) *Saccharomyces cerevisiae*	Na_2SO_3 w. continuous CO_2 sparge	Product concn of ≤ 230 g/L	165

"bisulfite-steered" process for glycerol and acetaldehyde production. The reported yields of products by this method were glycerol 25%, ethanol 17%, and acetaldehyde 11% [168–170]. At a 15¢/lb. price for sugar, the process was recommended as competitive, but difficulties in handling and recycling SO_2 and in product upgrading and separation of the coproducts, from which a 90% recovery of glycerol and 80% for acetaldehyde and ethanol was estimated, proved to be limiting.

REFERENCES

1. Thwaites, J. D. (Shell, Rotterdam), Chem. Ind. (London), 48(8–9): 508–509, 1969.
2. Anonymous, Glycerine by synthesis, Chem. Eng., 55(10): 100, 134–137, 1948.
3. Synthetic glycerine plant opened in Houston by Shell, Chem. Eng Progr. (AIChE), 44(10): 16, 1948.
4. Hightower, J. F., Glycerine from petroleum, Chem. Eng., 55(9): 96–98, 1948.
5. Kiddoo, G., Petrochemical processes (synthetic glycerine from propylene), Chem. Eng., 59(9): 164, 1952.
6. Faith, W. L., Keyes, D. B., and Clark, R. L., Industrial Chemicals, 3rd ed., John Wiley & Sons, New York, 1965 pp. 401–411.
7. Yerman, R. B. (Dow Chemicals). In Chemical and Process Technology Encyclopedia (D. J. Considine, ed.), McGraw Hill, New York, 1974, pp. 412–415.
8. Anonymous, Glycerol goes on a spree, Chem. Eng. News, Nov. 7, 1955, pp. 4734–4736.
9. Kern, J. C. In Encyclopedia of Chemical Technology, 2nd ed. (Kirk/Othmer, eds.), Wiley Interscience, New York, 1966, Vol. 10, pp. 619–631.
10. Anonymous, Glycerine supplies remain tight, Chem. Eng. News, July 31, 1967, pp. 48–50.
11. Anonymous, Picture of a new process, Chem. Week, March 24, 1956, pp. 44–46.
12. British Patent 536,428, 5/14/41, H. Dreyfus.
13. U.S. Patent 2,311,741, 2/23/43, H. Dreyfus (to Celanese Corp. of America).
14. U.S. Patent 2,318,032, 5/4/43, G. H. van de Griend and L. M. Peters (to Shell Development Co.).
15. U.S. Patent 2,378,104, 6/12/45 C. F. Reed (to Chas. L. Horn).
16. U.S. Patent 2,407,344, 9/10/46, C. F. Reed (to C. L. Horn).
17. Swedish Patent 119,077, 6/25/47, A. I. Hedlund (to Svenska Cellulosa Aktiebolaget).
18. U.S. Patent 2,605,293, 7/29/52, F. T. Tymstra (to Shell Development Co.).
19. Dutch Patent 83,256, 11/15/56, N. V. de Bataafsche Petroleum Maatscheppij.
20. French Patent 1,328,311, 5/31/63, Scientific Design Co., Inc.
21. British Patent 940,284, 10/30/63, Olin Matheison Chemical Corporation.
22. German Patent 1,156,774, 11/7/63, Halcon International, Inc.
23. Czechoslovakia Patent 104,347, 1/28/62, I. Ondrus and O. Lustik.
24. British Patent 971,633, 9/30/64, Standard Oil Co.
25. Polish Patent 50,594, 2/25/66, Politechinska Szcrecinska.

26. U.S.S.R. Patent 129,197, 11/17/66, Ya P. Choporov, R. D. Kruzhalov, T. F. Fel'd-
man, R. N. Rozanshtein, L. O. Ignatkova, N. A. Dubova and V. A. Fomina.
27. French Patent 1,557,588, 2/14/69, T. Reis.
28. Japanese Patent 73, 96,509, 12/19/73, K. Kobayaski, J. Maruyama, and H.
Yoshida, (to Nippon Soda Co., Ltd).
29. Czechoslovakia Patent 153,376, 5/15/74, V. Peterka.
30. Japanese Patent 76, 125,309, 11/1/76, N. Kimura, (to Osaka Soda Co., Ltd.).
31. U.S. Patent 4,053,525, 10/11/77, D. L. Saletan, R. S. Young, and W. R.
Pledger (to Shell Oil Co.).
32. French Patent 2,565,229, 12/6/85, N. Negato, H. Mori, K. Maki, and R. Ish-
ioka (to Showa Deako K.K.).
33. British Patent 540,360; 10/15/41, N. V. de Bataafsche Petroleum Maatschippij.
34. U.S. Patent 2,373,942, 4/17/45, I. Bergsteinsson (to Shell Development Co.).
35. U.S. Patent 2,376,257, 5/15/45, A. E. Lacomble (to Shell Development Co.).
36. U.S. Patent 2,451,485, 10/19/48, G. W. Hearne and M. L. Adams (to Shell Development Co.).
37. U.S. Patent 2,486,842, 11/1/49, G. W. Hearne and M. L. Adams (to Shell Development Co.).
38. U.S. Patent 2,606,932, 10/12/52, R. M. Cole, C. B. Dunn, and G. J. Pierotti
(to Shell Development Co.).
39. British Patent 619,014, 3/2/49, S. A. Ballard, H. deV. Finch, and E. A. Peterson (to N.V. de Bataafsche Petroleum Maatschippij).
40. U.S. Patent 2,731,502, 1/16/56, C. W. Smith (to Shell Development Co.).
41. U.S. Patent 2,414,385, 1/14/47, N. A. Milas (to Research Corp.).
42. U.S. Patent 2,500,599, 3/14/50, I. Bergsteinsson and T. W. Evans (to Shell Development Co.).
43. British Patent 654,764, 6/27/51, Distillers Co. and D. P. Young.
44. British Patent 725,375, 3/9/55, N. V. deBataafsche Petroleum Maatschippij.
45. British Patent 730,431, 5/25/55, N. V. deBataafsche Petroleum Maatschippij.
46. German Patent 907,774, 3/29/54, Badische Anilin & Soda-Fabrik Akt.-Ges.
47. U.S. Patent 2,739,173, 3/20/56, A. E. Corey and J. N. Cosby (to Allied Chemical & Dye Corp.).
48. U.S. Patent 2,754,325, 7/10/56, C. W. Smith (to Shell Development Co.).
49. Dutch Patent 83,775, 11/15/56, N. V. de Bataafsche Petroleum Maatschappij.
50. U.S. Patent 2,718,529, 9/20/55, C. W. Smith and R. T. Holm (to Shell Development Co.).
51. U.S.S.R. Patent 107,762, 12/25/57, P. G. Sargeav, L. M. Bukreeva, and A. G.
Polkoonikova.
52. British Patent 889,613, 2/21/62, Olin Mathieson Chemical Corp.
53. U.S. Patent 3,168,579, 2/2/65, G. A. Boswell and R. C. Morris (to Shell Oil Co.).
54. Polish Patent 48,251, 5/11/64, Inst. Ciezkiej Syntezy Organiczny.
55. U.S.S.R. Patent 166,009, 11/10/64, N. G. Markins, B. D. Krizhalov, and S. T.
Meshcheryakov.
56. Netherlands Patent 6,408,089, 1/18/65, Chemische Werke Huels A.-G.
57. Netherlands Patent 6,513,621, 5/10/66, Halcon International Inc.
58. French Patent 1,484,819, 6/16/67, Romanio Ministry of the Chemical Industry.

59. French Patent 1,509,277, 1/12/68, W. C. Fisher, S. M. Linder, R. L. Pelley, and H. P. Liao (to FMC Corp.).
60. French Patent 1,509,278, 1/12/68, H. P. Liao and L. A. Smith (to FMC Corp.).
61. French Patent 1,548,678, 12/6/68, J. Barthoux (to PROGIL, S.A.).
62. Polish Patent 57,650, 7/30/69, J. Beres, Z. Maciejewski, I. Szelejewska, and Z. Goral, (to Instytut Ciezkiej Syntezy Organicnej).
63. Japanese Patent 71, 42,884, 12/18/71, M. Tsutsumi and A. Yasui (Kuraray Co., Ltd.).
64. U.S.S.R. Patent 236,454, 4/11/72, V. F. Shvets, L. T. Kondrat'ev, and N. N. Lebedev (to Chem. Technol. Inst., Moscow).
65. U.S.S.R. Patent 322,973, 8/15/74, N. N. Lebedev, V. F. Shvets, L. T. Komdrat'ev (to Chemical-Technological Institut, Moscow).
66. Netherlands Patent 75, 04,156, 7/31/75, Halcon International.
67. U.S. Patent 3,954,815, 5/4/76, W. C. Fisher, S. M. Linder, R. L. Pelley, H. P. Liao (to FMC Corp.).
68. U.S.S.R. Patent 360,336, 11/28/72, M. I. Farberov, V. A. Podgornova, and L. V. Mel'nik (to Yaroslov Technological Institute).
69. U.S. Patent 3,846,478, 11/5/74, R. W. Cummins (to FMC Corp.).
70. Japanese Patent 77, 93,705, 8/6/77, Y. Kimoto and H. Masaki (to Daicel Ltd).
71. Bobylev, B. N., Mel'nik, L. V., Farberov, M. I., Subbotina, I. V., and Raevskii, B. M., Zh. Prikl. Khim. (Leningrad), 50(10): 2316 (Russ.), 1977.
72. U.S. Patent 4,049,724, 9/20/77, M. N. Sheng and W. A. Mameneskis (to Atlantic-Richfield Co.).
73. Japanese Patent 82, 77,636, 11/4/80, Sumitomo Chemical Co., Ltd.
74. Japanese Patent 86, 271,230, 12/1/86, T. Masuda, K. Asano, and S. Ando (to Mitsui Toatsu Chemicals, Inc.).
75. British Patent 655,237, 7/11/51, W. F. Gresham (to E. I. duPont de Nemours & Co.).
76. U.S. Patent 2,451,333, 10/12/48, W. F. Gresham and R. E. Brooks (to E. I. duPont de Nemours & Co.).
77. French Patent 1,335,323, 8/16/63, Shell Internationale Research Maatschoppij.
78. U.S. Patent 3,118,760, 1/28/64, R. W. Foreman and F. Veatch (to Standard Oil Co.).
79. U.S.S.R. Patent 486,811, 7/25/74, T. I. Poletneva, E. R. Leikin, T. V. Ipatova, V. I. Yakerson, E. Z. Golosman, V. S. Sobelevskii, A. I. Kreindel, and I. A. Marmaeva.
80. Japanese Patent 86, 271,230, 12/1/86, T. Masuda, T. Asano, and S. Ando (to Mitsui Toatsu Chemicals, Inc.).
81. Federonko, M., Chem. Pop., 41(6): 767 (Engl.), 1987.
82. German Patent 1,156,733, 11/7/63, Halcon International, Inc.
83. U.S. Patent 2,773,101, 12/4/56, C. W. Smith and R. T. Holm (to Shell Development Co.).
84. Canadian Patent 940,935, 1/24/74, N. S. Aprahamian (to Union Carbide Corp.).
85. German Patent 2,515,033, 10/9/75, N. Isogai, T. Okawa, and T. Takeda (to Mitsubishi Gas Chemical Co., Inc.).
86. Czechoslovokia Patent 220,530, 11/15/85, J. Hires.

87. Romanian Patent 64,299, 12/15/78, M. Georgescu and M. Mihailescu (to Institutul "Petrochim", Plo).
88. Lenth, C. W., and Du Puis, R. N., Ind. Eng. Chem., 37: 152, 1945.
89. Belgian Patent 635,207, 1/20/64, Atlas Chemical Industries, Inc.
90. Belgian Patent 635,208, 1/20/64, Atlas Chemical Industries, Inc.
91. U.S. Patent 3,396,199, 8/6/68, L. Kasehagen (to Atlas Chemical Industries, Inc.).
92. U.S. Patent 2,642,462, 6/15/53, L. Kasehagen (to Atlas Powder Co.).
93. U.S. Patent 3,691,100, 9/12/72, L. W. Wright (to Atlas Chemical Industries, Inc.).
94. U.S. Patent 2,381,316, 8/7/45, L. A. Stengel and F. E. Maple (to Commercial Solvents Corp.).
95. U.S. Patent 2,271,083, 1/27/42, E. J. Lorand (to Hercules Powder Co.).
96. U.S. Patent 2,335,731, 11/20/43, R. R. Bottoms.
97. British Patent 499,417, 1/24/39, Association of American Soap & Glycerine Producers, Inc.
98. U.S. Patent 2,282,603, 5/12/42, R. D. DuPuis (to Association of American Soap & Glycerine Producers, Inc.).
99. Swiss Patent 213,251, 5/1/41, Bombrini-Delfino.
100. U.S. Patent 3,030,429, 4/17/62, F. Conradin, G. Bertossa, and J. Giese (to Inventa A.-G fuer Forschung und Patent—verwertung).
101. Swiss Patent 423,738, 5/13/67, Inventa A.-G. fuer Forschung und Patentverwertung.
102. German Patent 1,004,157, 3/14/57, Fabwerke Hoechst Akt.-Ges. vorm Meister Lucius & Brüning.
103. U.S. Patent 3,471,380, 10/7/69, K. C. Hellwig and S. C. Schuman (to Hydrocarbon Research Inc.).
104. U.S. Patent 4,338,472, 7/6/82, A. K. Sirker (to Hydrocarbon Research, Inc.).
105. Netherlands Patent 81, 05,092, 8/16/82, Hydrocarbon Research, Inc.
106. Vasyunina, N. A., Balandin, A. A., Karzhev, V. I., Rabinovich, B. Ya, Chepigo, S. V., Grigoryan, E. S., Slutskin, R. L., Khim Prom., 82–86, 1962.
107. Japanese Patent 68, 18,887, 8/16/68, T. Nomura and K. Mishida.
108. British Patent 1,150,490, 4/30/69, Kyowa Chemicals Co., Ltd.
109. U.S.S.R. Patent 374,271, 3/20/73, N. A. Vasyunina, Ya M. Kovkin, and N. D. Zelinski (to Institute of Organic Chemistry).
110. Vasyunina, N. A., Filatova, T. N., Klabunovskii, E. I., Barysheva, G. S., Alekseev, S. B., and Blonsova, S., Protessy Uchastiem. Mol. Vodoroda, 111–124 (Russ.), 1973.
111. U.S.S.R. Patent 422,718, 4/5/74, N. E. Vishurevskii, V. I. Losik, R. L. Slutskin, and L. A. Osip'yan.
112. Stefoglo, E. F., and Ermokova, A., Khim. Prom. (Moscow), 5: 325 (Russ.), 1974.
113. Sokolov, N. M., Shtrom, M. I., and Zhavronkov, N. M., Khim. Prom. (Moscow), 7: 495 (Russ.), 1975.
114. U.S.S.R. Patent 245,754, 3/25/76, I. D. Rozhdestvenskaya, T. N. Fadeeva, and M. I. Kirsh (to all-Union Scientific-Research Institute of Protein Biosynthesis).

115. Gregor'ev, G. L., Popov, O. S., and Zakharova, V. S., V sb Khimiya i Khim. Tekhnol., 31 (Russ.), 1975.
116. German Patent 658,047, 4/27/38, K. Lüdecke.
117. German Patent 655,177, 1/11/38, Norddeutsche Hefeindustrie A.-G.
118. German Patent 664,575, 8/29/38, Norddeutsche Hefeindustrie A.-G.
119. British Patent 490,783, 8/22/38, Societe' industrielle de nouvequx appareils (S.I.N.A.).
120. French Patent 49,972, 9/29/39, Societe' industrielle de nouveaux appareils (S.I.N.A.).
121. French Patent 829,539, 6/29/38, C. Mariller.
122. German Patent 664,576, 8/29/38, Henkel & Cie G.m.b.H.
123. British Patent 505,733, 5/16/39, Deutsche Gold-und Silber-Scheideanstelt vormals Roessler (to the Distillers Co., Ltd.).
124. British Patent 488, 464, 7/7/38, H. Haehn.
125. U.S. Patent 2,189,793, 2/13/40, H. Haehn.
126. U.S. Patent 2,275,639, 3/10/42, N. M. Mnookin (to Speas Development Co.).
127. U.S. Patent 2,381,052, 8/7/45, H. M. Hodge (to United States Industrial Chemicals, Inc.).
128. Danish Patent 62,582, 8/21/44, Aktieselkebet Dansk Goerings-Industri.
129. German Patent 727,555, 10/1/42, I. G. Farbenind, A.-G.
130. British Patent 572,539, 10/12/45, The Distillers Co., Ltd.
131. U.S. Patent 2,428,766, 10/7/47, A. L. Shade (to Overly Biochemical Research Foundation).
132. U.S. Patent 2,388,840, 11/13/45, E. I. Fulmer, L. A. Underkofler, and R. J. Hickey (to Iowa State College Research Foundation).
133. U.S. Patent 2,416,745, 3/4/47, E. I. Fulmer, L. A. Underkofler, and R. J. Hickey (to Iowa State Coll. Res. Foundation).
134. U.S. Patent 2,386,381, 10/9/45, R. T. K. Cornwell (to Sylvania Industrial Corp.).
135. U.S. Patent 2,410,518, 11/5/46, C. A. Neuberg and I. S. Roberts.
136. British Patent 569,683, 6/5/45, H. M. Hodge (to United States Industrial Alcohol Co.).
137. U.S. Patent 2,422,453, 6/17/47, R. A. Walmesley (to Imperial Chemical Industries, Ltd.).
138. Canadian Patent 408,881, 11/24/42, H. M. Hodge (to United States Alcohol Co.).
139. U.S. Patent 2,430,170, 11/4/47, C. E. Grover.
140. U.S. Patent 2,432,032, 12/2/47, A. C. Neish, G. A. Ledingham, and A. C. Blackwood (to Honorary Advisory Council for Scientific and Industrial Research).
141. French Patent 865,691, 5/30/41, C. C. Cornee.
142. Japanese Patent 4498 ('50), 12/18/50, S. Yasuda.
143. U.S. Patent 2,571,210, 10/16/51, A. E. Craver.
144. French Patent 944,808, 4/15/49, G. R. Laborousse.
145. U.S. Patent 2,614,964, 10/21/52, B. T. Brooks (to Glycerine Corp. of America, Inc.).

146. U.S. Patent 2,680,703, 6/8/54, B. T. Brooks (to Glycerine Corp. of America, Inc.).
147. U.S. Patent 2,680,704, 6/8/54, K. Schneider (to Backhefe G.m.b.H.).
148. British Patent 739,703, 11/2/55, Backlefe G.m.b.H.
149. Belgian Patent 502,759, 8/16/51, Y. Miroluboff and T. Colpaert.
150. Belgian Patent 517,358, 6/1/53, Y. Miroluboff.
151. Spanish Patent 213,353, 9/2/54, J. P. Sala and T. G. Fuster.
152. British Patent 717,939, 11/2/54, K. J. C. Luckhurst (to Imperial Chemical Industries, Ltd.).
153. U.S.S.R. Patent 104,880, 3/25/57, A. S. Egerov, G. L. Visnevskaya, and A. I. Skirstymanskii.
154. U.S. Patent 2,772,206, 11/27/56, E. M. Frankel and S. L. Goldheim.
155. U.S. Patent 2,772,207, 11/27/56, E. M. Frankel and S. L. Goldheim.
156. U.S. Patent 2,793,981, 5/28/57, J. F. T. Spencer, J. M. Roxburgh, and H. R. Sallans (to National Research Council, Canada).
157. German Patent 1,022,180, 1/9/58, National Research Council, Canada.
158. Austrian Patent 198,715, 7/25/58, J. Dyr, M. Verner, and O. Zvacek.
159. British Patent 823,740, 11/18/59, Commercial Solvents (Great Britain), Ltd.
160. French Patent 1,148,274, 12/5/57, L. Basseguy.
161. U.S. Patent 2,986,495, 5/30/61, O. Hiroshi (to Noda Industrial Science Research Laboratory).
162. Malchenko, A. L., Kristul, F. P., and Skirstymonskii, A. L., Spirtomaya Prom., 28(4): 14, 1962.
163. U.S. Patent 3,198,843, 8/3/65, R. S. Barker (to Halcon International, Inc.).
164. Parekh, S. R., and Pandey, S. R., Biotechnol. Bioeng., 27(7): 1089 (Engl.), 1985.
165. Kalla, G. P., Naik, S. C., and Lashkari, V. Z., J. Ferment. Technol., 63(3): 231 (Engl.), 1985.
166. Van Ling, G., and Vluger, J. C., J. Appl. Chem. (London), 19(1): 43 (Engl.), 1969.
167. Van Ling, G., Driessen, A. J., Piet, A. C., and Vlugter, J. C., Ind. Eng. Chem. Prod. Res. Dev., 9(2): 210, 1970.
168. Harris, J. F., and Hajny, G. J., J. Biochem. Microbiol. Technol., 11(1): 9, 1960.
169. Button, D. K., Garver, J. C., and Hajny, G. J., Appl. Microbiol., 14(2): 292, 1966.
170. Hajny, G. J., Hendershot, W. F., and Peterson, W. H., Appl. Microbiol., 52(2): 5, 1960.

5

Chemical Reactions of Glycerine

Eric Jungermann

Jungermann Associates, Inc., Phoenix, Arizona

I. INTRODUCTION

Glycerine is the simplest example of a trihydric alcohol. It can be considered a derivative of propane and is most clearly described as 1,2,3–trihydroxypropane. It has the empirical formula $C_3H_8O_3$ and a molecular weight of 92.09. Glycerine is a viscous, colorless, odorless liquid with a sweet taste.

$$
\begin{array}{c}
H \\
| \\
H-C-O-H \\
| \\
H-C-O-H \\
| \\
H-C-O-H \\
| \\
H
\end{array}
$$

The glycerine molecule contains two primary and one secondary hydroxyl group and the three hydroxyl groups are on adjacent carbons. Because of the multiple hydroxyl groups and their positions on the carbon chain, glycerine has the potential to form more derivatives than is usual for an ordinary alcohol. Utilizing the reactivity of the hydroxyl groups, numerous derivatives can be prepared, including mono-, di-, and triesters and ethers. Oxidation can lead to numerous derivatives, such as glyceraldehyde, dihydroxy acetone, and glyceric aldehyde. Reactions involving hydroxyl groups at adjacent carbon atoms can result in breakage of the carbon–carbon bonds, as in the well-known analytical procedure with periodic acid, or by the condensation of two hydroxyl groups with another reagent, such as a ketone, to form heterocyclic derivatives. Many of these reactions find applications in the production of industrially important materials.

II. GENERAL REACTIONS OF THE HYDROXYL GROUP(S) OF GLYCERINE

When one or more of the hydroxyl groups of glycerine react with a variety of reagents, several isomers can be formed. When only one of the hydroxyl groups reacts, two isomers are formed:

$$
\begin{array}{cc}
\begin{array}{l}
H \\
| \\
H-C-X \\
| \\
H-C-OH \\
| \\
H-C-OH \\
| \\
H
\end{array}
&
\begin{array}{l}
H \\
| \\
H-C-OH \\
| \\
H-C-X \\
| \\
H-C-OH \\
| \\
H
\end{array}
\end{array}
$$

Alpha (α) isomer   Beta (β) isomer

where X = OOCR, Cl, OR, etc.

When two of the hydroxyl groups react, two isomers are possible:

$$
\begin{array}{cc}
\begin{array}{l}
H \\
| \\
H-C-X \\
| \\
H-C-X \\
| \\
H-C-OH \\
| \\
H
\end{array}
&
\begin{array}{l}
H \\
| \\
H-C-X \\
| \\
H-C-OH \\
| \\
H-C-X \\
| \\
H
\end{array}
\end{array}
$$

α, β diisomer α, α diisomer

When all three of the hydroxyl groups are reacted with a particular reagent only one isomer results:

$$
\begin{array}{c}
H \\
| \\
H-C-X \\
| \\
H-C-X \\
| \\
H-C-X \\
| \\
H
\end{array}
$$

In practice, it is difficult to produce one particular isomer, and mixtures of all of the above can usually be found, unless specific steps are undertaken to synthesize a particular isomer.

One example of how specific isomers can be synthesized is via the formation of trityl ethers. One or two hydroxyls of glycerine can be blocked by etherification with the bulky trityl group [1,2]. The remaining hydroxyls are then esterified with an acid chloride in a typical Schotten-Bauman reaction; next the trityl groups are removed by catalytic hydrogenation with a platinum catalyst in absolute alcohol at 40–50°C at 45 lbs pressure. These mild conditions minimize migration of the ester group:

$$
\begin{array}{lll}
CH_2OC(C_6H_5)_3 & CH_2OC(C_6H_5)_3 & CH_2OH \\
| & | & | \\
CHOH + 2RCOCl \rightarrow CHOCOR \underline{\quad} H^+ \longrightarrow & CHOCOR \\
| & | & | \\
CH_2OH & CH_2OCOR \quad Pt \text{ in EtOH} & CH_2OCOR
\end{array}
$$
$$
\text{1,2-diglyceride}
$$

$$
\begin{array}{l}
CH_2OH \\
| \\
CH-OH + (C_6H_5)_3CCl \\
| \\
CH_2OH
\end{array}
$$

$$
\begin{array}{lll}
CH_2OC(C_6H_5)_3 & CH_2OC(C_6H_5)_3 & CH_2OH \\
| & | & | \\
CHOH + RCOCl \rightarrow & CHOCOR \underline{\quad} H^+ \longrightarrow & CHOCOR \\
| & | & | \\
CH_2OC(C_6H_5)_3 & CH_2OC(C_6H_5)_3 & CH_2OH
\end{array}
$$
$$
\text{2-monoglyceride}
$$

Other methods for preparing pure mono- and diglycerides have been reviewed by Hartman [3].

A. Esters of Glycerine

1. Organic Acid Ester

Long-chain fatty acid esters of glycerine are the most common and most diverse group of glycerine derivatives. They occur naturally as triglycerides in animal and vegetable fats and oils. Small amounts of mono- and diesters are often found in these fats and oils due to partial hydrolysis of the triester.

Glycerine with its three hydroxyl groups can be esterified with any combination of fatty acids to form a number of fatty esters. Fatty monoglycerides are an important class of compounds that have found a wide range of applications in the cosmetic and food industry. They are discussed in greater detail in Chapter 12 of this volume. In general, they are prepared by the direct condensation of a fatty acid or a fat (triglyceride) with glycerine resulting in mixtures containing 40–60% monoglyceride, 30–45% diglycerides, and free glycerine. When a fat is used as the starting material and reacted with an equimolar quantity of glycerine, approximately half the hydroxyl groups are esterified:

$$
\begin{array}{ccccccc}
CH_2OOCR & & CH_2OH & & CH_2OOCR & & CH_2OOCR \\
| & & | & & | & & | \\
CHOOCR & + & CHOH & \rightleftharpoons & CHOH & + & CHOH \\
| & & | & & | & & | \\
CH_2OOCR & & CH_2OH & & CH_2OOCR & & CH_2OH
\end{array}
$$

The reactions are run at elevated temperature, which improves the miscibility of the fatty material with glycerine and increases the rate of reaction. Solvents, such as dioxane and phenol, have been recommended when a high monoglyceride content is desired [4,5]. The reactions can be run either batchwise or continuously, and a number of useful catalysts have been reported [6–8]. Monoglycerides with a purity of 90 + % have been prepared commercially by molecular distillation [9].

Acetins are the mono-, di-, and triacetates of glycerine formed by the reaction of glycerine with acetic acid, acetic anhydride, or ketene [10]. The choice of catalyst plays a role in directing the reaction to form one isomer predominantly. Monoacetin is a thick hygroscopic liquid used in tanning and in the manufacture of explosives. Diacetin is a hygroscopic liquid and is used as a plasticizer and softening agent. The triacetin is a cellulose plasticizer used in cigarette filters and as a binder in solid rocket fuels. In the cosmetic area, it has found use as a fixative in perfumes.

Esters of various lower molecular weight fatty acids have been reported. These include formates, oxalates, propionates, butyrates, and benzoates [11–13].

2. Mixed Organic Esters

The preparation of alkyd resins, which are polyester surface-coating resins, illustrates interesting examples of mixed esters of glycerine. Aklyds are synthesized by reacting polyfunctional alcohols and polyfunctional organic acids, often followed by reaction with a fatty acid. Among the alcohols used are glycerine, ethylene glycol, sorbitol, and pentaerythritol; examples of polyfunctional acids are phthalic and maleic anhydride and sebacic acid. Unsaturated or saturated monocarboxylic fatty acids are used as blocking and modifying agents.

In a typical reaction, glycerine and phthalic anhydride are heated to 200°C, when the carboxyl group esterifies readily with the primary groups of glycerine to form multifunctional resinoids:

$$
\begin{array}{c}
\text{CO} \\
\text{O} \\
\text{CO}
\end{array}
+
\begin{array}{c}
\text{H}_2\text{COH} \\
\text{HCOH} \\
\text{H}_2\text{COH}
\end{array}
\longrightarrow
\quad
\begin{array}{c}
\text{COOCH}_2\text{CHOHCH}_2\text{OH} \\
\\
\text{COOH}
\end{array}
\quad +
$$

$$
+
\quad
\begin{array}{c}
\text{COOH} \quad\quad\quad \text{HOOC} \\
\text{---COOCH}_2\text{CHOHCH}_2\text{OOC---}
\end{array}
$$

Because the resinoids are multifunctional, they can form either chain or space polymers. The danger of premature crosslinking and gelation can be minimized by blocking one of the hydroxyl groups of the glycerine with a monocarboxylic fatty acid:

$$
\begin{array}{c}
\text{---G---} \\
|
\end{array}
+ \text{FA} \quad\text{---------------}\quad
\begin{array}{c}
\text{---G---} \\
| \\
\text{FA}
\end{array}
$$

followed by:

$$
\begin{array}{c}
\text{---G---} \\
| \\
\text{FA}
\end{array}
+ \text{---P---} \quad\text{-----------}\quad
--\left[\begin{array}{c} \text{G---P} \\ | \\ \text{FA} \end{array}\right]_n -- + \text{H}_2\text{O}
$$

where G = glycerine, FA = fatty acid, P = phthalic

Since 18-carbon saturated or unsaturated fatty acids are commonly employed, the potential for internal plasticization is strong. By varying the amount and kind of modifying fatty acid, resin chemists can develop products with virtually limitless range and coating requirements.

Instead of using the fatty acids as shown in the previous example, alkyd resins can also be produced using drying oils. Glycerine is heated with the drying oil in the presence of metallic soaps at temperatures just high enough to promote ester interchange. Mono- and diglycerides are formed which are then condensed with phthalic anhydride or other bifunctional acids to form the resins [14].

Acetoglycerides are another important group of derivatives of mono- and diglycerides that have found practical commercial uses in the food industry [15,16]. They are prepared by esterification of a partial glyceride with acetic anhydride.

$$C_{17}H_{35}COOCH_2CHOHCH_2OH + (CH_3CO)_2O \underline{\hspace{1cm}} 110°C \longrightarrow$$
$$\underline{\hspace{1cm}} \longrightarrow C_{17}H_{35}COOCH_2CHOHCH_2OOCCH_3 + CH_3COOH$$

Di- and triglycerides containing the lactate residue are prepared by direct esterification of glycerine with mixtures of fatty acids and lactic acid [17]. In this reaction, lactic acid acts both as a coreactant and a solvent; this allows the use of somewhat lower temperature for esterification than is usual with glycerine and fatty acids alone. These lactic acid derivatives have found use as food emulsifiers. Other derivatives of partial glycerides that have been reported include the tartrates and citrates [18].

3. Mixed Organic–Inorganic Esters

Sulfated fatty monoglycerides are good examples of this category. These materials are excellent detergents and have been used in a hard water bar soap (Vel), shampoos, and in tooth paste fomulations. They were introduced into the United States in the early 1940s by Colgate-Palmolive Co. These materials are produced by reaction between a fat (coconut oil), glycerine, and sulfuric acid [19–21]:

$$
\begin{array}{lll}
H_2C{-}OOCR_a & H_2COH & H_2COOCR_{a,b,c} \\
| & | & | \\
HC{-}OOCR_b + 2\ HCOH - (1)H_2SO_4 \rightarrow (2)NaOH \rightarrow 3\ HCOH + 3\ H_2O \\
| & | & | \\
H_2C{-}OOCR_c & H_2COH & H_2COSO_3Na
\end{array}
$$

Fats are reacted with oleum, followed by treatment with glycerine–sulfuric acid. The sulfated monoglycerides are somewhat unstable, since the carboxylate linkage is sensitive to alkalies and the sulfate linkage is sensitive to acid hydrolysis.

4. Inorganic Esters

Glycerine forms esters of many inorganic acids such as hydrogen halides, sulfuric, phosphoric, nitrous, nitric, and boric acids. Some of these are used as intermediates in chemical reactions, others have distinct end uses, such as trinitroglycerine, the well-known drug and explosive.

Many esters of glycerine and hydrogen halides are known. They are called halohydrins (i.e., chlorhydrins, bromohydrins). Two possible monochlorhydrins are formed when glycerine is saturated with hydrogen chloride [22]:

$CH_2ClCHOHCH_2OH$ $CH_2OHCHClCH_2OH$

α-Monochlorhydrin   β-Monochlorhydrin

Alpha- and beta-dichlorhydrins, as well as monochlorhydrins are obtained when a mixture of glycerine and glacial acetic acid saturated with hydrogen chloride is boiled; the amount of hydrogen chloride used determines the yield of both mono- and dichlorhydrin:

$$CH_2ClCHOHCH_2Cl \qquad\qquad CH_2ClCHClCH_2OH$$
α-Dichlorhydrin   β-Dichlorhydrin

Reaction of dichlorhydrins with alkali results in the formation of epichlorhydrin:

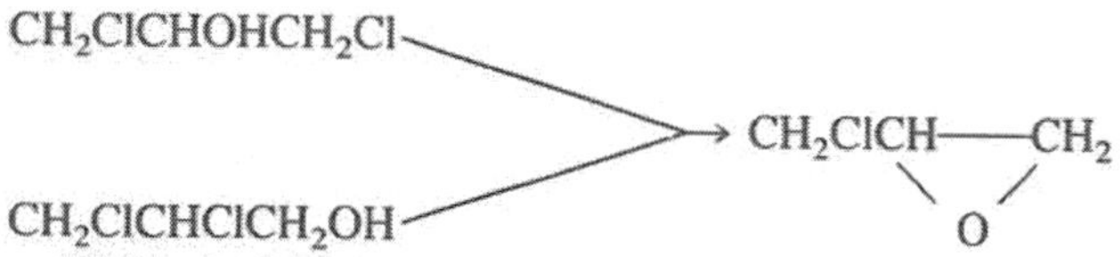

CH₂ClCHOHCH₂Cl
CH₂ClCHClCH₂OH → CH₂ClCH——CH₂ (epoxide, O bridge)
Dichlorhydrins   Epichlorhydrin

The epoxide ring in epichlorhydrin is readily opened by a variety of reagents; hence epichlorhydrin is an important intermediary in the synthesis of glycerine derivatives. For example, pure α-monochlorhydrin is obtained by the acid catalyzed hydrolysis of epichlorhydrin.

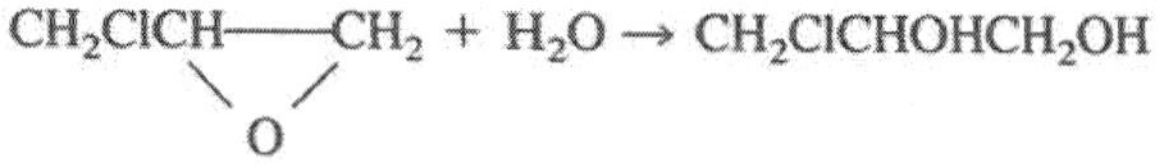

$$CH_2ClCH\text{——}CH_2 + H_2O \rightarrow CH_2ClCHOHCH_2OH$$

Reaction of the epichlorhydrin with sodium in ether yields glycidol, 2,3-epoxy-1-propanol:

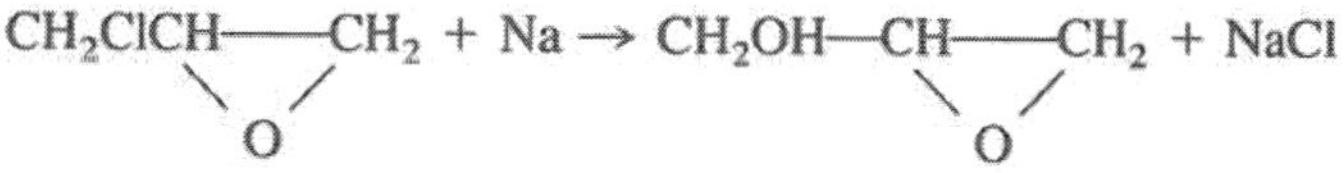

$$CH_2ClCH\text{——}CH_2 + Na \rightarrow CH_2OH\text{—}CH\text{——}CH_2 + NaCl$$

Reaction between hydrogen chloride gas and glycerine results in the formation of both alpha and beta dichlorhydrin. Excess hydrogen chloride gas, 2% acetic acid, and a temperature of 100–110°C are recommended by Gilman and Blatt [23]. Pure 1,3-dichlorhydrin is obtained by reacting epichlorhydrin with hydrochloric acid:

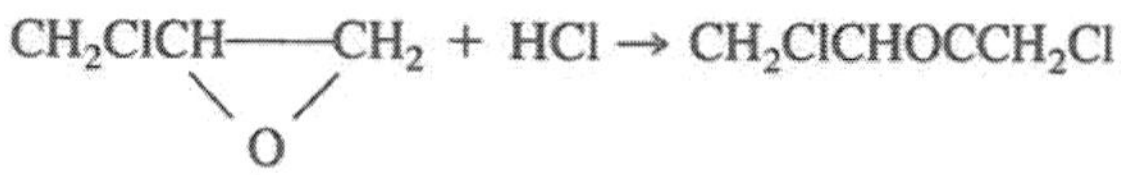

$$CH_2ClCH\text{——}CH_2 + HCl \rightarrow CH_2ClCHOCCH_2Cl$$

The trichlorohydrin can be prepared by the reaction between dichlorhydrins and thionyl chloride; alternately, it is prepared by the chlorination of propane and propylene.

Mono- and dibromohydrins have been prepared by the reaction between glycerine and hydrobromic acid or dry hydrogen bromide. They are colorless, viscous liquids, and occur in the various isomeric forms. Iodohydrins have not been made directly from glycerine; instead, they have been prepared from chlorhydrins by replacing the chloride with iodine [24].

Glycerine esters of nitric acid have been known for over a century. The trinitrate, nitroglycerine, was discovered in 1846; it has been used as a drug in the treatment of angina pectoris and athma, and as an explosive. The latter use was developed by Alfred Nobel in 1867, when he discovered that nitroglycerine, a liquid, can be absorbed into kieselguhr, producing dynamite. Dynamite per se is relatively stable, and requires detonators, such as mercury fulminate, or lead azide, to explode the whole mass. On explosion, nitroglycerine breaks down into oxygen, nitrogen, carbon dioxide, and water:

$$2C_3H_8(ONO_2)_3 = 3N_2 + 6CO_2 + O + 5H_2O$$

Glycerine can form mono-, di-, and trinitrates. The trinitrate is prepared by adding glycerine to a mixture of concentrated nitric and sulfuric acid at 10°C. The reaction is exothermic and requires cooling [22]. It is a colorless oil, and at room temperature, it is practically odorless however, when heated to 50°C, it has a characteristic odor.

Glycerine forms a trinitrite, $CH_2(ONO)$—$CH(ONO)$—$CH_2(ONO)$, when dry nitrous acid anhydride, N_2O_3 is passed through glycerine [25].

Glycerine forms esters with phosphoric and phosphorous acid and a large number of isomers are possible. The most important phosphate esters are the alpha and beta isomers of glyceryl monophosphate:

$$
\begin{array}{ll}
CH_2OH & CH_2OH \\
| & | \\
CH\text{—}OH & CH\text{—}O\text{—}PO(OH)_2 \\
| & | \\
CH_2\text{—}O\text{—}PO(OH)_2 & CH_2OH \\
\alpha\text{-Isomer} & \beta\text{-Isomer}
\end{array}
$$

When glycerine is esterified with phosphoric acid at 100°C, the alpha mono isomer is formed primarily. When esterification is carried out with sodium dihydrogen phosphate, the beta isomer is formed primarily. Relative amounts of the two isomers formed can be determined by titration with periodic acid which only reacts with the *alpha* isomer. Salts of glyceryl monophosphates, particularly the ammonium, calcium, iron, and sodium salts, are used in the manufacture of soft drinks and pharmaceuticals.

The glyceryl monophosphate ester structure is the backbone of phosphotides, an important group of naturally occurring substances. These are fatlike substances containing phosphorus and a basic component, such as choline or colamine. The best known phosphotide is lecithin which has the following structure:

$$CH_2OCOR_I$$
$$CH\text{---}OCOR_{II}$$
$$CH_2OPO_3CH_2CH_2N(CH_3)_3OH$$

R = fatty acid radicals, such as oleic, palmitic, stearic, etc.

When glycerine and boric acid are heated together an ester is formed which titrates as a monobasic acid [26]. This reaction can be represented as follows:

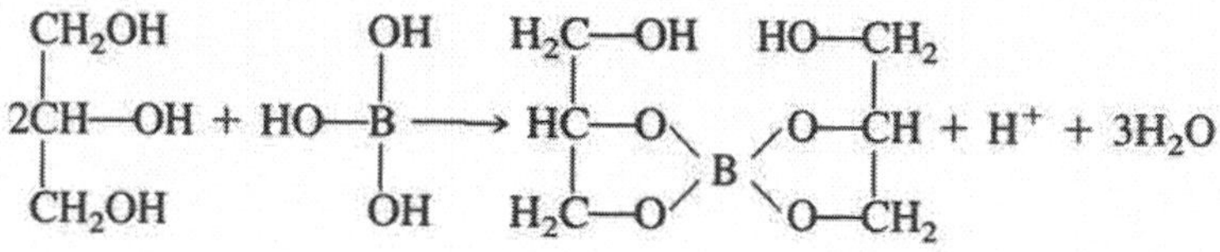

B. Ethers of Glycerine

Ethers of glycerine and fatty alcohols occur in natural products. Some typical examples of naturally occurring fatty ethers of glycerine are the alpha glyceryl monoethers of stearyl, oleyl, and cetyl alcohol; these ether–alcohols are called batyl, selachyl, and chimyl alcohol, respectively [27].

Glyceryl ethers are most readily prepared by reaction of the corresponding chlorhydrin with an alcohol or phenol in the presence of an alkali. The reaction proceeds via the epoxide:

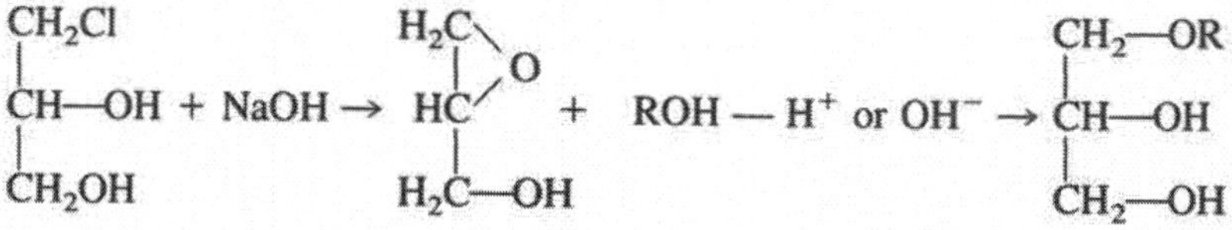

In this sequence, the ether group always goes to the alpha position.

Ethers of glycerine can also be formed directly from glycerine by reacting the sodium glyceroxide with alkyl or aryl halide or alkyl sulfate [28]. For example, glycerine trimethyl ether can be prepared by treating glycerine with dimethyl sulfate and sodium hydroxide at 100°C. Boiling monosodium glycerate with ethyl bromide in absolute alcohol yielded the alpha monomethyl ether of glycerine, and reaction between glycerine, 2 moles of ethyl bromide and potassium hydroxide at 100°C gives the alpha', alpha-diethyl ether of glycerine.

Glycerine reacts with ethylene oxide or propylene oxide to form polyethers.

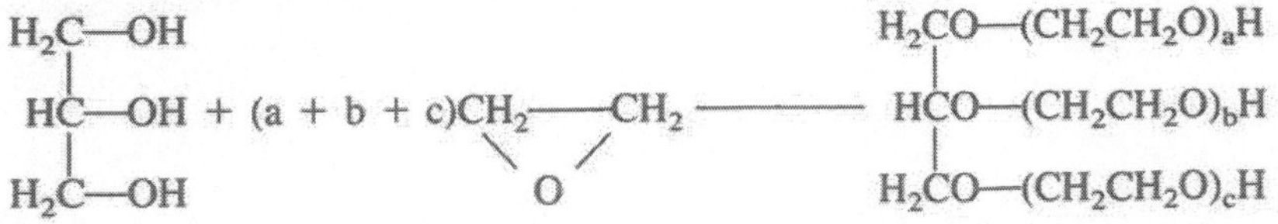

$$\begin{array}{ccc}
H_2C\!-\!OH & & H_2CO\!-\!(CH_2CH_2O)_aH \\
| & & | \\
HC\!-\!OH + (a + b + c)CH_2\!\!-\!\!-\!\!CH_2 \longrightarrow & HCO\!-\!(CH_2CH_2O)_bH \\
| & \diagdown\!\!\diagup & | \\
H_2C\!-\!OH & O & H_2CO\!-\!(CH_2CH_2O)_cH
\end{array}$$

By first adding the hydrophobic propoxy chain followed by the addition of hydrophillic ethoxy chains or vice versa "block copolymers" can be prepared [29]. When the ethoxy and propoxy chains are properly balanced, these block copolymers have surface-active properties. They also have been used as intermediates in the manufacture of some polymers.

C. Polymerization of Glycerine

Polymeric ethers are formed when glycerine reacts with itself to form polyglycerols via intermolecular dehydration

$$n \; HOCH_2CHOHCH_2OH \rightarrow HOCH_2CHOHCH_2O\!-\!(CH_2CHOHCH_2O)_{n-2}\!-\!CH_2CHOHCH_2OH$$

Polyglycerols are produced commercially by heating glycerine at temperatures in the range of 200–275°C at normal or reduced pressure in the presence of an alkaline catalyst, such as sodium or potassium hydroxide or sodium acetate. The reaction is usually carried out in an atmosphere of carbon dioxide or nitrogen [10, 30, 31]. Proper processing conditions can greatly affect the quality, color, and odor of the resultant polyglycerol. Babayan et al [32] have demonstrated the linear nature of polyglycerols, as shown above, but partial anhydrization can also occur, resulting in structures like the one shown below [33, 34]:

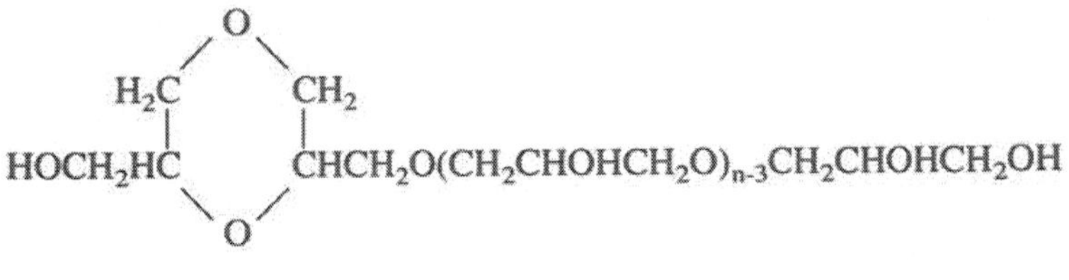

$$\begin{array}{c}
O \\
\diagup \quad \diagdown \\
H_2C \qquad\quad CH_2 \\
| \qquad\qquad | \\
HOCH_2HC \qquad\quad CHCH_2O(CH_2CHOHCH_2O)_{n-3}CH_2CHOHCH_2OH \\
\diagdown \qquad \diagup \\
O
\end{array}$$

In the commercial production of polyglycerols, products containing an average number of glycerine units ranging from 2 to 35 are obtained. Some unreacted glycerine usually stays in the mixture [35]. Vacuum stripping to reduce the free glycerine level is used; this also helps narrow the range of molecular weights in the resultant product. Viscosity, hydroxyl values, and refractive index are used to control the properties of the final polyglycerol. Polyglycerols range from viscous liquids for the lower molecular weight examples to solids. Viscosities and theoretical hydroxyl values, assuming no cyclization, are shown in Table 5.1.

Table 5.1 Viscosities of Polyglycerols

	M W	No. of OH groups	Viscosity CTKS @ 50°C	Hydroxyl values
Glycerine	92	3	45	1830
Diglycerol	166	4	287	1352
Triglycerol	240	5	647	1169
Tetraglycerol	314	6	1067	1071
Pentaglycerol	388	7	1408	1012
Hexaglycerol	462	8	1671	970
Heptaglycerol	536	9	2053	941
Octaglycerol	610	10	2292	920
Nonaglycerol	684	11	2817	903
Decaglycerol	758	12	3199	880
Pentadecaglycerol	1128	17	4893	846

The polyglycerols are soluble in water, alcohol, and many polar solvents. They have found use as plasticicers, emulsifiers, surfactants, lubricants, and as intermediates in the formation of polyglycerol esters (36). In analytical chemistry, polyglycerols have found use as liquid phases in gas chromatography [37].

The polyglycerols have been esterified with a wide variety of long and short chain fatty acids, including saturated, unsaturated, polyunsaturated, hydroxyacids, and others. They are prepared either as partial esters or can be completely esterified. Esterification with fatty acids is carried out at 200°C; some anhydride formation can occur at these temperatures. The reactions are usually base catalyzed. Variations in the molar ratios used will determine the extent of the formation of mono- and diesters.

Variations in the polyglycerol used and the nature of the fatty acid allows the preparation of a wide variety of materials with a wide range of properties. Polyglycerols have found extensive use in the food, cosmetic and pharmaceutical industry as surfactants, emulsifiers, lubricants gelling agents, humectants, adhesives, urethane intermediates, and textile fiber finishes [38–40].

D. Oxidation of Glycerine

Glycerine is quite stable in the presence of oxygen under normal conditions, but oxidizes in the presence of certain catalysts, such as iron or copper. Glycerine is readily oxidized by a variety of chemical and microbiological oxidants, as well as by electrolysis. In reactions with strong oxidants, such as potassium permanganate, glycerine is oxidized completely to carbon dioxide and water:

$$2C_3H_8O_3 + 7O_2 \rightarrow 6CO_2 + 8H_2O$$

Theoretically, glycerine can be oxidized to the 11 oxidation products shown below, while maintaining the original three-carbon chain. Partial oxidation is usually hard to control; the eleven oxidation products have been isolated, though more often they are prepared by indirect methods, rather than the controlled oxidation of glycerine.

CHO   CH₂OH   CHO   CHO
| | | |
CHOH   C=O   C=O   CHOH
| | | |
CH₂OH   CH₂OH   CH₂OH   CHO
Glyceraldehyde   Dihydroxy   Hydroxypyruvic   Tartronic
                         acetone   aldehyde   dialdehyde

CHO   COOH   COOH   COOH
| | | |
C=O   CHOH   C=O   CHOH
| | | |
CHO   CH₂OH   CH₂OH   CHO
Mesoxalic   Glyceric   Hydroxy   Tartronic
dialdehyde   acid   pyroaracemic   semialdehyde
                                                  acid

            COOH   COOH   COOH
            | | |
            CHOH   C=O   C=O
            | | |
            COOH   CHO   COOH
            Tartronic   Mesoxalic   Mesoxalic
            acid   semialdehyde   acid

Oxidation of glycerine with bromine and sodium carbonate results in the formation of a mixture of glyceraldehyde and dihydroxacetone; this mixture is often called glycerose. The aldehyde is the predominant component in this mixture. These compounds are the simplest aldose and ketose, and as such are the forerunners of the carbohydrates. Dihydroxyacetone, can also be formed by the microbiological oxidation of glycerine by the sorbose bacterium which exclusively attacks secondary hydroxyl groups [41]. Dihydroxyacetone is a skin-tanning agent and is used in cosmetic formulations designed for this purpose. Anodic oxidation of glycerol at a silver oxide electrode yields glyceric acid [42]. Glycerine is also converted into glyceric acid by the reaction of concentrated nitric acid.

Oxidation with sodium chromate and chromic acid in strongly acid solution results in the quantitative conversion of glycerine into carbon dioxide and water:

$$3CH_2OHCHOHCH_2OH + 7Cr_2O_7^= + 56H^+ \rightarrow 9CO_2 + 40\ H_2O + 14Cr^{+3}$$

This method is used for the analysis of glycerine, especially when it is present at low levels. This is discussed in more detail in Chapter 7. If the oxidizing solution is made alkaline, glyceric acid is the main reaction product, although this is not a clean or quantitative reaction.

Another technique for oxidizing glycerine and one that also has found application in analytical procedures for the determination of glycerine, is based on the use of periodic acid. Periodic acid reacts with compounds that contain hydroxyl groups on adjacent carbon atoms. In the case of glycerine, carbon bonds are broken on both sides of the central carbon, resulting in the formation of 2 moles of formaldehyde and one mole of formic acid:

$$CH_2OHCHOHCH_2OH + 2IO_4^- \rightarrow 2CH_2O + HCOOH + H_2O + 2IO_3^-$$

This is an important analytical technique for determining glycerine, or mixtures of glycerine, monoglycerides and diglycerides, and is discussed more fully in Chapter 7. Lead tetraacetate also oxidizes glycerine and other polyhydric alcohols with adjacent hydroxyl groups, but this reaction is not quantitative due to further reaction between the lead tetraacetate and the formic acid formed in the reaction [43].

E. Miscellaneous Reactions of Glycerine

Dehydration of glycerine results in the formation of acrolein, a clear liquid (bp 52°C) with a very pungent smell:

$$CH_2OHCHOHCH_2OH \underset{360°C}{\overset{Al_2O_3}{\longrightarrow}} \underset{acrolein}{CH_2 = CH - CHO} + 2\ H_2O$$

Acrolein can be formed in trace quantities when glycerine is stored in inadequately protected metal containers and exposed to elevated temperatures, resulting in very noticeable, strong pungent off-odors.

When glycerine is heated with oxalic acid, an intermediate glyceryl oxalate is formed, which on further heating breaks down into allyl alcohol and carbon dioxide:

$$\begin{array}{ccccc} CH_2OH & & \left[CH_2OH \right. & & CH_2OH \\ | & & | & & | \\ CHOH & + \begin{array}{c} COOH \\ | \\ COOH \end{array} \rightarrow & \left. \begin{array}{c} CH-O-CO \\ | \\ CH_2O-CO \end{array} \right] \rightarrow & \begin{array}{c} CH \\ || \\ CH \end{array} + 2CO_2 + 2H_2O \end{array}$$

Hydrogenation at pressures from 10–100 atmospheres and a temperature of 150–160°C over metal catalysts, such as nickel, platinum, or copper chromite primarily yields propylene glycol:

$$CH_2OHCHOHCH_2OH \xrightarrow[\text{Ni, 150°C}]{H_2} CH_3CHOHCH_2OH$$

Glycerine reacts with alkalies and certain metallic oxides forming glyceroxides. For example, litharge and glycerine react exothermically to form lead glyceroxides:

$$CH_2OHCHOHCH_2OH + PbO \rightarrow C_3H_5(OH)O_2Pb + H_2O$$

Monosodium glyceroxide is prepared by reacting metallic sodium in absolute alcohol; the disodium glyceroxide is obtained from the monoisomer by reaction at elevated temperature with an equivalent amount of sodium ethylate. Glycerine acetals can be prepared by the reaction of glycerine with an aldehyde or ketone. These are heterocyclic compounds displaying geometric and optical isomerism.

Isomer formation is a function of temperature; higher temperatures favor formation of the 5-membered ring, lower temperature leads primerily to a 6-membered ring [44]. Glycerine has been reported to react with acetylene in the presence of mercuric sulfate to form an acetal, α,β—ethylidene glycerine [28].

Many other reactions have been reported in the literature. This is not surprising for a chemical that has been around for more than 200 years. A thorough review can be found in the section on "Glycerine and Some Glycerine Derivatives" in Beilstein's *Handbook of Organic Chemistry* [28].

REFERENCES

1. Daubert, B. F., J. Am. Chem. Soc., 62: 1713, 1940.
2. Verkade, P. E., van der Lee, J., and Meerburg, W., Rec. Trav. Chim., 56: 613, 1937.
3. Hartman, L., Chem. Rev., 58: 559, 1958.
4. U.S. Patent 2,073,797, Hilditch, T. P. and Riggs, J. G. (to ICI, Ltd), 1937.
5. U.S. Patent 2,251,693, Richardson, A. S. and Eckey, E. W. (to P&G) 1941.

6. U.S. Patent 2,393,581, Arrowsmith, C. J. and Ross, J. (to Colgate-Palmolive Co.), 1945.
7. U.S. Patent 2,206,167, Edeler, A. and Richardson, A. S. (to P&G), 1940.
8. Goldsmith, H. A., Chem. Rev., 33: 257, 1943.
9. Nystrom, R. F. and Brown, W. G., J. Am. Chem. Soc., 70: 3738, 1948.
10. Miner, C. S. and Dalton, N. N., Glycerol, ACS Monograph 117, Reinhold Publishing Corp., New York, 1953.
11. U.S. Patent 2,405,936, Bartlett, E. P., (to E. I. du Pont de Nemours & Co.), 1946.
12. Gilchrist, P. A. and Schuette, H. A., J. Am. Chem. Soc., 53: 3480, 1931.
13. Fairbourne, A. and Foster, G. E., J. Chem. Soc., 127: 2749, 1925.
14. Kraft, W. M., Am Paint J., 41: 96, 1957.
15. Alfin-Slater R. B., Coleman, R. D., Feuge, R. O., and Altschul, A. M., J. Am. Oil Chem. Soc., 35: 122, 1958.
16. Zabik, M. E. and Dawson, L. E., Food Technol., 17: 87, 1963.
17. U.S. Patent 3,012,048, Shapiro, S. (to Armour & Co.), 1961.
18. U.S. Patent 2,813,032, Hall, L. A. (to to Griffith Lab. Inc.), 1957.
19. U.S. Patent 2,478,354, Bell, A. C. and Alsop, W. G., (to Colgate-Palmolive Co.), 1949.
20. U.S. Patent 2,868,812, Gray, F. W., (to Colgate-Palmolive Co.), 1959.
21. U.S. Patent 2,660,588, Gebhart, A. I. and Mitchell, J. E., (to Colgate-Palmolive Co.), 1953.
22. Karrer, P., Organic Chemistry, Elsevier Press, New York, 1938, Chap. 19.
23. Gilman, H. And Blatt, A. H., Organic Syntheses, Vol. I, Wiley and Sons, New York, 1941, p. 2992.
24. Glattfeld, J. W. E and Klass, R., J. Am. Chem. Soc., 55: 1114, 1933.
25. Masson, A., Berichte, 16: 1697, 1883.
26. Soine, T. O. and Wilson, C. O., Rogers Inorganic Pharmaceutical Chemistry, 11th ed. Lea & Febiger, Philadelphia, 1957.
27. Baer, E., Fischer, H. O. L., and Rubin, L. J., J. Biol. Chem., 170:337, 1947.
28. Beilstein's Handbook of Organic Chemistry, fourth edition
29. Schmolka, I. R. In Nonionic Surfactants M. J. Schick, ed.), Marcel Dekker, Inc., New York, 1967, Chap. 10.
30. U.S. Patent 1,917,257, Harris, B. R., 1933.
31. U.S. Patent 2,023,388, Harris, R. B., 1935.
32. Babyan, V. K., Kaufman, T. G., Lehman, H., and Tkaczuk, JSCC, 15: 473, 1964.
33. Summerbell, R. K., and Stephens, J. R., J. Am. Chem. Soc., 76: 6401, 1954.
34. Summerbell, R. K., and Poklacki, E. S., J. Am. Oil Chem. Soc., 39: 306, 1962.
35. Khim.-Farm. Zh. 23: 1137, 1988; CA 112(14)119754v.
36. JSCC (Jap), 22: 171, 1988 CA 112(24)223107x.
37. J. Chromatogr. 33: 411, 1968.
38. U.S. Patent 2,786,978, de Ren, W. J. F, and Gracht, V. D., (to Lever Brothers, Co.), 1957.
39. Mfg. Confect, 50: 36, 1970; CA 73(19)97606v.
40. Polym. Prep (Am. Chem Soc., Div Polym. Chem), 27: 7, 1986.

41. Blazejek, S. And Sobczak, E. Przem. Spozyw. 39: 11, 1985; CA
    106(13)100770b.
42. Kyriacou, D. and Tougas, T. P., J. Org. Chem., 52: 2318, 1987.
43. Hockett, R. C., Dienes, M. T., Fletcher, H. G., and Ramsden, H. E., J. Am.
    Chem. Soc., 66: 467, 1944.
44. Trister, S. M. and Hibbert, H., Can. J. Res., 14B: 415, 1936.

6

Physical Properties of Glycerine

S. R. Gregory

Humko Chemical Division, Witco Corporation, Memphis, Tennessee

Although the physical properties of glycerol have been studied for more than a century, our knowledge in the field is still far from complete. The preparation of pure anhydrous glycerol has always caused serious difficulties. Glycerol has a high affinity for moisture which has hindered the study of the anhydrous material. Progress continues to be made in understanding the physical properties, but in many cases an arbitrary choice has to be made when faced with the somewhat different results reported by earlier investigators. The tables included in this chapter are selections from the values published in the scientific literature and are by no means infallible.

I. GLYCEROL, GLYCERIN, 1,2,3-PROPANETRIOL

The simplest trihedric alcohol, when pure, is a colorless, odorless, viscous liquid with a sweet taste at ordinary room temperature.

II. SPECIFIC GRAVITY

It is important that specific gravity be accurately determined, since the study of all other properties is based on, or connected with it.

Measurement of specific gravity is the principal means of determining the glycerol content of distilled glycerin. For years specific gravity was best determined by the use of a pycnometer [1]. However, instruments are now available that measure the specific gravity by measuring the natural frequency of the sample in a measurement cell and the manufacturer reports an accuracy of $\pm 1 \times 10^{-5}$ g/cm^3 [2].

In 1927, Bosart and Snoddy [3] published their determination of the specific gravity of glycerol solutions, as measured under carefully controlled conditions at 15/15, 15.5/15.5, 20/20, and 25/25 $\pm$ 0.1°C. These figures are given in Table 6.1. They have been widely used in the industry.

Table 6.1 Specific Gravity and Percent Glycerol by Weight

Glycerol (%)	Apparent specific gravity (°C)				True specific gravity (°C)			
	15/15	15.5/15.5	20/20	25/25	15/15	15.5/15.5	20/20	25/25
100	1.26557	1.26532	1.26362	1.26201	1.26526	1.26501	1.26331	1.26170
99	1.26300	1.26275	1.26105	1.25945	1.26270	1.26245	1.26075	1.25910
98	1.26045	1.26020	1.25845	1.25685	1.26010	1.25985	1.25815	1.25655
97	1.25785	1.25760	1.25585	1.25425	1.25755	1.25730	1.25555	1.25395
96	1.25525	1.25500	1.25330	1.25165	1.25495	1.25470	1.25300	1.25140
95	1.25270	1.25245	1.25075	1.24910	1.25240	1.25215	1.25045	1.24880
94	1.25005	1.24980	1.24810	1.24645	1.24975	1.24950	1.24780	1.24615
93	1.24740	1.24715	1.24545	1.24380	1.24710	1.24685	1.24515	1.24350
92	1.24475	1.24450	1.24280	1.24115	1.24445	1.24420	1.24250	1.24085
91	1.24210	1.24185	1.24020	1.23850	1.24185	1.24155	1.23985	1.23825
90	1.23950	1.23920	1.23755	1.23585	1.23920	1.23895	1.23725	1.23500
89	1.23680	1.23655	1.23490	1.23320	1.23655	1.23625	1.23460	1.23295
88	1.23415	1.23390	1.23220	1.23055	1.23390	1.23360	1.23195	1.23025
87	1.23150	1.23120	1.22955	1.22790	1.23125	1.23095	1.22930	1.22760
86	1.22885	1.22855	1.22690	1.22520	1.22860	1.22830	1.22660	1.22495
85	1.22620	1.22590	1.22420	1.22255	1.22595	1.22565	1.22395	1.22230
84	1.22355	1.22325	1.22155	1.21990	1.22330	1.22300	1.22130	1.21965
83	1.22090	1.22055	1.21890	1.21720	1.22060	1.22030	1.21865	1.21695
82	1.21820	1.21790	1.21620	1.21455	1.21795	1.21765	1.21595	1.21430
81	1.21555	1.21525	1.21355	1.21190	1.21530	1.21500	1.21330	1.21165
80	1.21290	1.21260	1.21090	1.20925	1.21265	1.21235	1.21065	1.20900
79	1.21015	1.20985	1.20815	1.20655				
78	1.20740	1.20710	1.20540	1.20380				
77	1.20465	1.20440	1.20270	1.20110				
76	1.20190	1.20165	1.19995	1.19840				
75	1.19915	1.19890	1.19720	1.19565	1.19890	1.19865	1.19700	1.19540
74	1.19640	1.19615	1.19450	1.19295				
73	1.19365	1.19340	1.19175	1.19025				
72	1.19090	1.19079	1.18900	1.18755				
71	1.18815	1.18795	1.18630	1.18480				

Table 6.1 Specific Gravity and Percent Glycerol by Weight

Glycerol (%)	Apparent specific gravity (°C)				True specific gravity (°C)			
	15/15	15.5/15.5	20/20	25/25	15/15	15.5/15.5	20/20	25/25
70	1.18540	1.18520	1.18355	1.18210	1.18515	1.18495	1.18330	1.18185
69	1.18260	1.18240	1.18080	1.17935				
68	1.17985	1.17965	1.17805	1.17660				
67	1.17705	1.17685	1.17530	1.17385				
66	1.17430	1.17410	1.17255	1.17110				
65	1.17155	1.17130	1.16980	1.16835	1.17135	1.17110	1.16970	1.16815
64	1.16875	1.16855	1.16705	1.16560				
63	1.16600	1.16575	1.16430	1.16285				
62	1.16320	1.16300	1.16155	1.16010				
61	1.16045	1.16020	1.15875	1.15735				
60	1.15770	1.15745	1.15605	1.15460	1.15750	1.15725	1.15585	1.15445
59	1.15490	1.15465	1.15325	1.15185				
58	1.15210	1.15190	1.15050	1.14915				
57	1.14935	1.14910	1.14775	1.14640				
56	1.14655	1.14635	1.14500	1.14365				
55	1.14375	1.14355	1.14220	1.14090	1.14360	1.14340	1.14205	1.14075
54	1.14100	1.14080	1.13945	1.13815				
53	1.13820	1.13800	1.13670	1.13540				
52	1.13540	1.13525	1.13395	1.13265				
51	1.13265	1.13245	1.13120	1.12995				
50	1.12985	1.12970	1.12845	1.12720	1.12970	1.12955	1.12830	1.12705
49	1.12710	1.12695	1.12570	1.12450				
48	1.12440	1.12425	1.12300	1.12185				
47	1.12165	1.12150	1.12030	1.11915				
46	1.11890	1.11880	1.11760	1.11650				
45	1.11620	1.11605	1.11490	1.11380	1.11605	1.11595	1.11475	1.11365
44	1.11345	1.11335	1.11220	1.11115				
43	1.11075	1.11060	1.10950	1.10845				
42	1.10800	1.10790	1.10680	1.10575				
41	1.10525	1.10515	1.10410	1.10310				
40	1.10255	1.10245	1.10135	1.10040	1.10240	1.10235	1.10125	1.10030
39	1.09985	1.09975	1.09870	1.09775				
38	1.09715	1.09705	1.09605	1.09510				
37	1.09445	1.09435	1.09335	1.09245				
36	1.09175	1.09165	1.09070	1.08980				
35	1.08905	1.08895	1.08805	1.08715	1.08895	1.08885	1.08790	1.08705
34	1.08635	1.08625	1.08535	1.08445				
33	1.08365	1.08355	1.08270	1.08190				
32	1.08100	1.08085	1.08005	1.07925				
31	1.07830	1.07815	1.07735	1.07660				

Table 6.1 (*Continued*)

Glycerol (%)	Apparent specific gravity (°C)				True specific gravity (°C)			
	15/15	15.5/15.5	20/20	25/25	15/15	15.5/15.5	20/20	25/25
30	1.07560	1.07545	1.07470	1.07395	1.07550	1.07535	1.07460	1.07385
29	1.07295	1.07285	1.07210	1.07135				
28	1.07035	1.08025	1.06950	1.06880				
27	1.06770	1.06760	1.06690	1.06625				
26	1.06510	1.06500	1.06435	1.06370				
25	1.06250	1.06240	1.06175	1.06115	1.06240	1.06230	1.06165	1.06110
24	1.05985	1.05980	1.05915	1.05860				
23	1.05725	1.05715	1.05655	1.05605				
22	1.05460	1.05455	1.05400	1.05350				
21	1.05200	1.05105	1.05140	1.05095				
20	1.04935	1.04935	1.04880	1.04840	1.04930	1.04925	1.04875	1.04830
19	1.04685	1.04680	1.04630	1.04590				
18	1.04435	1.04430	1.04380	1.04345				
17	1.04180	1.04180	1.04135	1.04100				
16	1.03930	1.03925	1.03885	1.03850				
15	1.03675	1.03675	1.03635	1.03605	1.03670	1.06370	1.03630	1.03600
14	1.03425	1.03420	1.03390	1.03360				
13	1.03175	1.03170	1.03140	1.03110				
12	1.02920	1.02920	1.02890	1.02865				
11	1.02670	1.02665	1.02640	1.02620				
10	1.02415	1.02415	1.02395	1.02370	1.02415	1.02410	1.02390	1.02370
9	1.02175	1.02175	1.02155	1.02135				
8	1.01935	1.01930	1.01915	1.01900				
7	1.01690	1.01690	1.01675	1.01660				
6	1.01450	1.01450	1.01435	1.01425				
5	1.01210	1.01205	1.01195	1.01185	1.01205	1.01205	1.01195	1.01185
4	1.00965	1.00965	1.00955	1.00950				
3	1.00725	1.00725	1.00720	1.00710				
2	1.00485	1.00485	1.00480	1.00475				
1	1.00240	1.00240	1.00240	1.00235				

Source: From Ref. 5.

Another set of specific gravity determinations which has been employed in the industry are those determined by Timmerman and Hennaut-Roland [4], which show slight differences in the fourth and fifth decimal places. For 100% concentration at 15/4°C, the following true specific gravities were found:

Bosart and Snoddy	1.26415
Timmermans and Hennaut-Roland	1.26443

Table 6.2 has been calculated from Timmermans results, corrected to apparent specific gravity at 60/60°F = 1.26560. It shows concentrations corresponding to tabulated values of specific gravity at 60°F [5].

III. DENSITY

The density of glycerol solutions at various concentrations and temperature was calculated from their specific gravity data by Bosart and Snoddy [3], and are shown in Table 6.3. In a subsequent discussion [6] of these density values versus those appearing in the International Critical Table [7], it is suggested that the latter are unsatisfactory where accuracy in the fourth decimal place is necessary, particularly at or above 20°C.

The densities of glycerol–water covering a wider temperature range has been compiled from six different sources and is shown in Tables 6.4 and 6.5 [8].

The densities of glycerol solutions at low temperatures were determined by Green and Parke [9] and are shown in Table 6.6.

IV. BY WEIGHT VS. BY VOLUME

The correlation of the composition of glycerol solutions to achieve a definite percentage by weight of pure glycerol has been calculated and is given in Table 6.7.

Table 6.2 Specific Gravity (SG) of Aqueous Glycerin Solutions with Glycerol Concentration Between 95 and 100%

SG	0	1	2	3	4	5	6	7	8	9
1.252	94.72	94.76	94.80	94.84	94.88	94.92	94.96	95.00	95.03	95.07
1.253	94.11	95.15	95.19	95.22	95.26	95.30	95.34	95.38	95.42	95.46
1.254	95.49	95.53	95.57	95.61	95.65	95.69	95.73	95.77	95.80	95.84
1.255	95.88	95.92	95.96	96.00	96.04	96.08	96.11	96.15	96.19	96.23
1.256	96.27	96.31	96.35	96.39	96.42	96.46	96.50	96.54	96.58	96.62
1.257	96.66	96.69	96.73	96.77	96.81	96.85	96.89	96.93	96.97	97.00
1.258	97.04	97.08	97.12	97.16	97.20	97.24	97.28	97.32	97.36	97.40
1.259	97.44	97.48	97.52	97.56	97.60	97.64	97.68	97.72	97.76	97.80
1.260	97.84	97.88	97.92	97.96	98.00	98.04	98.07	98.10	98.14	98.18
1.261	98.22	98.26	98.30	98.34	98.38	98.41	98.45	98.49	98.53	98.57
1.262	98.61	98.65	98.69	98.72	98.76	98.80	98.84	98.88	98.92	98.96
1.263	99.00	99.03	99.07	99.11	99.15	99.19	99.23	99.27	99.31	99.34
1.264	99.38	99.42	99.46	99.50	99.54	99.58	99.61	99.65	99.69	99.73
1.265	99.77	99.81	99.85	99.89	99.93	99.86	100.00			

Source: From Ref. 5.

Table 6.3 Density and Percent of Glycerol

Glycerol (%)	Density at (°C)					Glycerol (%)	True specific gravity (°C)				
	15	15.5	20	25	30		15	15.5	20	25	30
100	1.26415	1.26381	1.26108	1.25802	1.25495	50	1.12870	1.12845	1.12630	1.12375	1.12110
99	1.26160	1.26125	1.25850	1.25545	1.25235	49	1.12600	1.12575	1.12360	1.12110	1.11845
98	1.25900	1.25865	1.25590	1.25290	1.24975	48	1.12325	1.12305	1.12090	1.11840	1.11580
97	1.25645	1.25610	1.25335	1.25030	1.24710	47	1.12055	1.12030	1.11820	1.11575	1.11320
96	1.25385	1.25350	1.25080	1.24770	1.24450	46	1.11780	1.11760	1.11550	1.11310	1.11055
95	1.25130	1.25095	1.24825	1.24515	1.24190	45	1.11510	1.11490	1.11280	1.11040	1.10795
94	1.24865	1.24830	1.24560	1.24250	1.23930	44	1.11235	1.11215	1.11010	1.10775	1.10530
93	1.24600	1.24565	1.24300	1.23985	1.23670	43	1.10960	1.10945	1.10740	1.10510	1.10265
92	1.24340	1.24305	1.24035	1.23725	1.23410	42	1.10690	1.10670	1.10470	1.10240	1.10005
91	1.24075	1.24040	1.23770	1.23460	1.23150	41	1.10415	1.10400	1.10200	1.09975	1.09740
90	1.23810	1.23775	1.23510	1.23200	1.22890	40	1.10145	1.10130	1.09930	1.09710	1.09475
89	1.23545	1.23510	1.23245	1.22935	1.22625	39	1.09875	1.09860	1.09665	1.09445	1.09215
88	1.23280	1.23245	1.22975	1.22665	1.22360	38	1.09605	1.09590	1.09400	1.09180	1.08955
87	1.23015	1.22980	1.22710	1.22400	1.22095	37	1.09340	1.09320	1.09135	1.08915	1.08690
86	1.22750	1.22710	1.22445	1.22135	1.21830	36	1.09070	1.09050	1.08865	1.08655	1.08430
85	1.22485	1.22445	1.22180	1.21870	1.21575	35	1.08800	1.08780	1.08600	1.08390	1.08165
84	1.22220	1.22180	1.21915	1.21605	1.21300	34	1.08530	1.08515	1.08335	1.08125	1.07905
83	1.21955	1.21915	1.21650	1.21340	1.21035	33	1.08265	1.08245	1.08070	1.07860	1.07645
82	1.21690	1.21650	1.21380	1.21075	1.20770	32	1.07995	1.07075	1.07800	1.07600	1.07380
81	1.21425	1.21385	1.21115	1.20810	1.20505	31	1.07725	1.07705	1.07535	1.07335	1.07120
80	1.21160	1.21120	1.20850	1.20545	1.20240	30	1.07455	1.07435	1.07270	1.07070	1.06855
79	1.20885	1.20845	1.20575	1.20275	1.19970	29	1.07105	1.07175	1.07010	1.06815	1.06605
78	1.20610	1.20570	1.20305	1.20005	1.19705	28	1.06935	1.06915	1.06755	1.06560	1.06355

77	1.20335	1.20300	1.20030	1.19735	1.19435	27	1.06670	1.06655	1.06495	1.06305	1.06105
76	1.20060	1.20025	1.19760	1.19465	1.19170	26	1.06410	1.06390	1.06240	1.06055	1.05855
75	1.19785	1.19750	1.19485	1.19195	1.18900	25	1.06150	1.06130	1.05980	1.05800	1.05605
74	1.19510	1.19480	1.19215	1.18925	1.18635	24	1.05995	1.05870	1.05720	1.05545	1.05350
73	1.19235	1.19205	1.18940	1.18650	1.18365	23	1.05625	1.05610	1.05465	1.05290	1.05100
72	1.18965	1.18930	1.18670	1.18380	1.18100	22	1.05365	1.05350	1.05205	1.05035	1.04850
71	1.18690	1.18655	1.18395	1.18110	1.17830	21	1.05100	1.05090	1.04950	1.04780	1.04600
70	1.18415	1.18385	1.18125	1.17840	1.17565	20	1.04840	1.04825	1.04690	1.04525	1.04350
69	1.18135	1.18105	1.17850	1.17565	1.17290	19	1.04590	1.04575	1.04440	1.04280	1.04105
68	1.17860	1.17830	1.17575	1.17295	1.17020	18	1.04335	1.04325	1.04195	1.04035	1.03860
67	1.17585	1.17555	1.17300	1.17020	1.16745	17	1.04085	1.04075	1.03945	1.03790	1.03615
66	1.17305	1.17257	1.17025	1.16745	1.16470	16	1.03835	1.03825	1.03605	1.03545	1.03370
65	1.17030	1.17000	1.16750	1.16475	1.16195	15	1.03580	1.03570	1.03450	1.03300	1.03130
64	1.16755	1.16725	1.16475	1.16200	1.15925	14	1.03330	1.03320	1.03200	1.03055	1.02885
63	1.16480	1.16445	1.16205	1.15925	1.15650	13	1.03080	1.13070	1.02955	1.02805	1.02640
62	1.16200	1.16170	1.15930	1.15655	1.15375	12	1.02830	1.02820	1.02705	1.02560	1.02395
60	1.15650	1.15615	1.15380	1.15105	1.14830	10	1.02325	1.02315	1.02210	1.02070	1.01905
59	1.15370	1.15340	1.15105	1.14835	1.14555	9	1.02085	1.02075	1.01970	1.01835	1.01670
58	1.15095	1.15065	1.14830	1.14560	1.14285	8	1.01840	1.01835	1.01730	1.01600	1.01440
57	1.14815	1.14785	1.14555	1.14285	1.14010	7	1.01600	1.01590	1.01495	1.01360	1.01205
56	1.14535	1.14510	1.14280	1.14015	1.13740	6	1.01360	1.01350	1.01255	1.01125	1.00970
55	1.14260	1.14230	1.14005	1.13740	1.13470	5	1.01120	1.01110	1.01015	1.00890	1.00735
54	1.13980	1.13955	1.13730	1.13475	1.13195	4	1.00875	1.00870	1.00780	1.00655	1.00505
53	1.13705	1.13680	1.13455	1.13195	1.12925	3	1.00635	1.00630	1.00540	1.00415	1.00270
52	1.13425	1.13400	1.13180	1.12920	1.12650	2	1.00395	1.00385	1.00300	1.00180	1.00035
51	1.13150	1.13125	1.12905	1.12650	1.12380	1	1.00155	1.00145	1.00060	0.99945	0.99800
						0	0.99913	0.99905	0.99823	0.99708	0.99568

Source: From Ref. 5.

Table 6.4   100 Percent Glycerol

Temp. (°C)	Density (g/ml)	Temp. (°C)	Density (g/ml)	Temp. (°C)	Density (g/ml)
0	1.27269	75.5	1.2256	180	1.14864
10	1.26699	99.5	1.2097	200	1.13178
15	1.26443	110	1.20178	220	1.11493
20	1.26134	120	1.19446	240	1.09857
30	1.25512	130	1.18721	260	1.08268
40	1.24896	140	1.17951	280	1.06725
54	1.2397	160	1.16440	290	1.05969

Source: From Ref. 5.

The dilution of glycerol to a specific concentration from a solution of a given percentage (by weight or by volume) can be calculated from the following formula [10]:

$$x = 100 \ (a-b)/b \tag{1}$$

$$y = 100 \ [(a-b)/b] \ D_t^t \tag{2}$$

$$K = 100 - [V(D_t^t - 1) + 100 \ d_t^t] \tag{3}$$

in which

x = parts by wt of water to be added to 100 parts of glycerol
y = parts by volume of water to be added to 100 volumes of glycerol
K = % contraction based on the final volume
a = % (wt or vol) of glycerol in the original material
b = % (wt or vol) of glycerol in the desired concentration
D_t^t = density of the original material
d_t^t = density of the desired concentration
V = vol. of the original material

V. REFRACTIVE INDEX

The refractive index can be used to measure the concentration of glycerol. It is easily measured and sensitive to dilution with water. However, these results are not quite as precise as those obtained from specific gravity. The refractive index of 100% glycerol is $(N^{20/D})$ 1.47399.

The determination of the refractive index of glycerol solutions have been made by a number of investigators including Hoyt [11] who prepared samples from 5 to 100% concentration as determined by Bosart and Snoddy. Refractive index values at 20 ± 0.1°C were plotted against concentration and found to

Table 6.5 Density of Glycerol–Water

% Wt. glycerol Temp. (°C)	Density (g/ml) (in vacuo)									
	90	80	70	60	50	40	30	20	10	0
0	1.24683	1.21962	1.19200	1.16349	1.13486	1.10667	1.07892	1.05161	1.02517	0.99987
10	1.24124	1.12140	1.18701	1.15909	1.13101	1.10336	1.07623	1.04951	1.02414	0.99973
20	1.23510	1.20850	1.18125	1.15380	1.12630	1.09930	1.07270	1.04690	1.02210	0.99823
30	1.22865	1.20231	1.17519	1.14822	1.12096	1.09452	1.06856	1.04347	1.01916	0.99567
40	1.22214	1.19606	1.16920	1.14247	1.11534	1.08924	1.06371	1.03945	1.01552	0.99224
50	1.2158	1.18999	1.16338	1.13655	1.10945	1.00380	1.05849	1.03505	1.01131	0.98807
60	1.20922	1.18352	1.15683	1.13015	1.1034	1.07800	1.05291	1.02958	1.00625	0.98324
70	1.20269	1.17679	1.15036	1.12382	1.09723	1.07184	1.04729	1.02386	1.00065	0.97781
80	1.19611	1.16990	1.14384	1.11745	1.09079	1.06564	1.04093	1.01752	0.99463	0.97183
90	1.18961	1.16332	1.13730	1.11094	1.08423	1.05901	1.0343	1.01097	0.98840	0.96534
100	1.18273	1.15604	1.13018	1.10388	1.07733	1.05217	1.02735	1.00392	0.98187	0.95838
Density at bp	1.15479	1.14004	1.12030	1.09763	1.07295	1.04947	1.02539	1.00277	0.98121	0.95838
bp	138	121	113.3	109	106	104	102.8	101.8	100.9	100

Source: From Ref. 5.

Table 6.6 Density and Percent of Glycerol at Low Temperatures

Glycerol (%)	F.p. (°C)	Temperature (°C)				
		−5	−10	−20	−30	−40
10	−1.6	—	—	—	—	—
20	−4.8	—	—	—	—	—
30	−9.5	1.0810	—	—	—	—
40	−15.4	1.1096	1.1109	—	—	—
50	−23.0	1.1397	1.1407	1.1450	—	—
60	−34.7	1.1663	1.1685	1.1732	1.1787	—
60.7	−46.7	1.1860	1.1889	1.1945	1.1985	1.2034
70	−38.5	1.1954	1.1993	1.2038	1.2079	—
80	−20.3	1.2210	1.2255	1.2305	—	—
90	−1.6	—	—	—	—	—

Source: From Ref. 5.

fall along a curve which could be divided into three sections: a curve from 0 to 44%, a straight line from 45 to 79%, and a curve from 80 to 100% concentration. The equations for these lines are:

1. Curve from 0 to 44% glycerol

$$y = 1.3303 + 0.001124x$$
$$+ 0.00000605x^2$$
$$- 0.000000555x^2$$

2. Straight line from 45 to 79%

$$y = 0.00149x + 1.32359$$

3. Curve from 80 to 100%

$$y = 0.90799 + 0.0154x$$
$$- 0.000155x^2$$
$$+ 0.000000576x^3$$

With these equations refractive indices shown in Table 6.8 were calculated. These calculated values deviated from the observed values only by 0.00011 and make possible the determination of glycerol concentration with an accuracy of at least 0.1% [5].

VI. BOILING POINT

Much attention has been paid to the determination of the boiling points of glycerol and its solutions under various pressure conditions. The importance of

Table 6.7 Comparison of Composition of Glycerol Solutions by Weight and by Volume

% wt of Pure glycerol	Equivalent % wt of aqueous glycerol of given concentration				Equivalent % vol of aqueous glycerol of given concentration				
	99%	95%	75%	60%	100%	99%	95%	75%	60%
0.00	0.00	0.00	0.00	0.00	0.00	0.00	0.00	0.00	0.00
5.00	5.05	5.26	6.67	8.33	4.00	4.05	4.25	5.63	7.28
10.00	10.10	10.53	13.33	17.78	9.10	8.19	8.61	11.38	14.74
15.00	15.15	15.79	20.00	25.00	12.29	12.40	13.07	17.29	22.38
20.00	20.20	21.05	26.67	33.33	16.58	16.78	17.63	23.34	30.21
25.00	25.25	26.31	33.33	41.67	20.98	21.24	22.31	29.53	38.23
30.00	30.30	31.58	40.00	50.00	25.49	25.80	27.10	35.88	46.45
35.00	35.35	36.84	46.67	58.33	30.11	30.48	32.02	42.38	54.87
40.00	40.40	42.10	53.33	66.67	34.84	35.26	37.04	49.03	63.49
45.00	45.45	47.36	60.00	75.00	39.67	40.16	42.18	55.85	72.31
50.00	50.50	52.63	66.67	83.33	44.63	45.17	47.45	62.81	81.32
55.00	55.55	57.89	73.33	91.67	49.68	50.30	52.83	69.94	90.56
60.00	60.60	63.16	80.00	100.00	54.86	55.54	58.34	77.23	100.00
65.00	65.65	68.42	86.67		60.16	60.89	63.97	84.67	
66.67	67.34	70.21	88.89		61.95	62.71	65.88	87.18	
70.00	70.70	73.68	93.33		65.56	66.35	69.71	92.25	
75.00	75.75	78.9	100.00		71.07	71.92	75.57	100.00	
80.00	80.80	84.21			76.69	77.59	81.55		
85.00	85.85	89.47			82.39	83.34	87.61		
90.00	90.90	94.74			88.20	90.20	93.78		
95.00	95.95	100.00			94.05	95.16	100.00		
98.00	98.98				97.60	98.78			
99.00	100.00				98.81	100.00			
100.00					100.00				

Source: From Ref. 5.

123

Table 6.8   Refractive Index (RI) of Glycerol—Water Solution at 20.0°C

Glycerol (% wt)	RI 20 n D	Difference for 1%	Glycerol (% wt)	RI 20 n D	Difference for 1%
100	1.47399	0.00165	50	1.39809	0.00149
99	1.47234	0.00163	49	1.39660	0.00147
98	1.47071	0.00161	48	1.39513	0.00145
97	1.46909	0.00157	47	1.39368	0.00141
96	1.46752	0.00154	46	1.39227	0.00138
95	1.46597	0.00154	45	1.39089	0.00136
94	1.46443	0.00153	44	1.38953	0.00135
93	1.46290	0.00151	43	1.38818	0.00135
92	1.46139	0.00150	42	1.38683	0.00135
91	1.45989	0.00150	41	1.38548	0.00135
90	1.45839	0.00150	40	1.38413	0.00135
89	1.45689	0.00150	39	1.38278	0.00135
88	1.45539	0.00150	38	1.38143	0.00135
87	1.45389	0.00152	37	1.38008	0.00134
86	1.45237	0.00152	36	1.37874	0.00134
85	1.45085	0.00155	35	1.37740	0.00134
84	1.44930	0.00156	34	1.37606	0.00134
83	1.44770	0.00160	33	1.37472	0.00134
82	1.44612	0.00162	32	1.37338	0.00134
81	1.44450	0.00160	31	1.37204	0.00134
80	1.44290	0.00155	30	1.37070	0.00134
79	1.44135	0.00153	29	1.36936	0.00134
78	1.43982	0.00150	28	1.36802	0.00133
77	1.43832	0.00149	27	1.36669	0.00133
76	1.43683	0.00149	26	1.36536	0,99132
75	1.43534	0.00149	25	1.36404	0.00132
74	1.43385	0.00149	24	1.36272	0.00131
73	1.43236	0.00149	23	1.36141	0.00131
72	1.43087	0.00149	22	1.36010	0.00131
71	1.42930	0.00149	21	1.35879	0.00130
70	1.42780	0.00149	20	1.35749	0.00130
69	1.42640	0.00149	19	1.35619	0.00129
68	1.42491	0.00149	18	1.35490	0.00129
67	1.42342	0.00148	17	1.35361	0.00128
66	1.42193	0.00149	16	1.35233	0.00127
65	1,42944	0.00149	15	1.35106	0.00126
64	1.41895	0.00149	14	1.34980	0.00126
63	1.41746	0.00149	13	1.34854	0.00125

Table 6.8 (*Continued*)

Glycerol (% wt)	RI 20 n D	Difference for 1%	Glycerol (% wt)	RI 20 n D	Difference for 1%
62	1.41597	0.00149	12	1.34729	0.00125
61	1.41448	0.00149	11	1.34604	0.00123
60	1.41299	0.00149	10	1.34481	0.00122
59	1.41150	0.00149	9	1.34359	0.00121
58	1.41001	0.00149	8	1.34238	0.00120
57	1.40852	0.00149	7	1.34118	0.00119
56	1.40693	0.00149	6	1,33999	0.00119
55	1.40554	0.00149	5	1.33880	0.00118
54	1.40405	0.00149	4	1.33762	0.00117
53	1.40256	0.00149	3	1.33645	0.00115
52	1.40107	0.00149	2	1.33530	0.00114
51	1.39958	0.00149	1	1.33416	0.00113
			0	1.33303	—

Source: From Ref. 5.

such data from the scientific and technical view point will be readily appreciated. Rather surprisingly, recent research tends to confirm the basic reliability of Gerlach's results, although the more refined methods available today permitted a number of marginal adjustments [12]. Boiling points under reduced pressure were calculated by von Rechenberg [13] and are shown in Table 6.9. A more detailed calculation of the increase in boiling point of glycerol-water solutions, including such solutions saturated with salt, at reduced pressures, has been drawn up [14]. It was found that Duhring's rule applies to these solutions, enabling boiling points to be calculated from the equation

$$\frac{T_1}{t_1} - \frac{T_2}{t_2} = K$$

Where T_1 is the boiling point of a given substance at pressure P_1; T_2 is its boiling point at another pressure P_2 and t_1 and t_2 and boiling points of a second similar liquid at the same pressure. K is a constant. The boiling points of glycerol-water solutions as calculated by this rule are given in Table 6.10.

VII. FREEZING POINT

The freezing point of pure glycerol has been reported at values closely around 18°C, with 18.17°C as perhaps the most precise.

Table 6.9   Boiling Points of Pure Glycerol at Various Pressures

mmHg		mmHg	
800	292.01	60	208.40
760	290.00	50	203.62
700	286.79	40	197.96
600	280.91	30	190.87
500	274.23	20	181.34
400	266.20	15	179.86
300	256.32	10	166.11
200	243.16	8	161.49
100	222.41	6	155.69
90	219.44	5	152.03
80	216.17	4	147.87
70	212.52		

Source: From Ref. 13.

Measurements of the freezing points of glycerol solutions have been determined by the U.S. Bureau of Standards [15]. Values between 10% and 60% glycerol are shown in Table 6.11. Glycerol 66.7% (by weight) and 33.3% water form a eutectic mixture which has a freezing point generally reported as -46.5°C, although Ross [16] gives the figure of -43.5°C in a more recent study.

VIII.   VISCOSITY

Many measurements of the viscosity of glycerol and glycerol solutions have been made, using different types of viscometers. Consequently, there is some disagreement in data from various sources. Precise work done by Sheeley covers the entire range of concentration in more than 100 measurements, and has been commended as suitable for the calibration of viscometers in the 20°C to 30°C temperature range [17]. A more general determination covering 0–100% concentration and 0 to 100°C temperature was made by Segur and Oberstar [18] and is given in Table 6.12.

The viscosity of glycerol solutions below 0°C was determined by Green and Parke [9], who used Ostwald viscometers except with 66.7% glycerol at−40°C, in which case a falling ball viscometer was used. Values are shown in Table 6.13. Viscosities of supercooled glycerol have also been determined by the falling ball viscometer down to -40°C and are given in Table 6.14. At −89°C, glycerol becomes hard, its viscosity having been calculated as 10^{13} poises [19].

The viscosity of glycerol at high temperature 80–167°C has also been derived and is given in Table 6.15 [20].

Table 6.10 Boiling Points of Glycerol-Water Solutions (Calculated from Duhring Lines)

Pressure (mm)	BP of water (°C)	Water (%) Glycerol (%)	Boiling Points (°C)									
			90 / 10	80 / 20	70 / 30	60 / 40	50 / 50	40 / 60	30 / 70	20 / 80	10 / 90	4.36 / 95.64
760.00	100		100.7	101.6	102.9	104.5	106.7	109.6	110.4	121.5	139.7	175.8
525.80	90		90.6	91.5	92.8	94.2	96.3	99.3	103.5	110.3	127.8	161.1
355.10	80		80.5	81.4	82.6	84.0	86.0	88.8	92.8	99.3	116.0	146.5
233.53	70		70.4	71.2	72.4	73.7	75.6	78.5	82.2	88.3	104.0	132.1
149.19	60		60.3	61.0	62.2	63.5	65.5	68.1	71.5	77.3	92.0	117.6
92.30	50		50.2	5.9	52.1	53.4	55.2	57.6	61.0	66.2	80.1	103.1

Source: From Ref. 14.

Table 6.11 Freezing Points of Glycerol-Water Solutions (°C)

Glycerol (% wt)	Lane	Bureau of Standards	Olsen, Brunjes, and Olsen	Feldman and Hahlstrom	Spangler and Daves
10	−1.6	−1.7	−2.3	−1.9	−1.99
20	−4.8	−4.8	−5.5	−5.4	−5.21
30	−9.5	−9.4	−9.8	−9.7	−9.92
35	−12.2	−12.3	−12.4	—	−12.65
40	−15.4	−15.6	−15.7	−15.4	−15.93
45	−18.8	−19.4	−18.6	—	−19.90
50	−23.0	−25.8	−28.8	−23.6	−24.55
55	−28.2	—	—	—	−30.10
60	−34.7	—	−37.2	−35.5	−37.90

Source: From Ref. 5.

Viscosity of glycerol rises with increasing pressure [21]. Generally, electrolytes dissolved in glycerol increases its viscosity, but a few have an opposite effect. Ammonium bromide, and iodide, rubidium chloride, bromide and iodide, and cesium chloride and nitrate belong to this group [22,23]. Potassium iodide also reduces the viscosity of glycerol [24].

Mixtures of glycerol-alcohol, glycerol-alcohol-water, and sugar solutions in glycerol have been studied with respect to viscosity [25]. Viscosities of glycerol-alcohol mixtures are shown in Table 6.16.

IX. VOLUMETRIC CONTRACTION

When glycerol and water are mixed, there is a rise in temperature and contraction in volume. This contraction was measured by Gerlach and is shown in Table 6.17.

X. VAPOR PRESSURE

Glycerol has a lower vapor pressure than would be expected from its molecular weight, as a result of the molecular association characteristic of alcohols.

The vapor pressure of 100% glycerol is below 0.001 (mmHg) at room temperature, and below 0.2 mm at 100°C. Vapor pressures of pure glycerol calculated by extrapolation from partial pressures of solutions to the point where the partial pressure equals the total pressure are estimated to be accurate to about 1% [26]. They are shown in Table 6.18.

More interest has centered on the vapor pressure of aqueous solutions of glycerin. Glycerol causes a larger reduction in the vapor pressure of the water

Table 6.12 Viscosity of Aqueous Glycerol Solutions Centipoises

Glycerol (% wt)	Temperatures (°C)										
	0	10	20	30	40	50	60	70	80	90	100
0	1.792	1.308	1.005	0.8007	0.6560	0.5494	0.4688	0.4061	0.3565	0.3165	0.2838
10	2.44	1.74	1.31	1.03	0.826	0.680	0.565	0.500	—	—	—
20	3.44	2.41	1.76	1.35	1.07	0.879	0.731	0.635	—	—	—
30	5.14	3.49	2.50	1.87	1.46	1.16	0.956	0.816	0.690	—	—
40	8.25	5.37	3.72	2.72	2.07	1.62	1.30	1.09	0.918	0.763	0.668
50	14.6	9.01	6.00	4.21	3.10	2.37	1.86	1.53	1.25	1.05	0.910
60	29.9	17.4	10.8	7.19	5.08	3.76	2.85	2.29	1.84	1.52	1.28
65	45.7	25.3	15.2	9.85	6.80	4.89	3.66	2.91	2.28	1.86	1.55
67	55.5	29.9	17.7	11.3	7.73	5.50	4.09	3.23	2.50	2.03	1.68
70	76	38.8	22.5	14.1	9.40	6.61	4.86	3.78	2.90	2.34	1.93
75	132	65.2	35.5	21.2	13.6	9.25	6.61	5.01	3.80	3.00	2.43
80	255	116	60.1	33.9	20.8	13.6	9.42	6.94	5.13	4.03	3.18
85	540	223	109	58	33.5	21.6	14.2	10.0	7.28	5.52	4.24
90	1310	498	219	109	60.1	35.5	22.5	15.5	11.0	7.93	6.00
91	1590	592	259	127	68.1	39.8	25.1	17.1	11.9	8.62	6.40
92	1950	729	310	147	78.3	44.8	28.0	19.0	13.1	9.46	6.82
93	2400	860	367	172	89	51.5	31.6	21.2	14.4	10.3	7.54
94	2930	1040	437	202	105	58.4	35.4	23.6	15.8	11.2	8.19
95	3690	1270	523	237	121	67.0	39.9	26.4	17.5	12.4	9.08
96	4600	1580	624	281	142	77.8	45.4	29.7	19.6	13.6	10.1
97	5770	1950	765	340	166	88.9	51.9	33.6	21.9	15.1	10.9
98	7370	2460	939	409	196	104	59.8	38.5	24.8	17.0	12.2
99	9420	3090	1150	500	235	122	69.1	43.6	27.8	19.0	13.3
100	12070	3900	1410	612	284	142	81.3	50.6	31.9	21.3	14.8

Source: From Ref. 18.

129

Table 6.13   Viscosity of Glycerol-Water Solutions at Low Temperatures in Centipoises

Glycerol (% wt)	F.p. (°C)	−5°C	−10°C	−20°C	−30°C	−40°C
10	−1.6	—	—	—	—	—
20	−4.8	—	—	—	—	—
30	−9.5	6.5	—	—	—	—
40	−15.4	10.3	14.4	—	—	—
50	−23.0	18.8	24.4	48.1	—	—
60	−34.7	41.6	59.1	108.0	244.0	—
66.7	−46.5	74.7	113.0	289.0	631.0	1398.0
70	−38.5	110.0	151.0	394.0	1046.0	—
80	−20.3	419.0	683.0	1600.0	—	—
90	−1.6	—	—	—	—	—

Source: From Ref. 8.

Table 6.14   Viscosity of Supercooled Glycerol

Glycerol (°C)	+0.5% Water (Poises)	Glycerol (°C)	Anhydrous (Poises)
−40.9	45000	−42.0	67100
−37.0	22000	−41.8	63000
−34.0	10600	−36.3	21700
−34.0	10800	−36.0	20500
−30.0	5220	−28.9	5360
−26.0	2830	−25.0	2600
−13.9	407	−25.0	2640
−13.9	405	−20.0	1340
−6.1	142	−19.5	1230
−6.1	144	−15.4	665
−6.0	137	−10.8	357
+2.0	53.8	−10.8	352
+1.8	56.5	−4.2	148
+1.7	62.4	−4.2	149
+6.3	34.4		
+6.3	35.7		
+9.1	26.7		
+9.2	26.0		
+12.8	18.4		
+12.8	18.3		

Source: From Ref. 5.

Table 6.15 The Viscosity of 100% Glycerol at High Temperatures

°C	80	90	100	110	120	130	150	158	167
Centipoise	32.18	2	14.60	10.48	7.797	5.986	3.823	3.282	2.806

Source: From Ref. 20.

Table 6.16 Viscosity of Mixtures of Glycerol with Methyl Alcohol and with Alcohol Centipoises

Methyl alcohol			Ethyl alcohol		
(% wt)	25°C	45°C	(% wt)	25°C	45°C
100.0	875	311	0	875	311
9.8	375	149	9.9	504	187.1
24.8	97.7	46.55	24.8	167	77.6
49.4	14.38	9.77	49.1	24.58	21.8
74.3	2.75	2.53	73.6	6.17	1.25
87.8	1.23	1.22	100.0	1.23	1.25
100.0	0.63	0.71	—	—	—

Source: From Ref. 25.

Table 6.17 Volumetric Contractions of Glycerol and Water When Mixed at 20°C (Gerlach)

100 Parts out of soln contains:		Average hypothetical Vol of soln $\dfrac{A}{1.262} + B$	Actual vol. of soln 100 $\dfrac{100}{Sp.\ Gr.}$	$\dfrac{D \times 100}{C}$	$100 - E$
Glycerol A	Water B	C	D	E	F
100	0	79.240	79.240	100.0	0.0
90	10	81.315	80.906	99.497	0.503
80	20	83.392	82.713	99.186	0.814
70	30	85.468	84.603	98.987	1.013
60	40	87.544	86.580	98.899	1.101
50	50	89.620	88.652	98.920	1.080
40	60	91.696	90.826	99.052	0.948
30	70	93.772	93.110	99.294	0.705
20	80	95.885	95.420	99.515	0.485
10	90	97.924	97.704	99.775	0.225
0	100	100.0	100.0	100.0	0.0

Source: From Ref. 26.

Table 6.18   Vapor Pressure (VP) of Pure Glycerol

Temp. (°C)	VP (mmHg)	Temp. (°C)	VP (mmHg)
200	46.0	120	0.74
190	30.3	110	0.385
180	19.3	100	0.195
170	12.1	90	0.093
160	7.4	80	0.041
150	4.30	70	0.017
140	2.43	60	0.0067
130	1.35	50	0.0025

Source: From Ref. 26.

than can be accounted for by its molar concentration, due to the formation of hydrates. This effect was measured at 70°C by Perman and Price [27] and expressed in terms of apparent molecular weight. A 45% glycerol solution has a "vapor pressure-lowering" effect on water equivalent to a compound having an 86.9 molecular weight. Table 6.19 gives the effect at other concentrations.

Vapor pressures for aqueous glycerol solutions can be calculated according to Duhring's rule. This rule states that if a solution and water have the same vapor pressure, P_1, at temperatures t_1 and t_2, respectively, and have vapor pressure, p_2, at temperatures t_3 and t_4, respectively, then:

$$(t_1 - t_3/(t_2 - t_4) = K$$

Values of K (Duhring's constant) have been calculated for several glycerol concentrations and relative vapor pressures are given in Table 6.20. Measurements made by a manometer technique by Campbell [28] gives comparable values at 70°C in Table 6.21.

Table 6.19   Vapor Presure of Glycerol–Water at 70°C and Apparent Molecular Weight of Glycerol

Moles glycerol/ moles water	% Wt of Glycerol	Apparent mol. wt.	Vapor pressure (mmHg)
0.0259	11.69	189.6	230.9
0.0266	11.97	126.3	229.3
0.0540	21.63	96.4	222.3
0.1602	45.02	86.9	199.9
0.3079	61.15	81.4	173.4
0.6694	77.38	74.7	128.1
1.4390	88.03	71.0	79.8

Source: From Ref. 27.

Table 6.20 Duhring's Constants and Calculated Relative Vapor Pressure [a] of Glycerol Solutions from 0 to 70°C

Glycerol (%)	Duhring's constant	Relative vapor pressure			
		(0°)	(20°)	(40°)	(70°)
90	1.194	0.324	0.320	0.309	0.303
80	1.106	0.496	0.517	0.509	0.505
60	1.040	0.700	0.738	0.738	0.744
40	1.022	0.822	0.865	0.866	0.862
20	1.014	0.900	0.942	0.939	0.931

[a] Relative vapor pressure $= \dfrac{\text{vapor pressure of solution}}{\text{vapor pressure of water}}$

Source: From Ref. 5.

A liquid evaporates into a vacuum at a rate equal to that predicted by kinetic theory multiplied by $\times$ called the evaporation coefficient. A new technique has been developed to measure $\times$ and with pure glycerol it was found to be equal to 0.05 [29].

XI. SPECIFIC HEAT

The specific heat of pure glycerol has been determined by several workers with values in the range of 0.575 to 0.5795 cal/°C/g at 26°C [30,31].

The specific heat of glycerol–water mixtures at temperatures from 0° to -31°C was thoroughly explored by Gucker and Marsh [32], and is given in Table 6.22.

Table 6.21 Vapor Pressure (VP) of Water in Contact with Glycerol at 70°C

Water % wt	Mole fraction	Observed pressure (mm)	Ratio observed pressure to pressure of pure water
12.10	0.410	79.8	0.341
25.11	0.599	128.1	0.548
38.75	0.764	173.4	0.741
54.97	0.862	199.9	0.855
88.00	0.974	229.3	0.980
89.26	0.977	230.9	0.987
100.00	1.000	233.8	1.00

Source: From Ref. 28.

Table 6.22 Specific Heats (cal/g/°C or Btu/lb/°F) of Glycerol-Water Mixtures

°F	°C	Glycerol								
		25%	30%	35%	40%	45%	50%	55%	60%	65%
35	1.7	0.88	0.87	0.86	0.84	0.82	0.80	0.77	0.74	0.71
30	−1.1	0.88	0.86	0.85	0.83	0.81	0.79	0.76	0.73	0.70
25	−3.9	0.87	0.86	0.84	0.82	0.80	0.78	0.75	0.72	0.69
20	−6.7	0.86	0.85	0.83	0.82	0.79	0.77	0.74	0.71	0.68
19[a]	−7.2	6.8[a]	—	—	—	—	—	—	—	—
15[a]	−9.4	4.1	4.8[a]	0.82	0.80	0.78	0.76	0.73	0.70	0.67
10.4[a]	−12.0	—	—	3.7[a]	—	—	—	—	—	—
10	−12.2	2.7	3.2	3.6	0.80	0.78	0.75	0.72	0.69	0.66
5	−15.0	2.1	2.4	2.7	0.79	0.77	0.74	0.71	0.67	0.65
4.6[a]	−15.2	—	—	—	2.9[a]	—	—	—	—	—
0	−17.8	1.7	1.9	2.1	2.4	0.76	0.73	0.70	0.66	0.63
−1.8[a]	−18.8	—	—	—	—	2.9[a]	—	—	—	—
−5	−20.6	1.4	1.6	1.8	2.0	2.2	0.72	0.69	0.65	0.62
−9.6	−23.1	—	—	—	—	—	2.0[a]	—	—	—
−10	−23.3	1.2	1.4	1.6	1.7	1.9	2.0	0.68	0.64	0.61
−15	−26.1	1.1	1.2	1.3	1.5	1.6	1.7	0.67	0.63	0.60
−18.9[a]	−28.3	—	—	—	—	—	—	1.7[a]	—	—
−20	−28.9	1.0	1.1	1.2	1.3	1.4	1.5	1.6	0.62	0.59
−25	−31.7	0.9	1.0	1.1	1.2	1.25	1.3	1.4	0.61	0.58

[a]Estimated freezing point, and maximum specific heat for mixtures of this composition. The horizontal lines in each column mark the lower limit of the one-phase system.

Source: From Ref. 32.

XII. HEAT OF SOLUTION

When glycerol is dissolved in water there is a slight rise in temperature. The maximum amount of heat (though not the greatest rise in temperature) is obtained when the glycerol is dissolved in a large excess of water. If the glycerol is not anhydrous some of its heat of solution will have been dissipated and so correspondingly smaller amounts of heat will be produced by further dilution. Such measurements made by Fricke [33] are given in Table 6.23.

Glycerol has a negative heat of solution in methyl alcohol and ethyl alcohol. The following molar values were reported by Kalossowsky [34].

Methyl alcohol, 1:50 mol −370 cal
Water + methyl alcohol, 1:(43 + 43) mol −417 cal
Water + ethyl alcohol, 1:(200 + 128) mol −337 cal
Water + ethyl alcohol, 1:(150 + 25) mol ± 0 cal

XIII. DIFFUSION

Theoretically the diffusion of a nonelectrolyte into a solvent should be inversely proportional to the viscosity of the solvent. However, when glycerol is the diffusing substance, the values obtained in water and alcohol are not constant. The diffusion constant of water into glycerol is 1.33×10^{-7} cm^2/s at 20.08°C [35].

According to other work [36] the diffusion coefficient is independent of concentration up to 70% glycerol, after which it rises. The increased mobility at very high glycerol concentrations is attributed to a decreased hydration of the glycerol molecules under these conditions.

Table 6.23 Molar Heat of Solution of Glycerol

Moles water/ mole glycerol	% Water	Molar heat of solution cal
0.000	0.00	1381
0.164	3.11	1329
0.304	5.62	1261
0.684	11.81	1140
1.633	24.21	962
2.866	35.93	788
3.711	42.07	705
4.564	47.17	615

Source: From Ref. 33.

Table 6.24   Heat of Vaporization of Glycerol

Temp. (°C)	L (cal/mol)	Temp. (°C)	L (cal/mol)
195	18,170	115	19,530
185	18,780	105	19,300
175	18,610	95	19,910
165	18,740	85	20,840
155	18,740	75	21,170
145	19,810	65	21,120
135	19,430	55	21,060
125	18,925		

Source: From Ref. 38.

XIV.   HEAT OF FUSION

The heat of fusion is reported as 47.5 cal/g and 47.9 cal/g [30,37].

XV.   HEAT OF VAPORIZATION

The heat of vaporization of glycerol is 21,060 cal/mol at 55°C and 18,170 at 195°C. The intermediate values have been determined by Stedman [38] and are given in Table 6.24.

XVI.   HEAT OF FORMATION

Figures published in the literature include; 159.80 Kcal at 15°C [39], and 158.60 Kcal at 15°C [40].

XVII.   HEAT OF COMBUSTION

The heat of combustion at 20°C is 397.0 kg calories per mole [41].

XVIII.   FLASH AND FIRE POINT

The flash and fire point of 99% glycerol are 177°C and 204°C, respectively, when determined in a Cleveland open cup, using the procedure described in Test D92–85 of the American Society for Testing Materials [42]. When aqueous glycerol is tested it will not flash until enough water has evaporated to bring the glycerol concentration to about 97.5% by weight. It will then flash at 190°C. This higher flash point may be due to the water vapor accompanying the glycerol vapor. Continuation of the test to the fire point of 204°C results in further concentration of the glycerol to about 98.6 by weight.

XIX. AUTO IGNITION TEMPERATURE

The auto ignition temperature of glycerol is 523°C on platinum, 429°C on glass, and 412°C in oxygen at 1 atm [43].

XX. DISSOCIATION CONSTANT

The dissociation constant of glycerol as a weak acid is 0.07×10^{-13} [44].

XXI. MAGNETIC SUSCEPTIBILITY

The unit of magnetic susceptibility of glycerol ($H \times 10^6$) at 22°C is -0.64 [45].

XXII. CONDUCTIVITY

At ordinary temperatures the conductivity of glycerol is 5×10^{-8} reciprocal ohms [23]. The specific conductivity of glycerol at temperatures from 0 to 21°C is given in Table 6.25 [46].

The electrical conductivity of glycerol has been determined for solutions of from 1 to 94% glycerol, as shown in Table 6.26.

XXIII. SURFACE TENSION

The surface tension of aqueous glycerol at temperatures from 17 to 90°C were determined by Muller and Sitzber [47] and the figures are given in Table 6.27.

Surface tension figures for 100% glycerol at temperatures 104°C to 202°C are given in Table 6.28 [12].

XXIV. COMPRESSIBILITY

Glycerol is one of the least compressible liquids with a response about half that of water. The compressibility of anhydrous glycerol is given as 21.1×10^{-6} cc/atm/cm^2 at 28.5°C. Of 18 organic liquids tested, using pressures up to 12,000 kg/cm^2 at 0, 50, and 95°C glycerol was the least compressible. Figures on the relative volumes of water and pure glycerol under pressure are given in Table 6.29 [48].

Table 6.25 Specific Conductivity, N, of Glycerol

°C	0.0	5.9	11.7	14.8	16.0	17.6	20.0	21.3
$N \times 10^8$	2.2	3.6	5.6	7.8	8.4	9.6	10.0	12.3

Source: From Ref. 46.

Table 6.26 Electrical Conductivity of Glycerol-Water Solutions

Glycerol content (wt %)	$\times 0.10^6$ ohm^{-1} cm^{-1}	Glycerol content (wt %)	$\times 0.10^6$ ohm^{-1} cm^{-1}	Glycerol content (wt %)	$\times 0.10^6$ ohm^{-1} cm^{-1}	Glycerol content (wt %)	$\times 0.10^6$ ohm^{-1} cm^{-1}
1	10.30	25	13.75	49	11.65	73	6.40
2	10.55	26	13.70	50	11.50	74	6.10
3	10.80	27	13.60	51	11.40	75	5.80
4	11.05	28	13.53	52	11.30	76	5.50
5	11.30	29	13.47	53	11.20	77	5.20
6	11.55	30	13.40	54	11.05	78	4.90
7	11.80	31	13.35	55	10.82	79	4.55
8	12.05	32	13.30	56	10.70	80	4.25
9	12.30	33	13.25	57	10.50	81	4.00
10	12.55	34	13.20	58	10.30	82	3.75
11	12.80	35	13.10	59	10.15	93	3.50
12	13.05	36	13.00	60	10.00	84	3.25
13	13.30	37	12.85	61	9.75	85	3.00
14	13.50	38	12.75	62	9.50	86	2.75
15	13.70	39	12.65	63	9.25	87	2.50
16	13.90	40	12.60	64	9.00	88	2.25
17	14.05	41	12.50	65	8.75	89	2.00
18	14.10	42	12.40	66	8.50	90	1.80
19	14.05	43	12.30	67	8.25	91	1.60
20	14.00	44	12.20	68	8.00	92	1.40
21	13.95	45	12.10	69	7.65	93	1.20
22	13.90	46	12.00	70	7.30	94	1.00
23	13.85	47	11.90	71	7.00		
24	13.80	48	11.80	72	6.70		

Source: From Ref. 46.

XXV. ADIABATIC EXPANSION

The change in temperature by adiabatic expansion of glycerol was determined by Pushin and Grebenshchikov [49]. Their data are shown in Table 6.30.

XXVI. THERMAL EXPANSION

The expansion of glycerol with increasing temperature may be measured by its change in volume or change in density. The values at temperature intervals above 15.5°C were determined by Comey and Backus [50] and shown in Table 6.31. Values calculated by Bosart and Snoddy are given in Table 6.32. Gerlach [51] made direct expansion measurements with a dilatometer to show

how volume increases with temperature up to the boiling point. Values are given in Table 6.33.

XXVII. SOUND TRANSMISSION

A number of researchers have studied the velocity of sound in pure glycerol. The results vary but the most generally accepted value is 1.92×10^5 cm/s at 20°C. Figures for velocities at varying temperatures and concentrations are given in Table 6.34 [52].

XXVIII. CRYOSCOPIC CONSTANT

The cryoscopic constant is defined as the freezing point depression per mol of solute (water) in 1000 g of solvent (glycerol). This was determined by Pushin and Glagoleva [53] and their results are given in Table 6.35.

XXIX. DIELECTRIC CONSTANT

The dielectric constant of glycerol–water solutions at 25°C was determined by Albright [54]. The measurements were made with a current having a frequency of 0.57×10^6 cycles/s. Values are given in Table 6.36.

The effect of temperature on the dielectric constant of pure glycerol and also glycerol–water solutions within the range 20–100°C is given in Table 6.37.

Table 6.27   Surface Tension of Glycerol (dynes/cm)

Temp. (°C)	Glycerol (wt %)					
	99.19	81.89	61.44	39.31	20.20	0.0
17	—	65.41	—	—	—	—
18	62.47	—	—	69.86	71.13	—
20	—	65.26	67.64	—	70.93	71.68
30	62.08	64.66	66.68	68.42	69.49	70.25
40	61.53	63.93	65.71	67.18	68.02	68.68
50	61.05	63.05	64.67	65.86	66.79	67.05
60	60.34	62.11	63.59	64.55	65.23	65.50
70	59.36	61.11	62.39	63.09	63.73	63.94
80	58.72	60.07	61.21	61.62	62.01	62.16
90	57.85	59.02	59.92	60.13	60.48	60.51

Source: From Ref. 47.

Table 6.28   Surface Tension of Pure Glycerol

Temp. (°C)	M (dynes/cm)	Temp. (°C)	M (dynes/cm)
104.1	55.7	171	48.9
121	54.9	184.5	47.5
130	54.1	202	45.3
151	51.1		

Source: From Ref. 12.

These measurements were made by the resonance method at a wave length of 150 meters, which corresponds to a frequency of 2×10^6 cycles/s [55].

XXX. CRYSTALLINE GLYCEROL

In spite of the tendency of glycerol to supercool, it can be crystallized by the use of seed crystals, or by cooling to -50°C or below and then slowly warming to about 0°C. Approximately 24 hours are required for crystallization to be complete.

Table 6.29   Compressibility of Water and Glycerol

Pressure	Water			Pure glycerol			50% Glycerol
kg/cm²	0°C	50°C	95°C	0°C	50°C	95°C	30°C
0	1.0000	—	—	1.0000	1.0266	—	1.0000
500	0.9771	—	—	0.9900	1.0136	—	—
1000	0.9567	0.9741	0.9984	0.9806	1.0025	1.0240	0.9674
1500	0.9396	0.9582	0.9812	0.9721	0.9930	1.0125	—
2000	0.9248	0.9439	0.9661	0.9641	0.9843	1.0024	0.9416
3000	0.8996	0.9201	0.9409	0.9501	0.9688	0.9853	0.9199
4000	0.8795	0.8997	0.9194	0.9373	0.9548	0.9700	0.9007
5000	0.8626	0.8824	0.9009	0.9264	0.9423	0.9565	0.8842
6000	—	0.8668	0.8849	0.9157	0.9310	0.9447	0.8701
7000	—	0.8530	0.8705	0.9057	0.9211	0.9342	0.8566
8000	—	0.8407	0.8577	0.8958	0.9121	—	0.8447
9000	—	0.8296	0.8461	0.8867	0.9036	—	0.8342
10000	—	0.8192	0.8352	0.8783	0.8955	—	0.8241
11000	—	—	0.8256	0.8712	0.8879	—	0.8149
12000	—	—	—	0.8648	0.8800	—	—

Source: From Ref. 48.

Table 6.30 Adiabatic Expansion of Glycerol and Water

Pressure, p (kg/cm^2)	dt/dp						
	1	500	1000	1500	2000	2500	3000
Water at 0°C	−0.00130	−0.00020	0.0064	0.00116	0.00150	0.00173	0.00189
Water at 80°C.	0.00490	0.00468	0.00445	0.00423	0.00406	0.00392	0.00382
Glycerol at 25°C	0.00437	0.00407	0.00380	0.00352	0.00327	0.00308	0.00294
Glycerol at 98.2°C	0.00625	0.00570	0.00520	0.00475	0.00441	—	—

dt/dp = change of temp./change of pressure of 1 kg/cm^2 at a pressure p.
Source: From Ref. 49.

Table 6.31 Coefficient of Expansion (Gravimetric) of High Gravity Glycerol of Specific Gravity 1.254 to 1.264 at 15.5/15.5°C

Temp. interval	Coefficient
15.5–20	0.000612
15.5–25	0.000617
15.5–30	0.000622

Source: From Ref. 50.

XXXI. VITREOUS GLYCEROL

In the temperature range of −70 to −110°C, the already highly viscous supercooled (amorphous) glycerol passes into the vitreous state. Usually, the change occurs at −83°C with a sudden lowering of the specific heat.

XXXII. BINARY AND TERNARY MIXTURES

The boiling points of various glycerol mixtures, both azeotropic and nonazeotropic, has been compiled and is shown in Table 6.38.

Table 6.32 Coefficient of Thermal Expansion (Gravimetric) of Mixture of Glycerol and Water

Glycerol (%)	Change in S G/degree		
	15–20°C	15–25°C	20–25°C
100	0.000615	0.000615	0.000610
97.5	0.000620	0.000615	0.000605
95	0.000615	0.000615	0.000615
90	0.000610	0.000615	0.000620
80	0.000620	0.000615	0.000610
70	0.000580	0.000570	0.000565
60	0.000540	0.000545	0.000550
50	0.000485	0.000495	0.000510
40	0.000430	0.000435	0.000445
30	0.000370	0.000385	0.000400
20	0.000300	0.000315	0.000325
10	0.000230	0.000255	0.000280
Water	0.000180	0.000205	0.000230

Source: From Ref. 5.

Table 6.33   Thermal Expansion of Pure Glycerol

°C Vol	°C Vol	°C Vol
0 = 10,000	100 = 10,530	200 = 11,245
10 = 10,045	110 = 10,590	210 = 11,330
20 = 10,090	120 = 10,655	220 = 11,415
30 = 10,140	130 = 10,720	230 = 11,500
40 = 10,190	140 = 10,790	240 = 11,585
50 = 10,240	150 = 10,860	250 = 11,670
60 = 10,295	160 = 10,930	260 = 11,755
70 = 10,350	170 = 11,005	270 = 11,840
80 = 10,410	180 = 11,080	280 = 11,925
90 = 10,470	190 = 11,160	290 = 12,010

Source: From Ref. 52.

XXXIII.   SOLUBILITY

Glycerol will dissolve a large number of organic and inorganic compounds and will be miscible with many other substances. Thus, glycerol will be completely miscible with most of the lower aliphatic alcohols, phenol, ethylene, propylene, and trialkyl glycols, some glycol ethers, but only partially, or not at all, with others.

The introduction of hydroxyl and amine groups into aliphatic and aromatic hydrocarbons increases their miscibility with glycerol, while the introduction of alkyl groups decreases their miscibility. Miscibility data for a number of compounds with glycerol are given in Tables 6.39 and 6.40.

Table 6.34   Adiabatic Compressibility of Glycerol; Velocity of Sound in Glycerol

Temp. (°C)	Bhagavantum and Joga Rao			Freyer et al.		
	Density	Velocity of sound (m/s)	Compressibility (atm/cc × 10^{-6})	Density	Velocity of sound (m/s)	Compressibility (atm/cc × 10^{-6})
10	—	—	—	1.2671	1941.5	21.2
20	—	—	—	1.2613	1923	21.7
28.5	1.2562	1957	21.1	—	—	—
30	—	—	—	1.2553	1905	22.3
40	1.2490	1926	21.9	1.2491	1886.5	22.8
50	1.2426	1895	22.7	1.2427	1868.5	23.4
60	1.2360	1862	23.6	—	—	—
70	1.2292	1840	24.3	—	—	—

Source: From Ref. 52.

Table 6.35 Cryoscopic Constant of Glycerol

Glycerol (g)	Water (g)	t (°C)[a]	K[b]
11.4474	0.0435	0.690	3.27
10.7008	0.0610	1.116	3.52
11.4474	0.1297	2.326	3.69

[a] t = depression of the freezing point.
[b] K = cryoscopic constant; the freezing point depression per mole of solute (water) in 1000 g of solvent (glycerol).
Source: From Ref. 53.

Table 6.36 Dielectric Constant, e, of Glycerol-Water Solutions at 25°C (Current Frequency = 0.57×10^6 cycles/s)

Glycerol (% wt)		Glycerol (% wt)	
0.00	78.48	60.15	62.38
9.88	75.98	70.00	58.52
20.33	73.86	79.86	54.08
30.19	71.44	90.42	48.66
39.67	68.93	100.00	42.48
50.23	65.72		

Source: From Ref. 54.

Table 6.37 Dielectric Constant, e, of Glycerol-Water Solutions (Current Frequency = 2×10^6 cycles/s)

Glycerol (% wt)	20°C	25°C[a]	40°C	60°C	80°C	100°C
0	80.37	78.5	73.12	66.62	60.58	55.10
10	77.55	75.7	70.41	63.98	58.31	—
20	74.72	72.9	67.70	61.56	56.01	—
30	71.77	70.0	64.87	58.97	53.65	—
40	68.76	67.1	62.03	56.24	51.17	—
50	65.63	64.0	59.55	53.36	48.52	—
60	62.03	60.0	55.48	50.17	45.39	41.08
70	57.06	55.6	51.41	46.33	41.90	38.07
80	52.27	50.6	46.92	42.32	38.30	34.70
90	46.98	45.5	42.26	38.19	34.47	31.34
100	41.14	40.1	37.30	33.82	30.63	27.88

[a] The data for 25°C were obtained by interpolation.
Source: From Ref. 55.

Table 6.38   Binary Systems Containing Glycerol Azeotropic and Nonazeotropic

	Second component		Azeotropic data	
Formula	Name	B P (°C)	B P (°C)	(wt %) Glycerol
H_2O	Water	100	Nonazeotropic	
$C_6H_4Br_2$	p-Dibromobenzene	220.25	217.1	10
$C_6H_5NO_2$	Nitrobenzene	210.75	Nonazeotropic	
$C_6H_6O_2$	Pyrocatechol	232.9	Nonazeotropic	
$C_6H_6O_2$	Resorcinol	281.4	Nonazeotropic	
$C_7H_7NO_2$	m-Nitrotoluene	230.8	229.5	32
$C_7H_7NO_2$	o-Nitrotoluene	221.85	220.8	8
$C_7H_7NO_2$	p-Nitrotoluene	239.0	235.7	17
C_7H_8	Toluene	110.75	Nonazeotropic	
C_6H_8O	o-Cresol	191.1	Nonazeotropic	
C_7H_8O	p-Cresol	201.7	Nonazeotropic	
C_8H_8	Styrene	145.8	Nonazeotropic	
$C_8H_8O_2$	Benzyl formate	202.3	Nonazeotropic	
$C_8H_8O_2$	Methyl benzoate	199.45	Nonazeotropic	
$C_8H_8O_3$	Methyl salicylate	222.35	221.4	7.5
C_8H_{10}	m-Xylene	139.0	Nonazeotropic	
C_8H_{10}	o-Xylene	143.6	Nonazeotropic	
$C_8H_{10}O$	Phenethyl alcohol	219.4	Nonazeotropic	
$C_8H_{10}O_2$	m-Dimethoxybenzene	214.7	212.5	7
$C_9H_{10}O$	p-Methylacetophenone	226.35	Nonazeotropic	
$C_9H_{10}O_2$	Benzyl acetate	214.9	Nonazeotropic	
$C_9H_{10}O_2$	Ethyl benzoate	212.6	Nonazeotropic	
$C_9H_{10}O_3$	Ethyl salicylate	233.7	230.5	10.3
C_9H_{12}	Mesitylene	164.6	Nonazeotropic	
C_9H_{12}	Propyl benzene	158.8	Nonzaeotropic	
$C_{10}H_8$	Naphthalene	218.05	215.2	10
$C_{10}H_{10}O_2$	Isosafrole	252.0	243.8	16
$C_{10}H_{10}O_2$	Safrole	235.9	231.3	14.5
$C_{10}H_{12}O$	Estragole	215.6	213.5	7.5
$C_{10}H_{12}O_2$	Ethyl α-toluene	228.75	228.6	7
$C_{10}H_{12}O_2$	Propyl benzoate	230.85	228.8	8
$C_{10}H_{12}O_2$	Eugenol	252.7	251.0	14
$C_{10}H_{14}O$	Carvone	231.0	230.85	3
$C_{10}H_{14}O$	Thymol	232.8	Nonazeotropic	
$C_{10}H_{14}O_2$	m-Diethoxybenzene	235.4	231.0	13

Table 6.38   (*Continued*)

	Second component	Azeotropic data		
Formula	Name	B P (°C)	B P (°C)	(wt %) Glycerol
$C_{10}H_{16}$	Camphene	159.6	Nonazeotropic	
$C_{10}H_{16}$	*d*-Limonene	177.8	177.7	1
$C_{10}H_{16}$	α-Pinene	155.8	Nonazeotropic	
$C_{10}H_{16}$	Thymene	179.7	179.6	1
$C_{11}H_{10}$	*1*-Methylnaphthalene	244.9	237.25	18
$C_{11}H_{10}$	2-Methylnaphthalene	241.15	233.7	16.5
$C_{11}H_{14}O_2$	1-Allyl-3,4-dimethyoxybenzene	255.0	248.3	18
$C_{11}H_{14}O_2$	1,2-Dimethoxy-4-Propenylbenzene	270.5	258.4	25
$C_{11}H_{14}O_2$	Butyl benzoate	249.8	243.0	17
$C_{11}H_{14}O_2$	Isobutyl benzoate	241.9	237.4	14
$C_{11}H_{20}O$	Methyl terpineol ether	216.2	214.0	8
$C_{12}H_{10}$	Acenaphthene	217.9	259.1	29
$C_{12}H_{10}$	Biphenyl	254.9	243.8	55
$C_{12}H_{10}O$	Phenyl ether	257.7	246.3	22
$C_{12}H_{16}$	Isoamyl benzoate	262.05	251.6	22
$C_{12}H_{16}O_3$	Isoamyl salicylate	279.0	267.0	—
$C_{12}H_{18}$	1,3,5-Triethylbenzene	215.5	212.9	8
$C_{12}H_{20}O_2$	Bornyl acetate	227.7	226.0	10
$C_{13}H_{10}O_2$	Phenyl benzoate	315.0	279.0	55
$C_{13}H_{12}$	Diphenyl methane	265.6	250.8	27
$C_{14}H_{12}O_2$	Benzyl benzoate	324.0	282.5	—
$C_{14}H_{14}$	1,2-Diphenylethane	284.0	261.3	32

Source: From Ref. 5.

The solubility of gases in glycerol, as in all liquids, depends on both temperature and pressure. The solubility of hydrogen and nitrogen in aqueous glycerol solutions at 25°C was measured by Drucker and Moles [56]. Solubilities of nitrogen and carbon dioxide at 15°C were determined by Hammel [57]. These results are combined in Table 6.41.

In many cases the combination of two miscible compounds may lead to a mixture immiscible with glycerol. Temperature and the relative proportions of the ingredients can be important factors. Table 42 shows the solubility of a large number of organic and inorganic substances.

XXXIV.   THERMAL CONDUCTIVITY

The thermal conductivity of glycerol solutions increases with rising temperatures and increase in water content. The rate of change being linear. Hence the

Table 6.39 Miscibility of Organic Solvents with Glycerol

Acetone I	Diethylenetriamine M	α-Methylbenzylamine M
Isoamyl acetate I	Diethyl formamide M	α-Methylbenzyldiethanolamine M
n-Amyl cyanate I	Di (2-ethylhexyl) amine I	
Anisaldehyde I	Diisopropylamine M	α-Methylbenzyldimethylamine I
Benzene I	Di-n-propyl aniline I	α-Methylbenzylethanolamine M
Benzyl ether I	Ethyl alcohol M	2-Methyl-5-ethylpyridine M
Chloroform I	Ethyl chloracetate I	Methyl isopropyl ketone I
Cinnamaldehyde I	Ethyl cinnamate I	4-methyl-n-valeric acid I
o-Cresol M	Ethyl ether I	0-Phenetidine I
Di-n-amylamine S	Ethyl phenyacetate I	2-Phenylethylamine M
Di-n-butylamine S	3-Heptanol I	Isopropanolamine M
Diisobutyl ketone I	n-Heptyl acetate I	Pyridine M
Diethyl acetic acid I	n-Hexyl ether I	Salicylaldehyde I
2,6,8 Trimethyl 4-nonanone I	Triethylenetetramine M	Tri-n-butyl phosphate I

Source: Ind. Eng. Chem. 31, 1491–1494 (1959).
M = miscible; I = immiscible; S = partially miscible.

Table 6.40 Miscibility of Various Substances with Glycerol of $n_D^{24.8} = 1.4634$

Acetone		M-Toluidine		Acetophenone		Methyl ethyl	
(% wt)[a]	°C[b]	(% wt)[a]	(°C)[b]	(% wt)[a]	(°C)[b]	(% wt)[a]	(°C)[b]
89.61	40.0	83.23	89.0	97.13	90.5	92.14	55.5
86.93	58.5	79.23	102.0	95.30	113.5	86.55	118.5
76.96	81.3	71.58	113.5	83.42	162.5	74.78	150.0
67.42	91.7	64.30	119.4	75.07	175.5	67.14	161.5
64.47	93.5	58.68	120.5	61.90	183.6	60.25	164.5
57.25	95.5	53.10	120.2	53.32	185.5	53.84	164.5
56.59	95.5	45.68	119.5	51.13	185.4	41.27	163.2
55.34	95.6	36.87	117.5	42.00	185.0	36.17	162.5
53.07	95.7	17.71	88.5	34.62	184.0	26.75	155.5
51.28	95.6			21.14	174.5	13.21	128.5
48.43	95.5	83.62	33.4	15.88	164.0	10.73	116.5
46.31	95.3	66.04	7.8	8.86	136.5	4.00	37.5
45.75	95.3	62.86	7.0	4.38	97.5		
44.67	95.2	48.80	6.7				
34.74	90.9	40.99	8.2				
29.24	85.3	31.40	9.2				
26.58	81.3	21.68	14.2				
20.44	66.6	18.72	16.8				
15.77	44.8	13.99	23 & 69				
10.90	9.5	10.96	no separation				

Table 6.40 *(Continued)*

Isoamyl Alcohol (% wt)[a]	(°C)[b]	Benzaldehyde (% wt)[a]	(°C)[b]	Salicylaldehyde (% wt)[a]	(°C)[b]	O-Toluidine (% wt)[a]	(°C)[b]	Anisole (% wt)[a]	(°C)[b]
84.26	12.5	97.02	85.5	95.60	106.5	92.20	100.0	90.12	230.5
76.21	36.8	94.54	107.5	91.38	143.5	86.14	130.0	78.80	263.5
62.40	61.4	90.10	127.5	77.02	170.5	73.42	150.0	69.46	273.5
54.41	69.3	77.13	152.5	58.67	176.5	63.28	154.0	53.41	275.5
46.16	73.0	62.30	159.5	52.22	176.6	52.53	154.4	44.02	274.5
36.79	74.1	55.29	160.7	48.32	176.5	46.59	154.0	27.68	250.5
31.90	74.2	49.22	160.3	41.82	175.5	40.97	153.0	11.29	185.5
27.62	73.7	26.63	144.5	26.54	165.5	32.04	150.0	6.07	161.5
19.20	71.5	23.87	140.0	18.30	148.5	20.96	137.0		
13.97	66.5	12.42	123.5	5.36	91.5	12.42	99.2		
10.35	58.0	7.74	103.5						
5.05	21.5	4.53	67.5						

Monomethlylaniline (% wt)[a]	(°C)[b]	Dimethylaniline (% wt)[a]	(°C)[b]	Pyrocatechol monoethyl ether (% wt)[a]	(°C)[b]	σ-Anisidine (% wt)[a]	(°C)[b]
89.50	197.5	92.40	197.5	79.37	183.0	73.09	142.5
73.50	220.0	86.00	245.0	63.71	192.0	61.25	145.0
66.42	223.0	68.02	282.0	48.68	192.8	51.69	144.5
59.48	224.5	58.54	286.0	44.42	192.9	43.57	143.0
51.56	223.5	50.06	287.0	35.30	191.0	34.25	141.0
40.60	222.5	35.68	284.0	20.05	172.5		
30.26	219.0	21.71	273.0				
14.60	190.5	9.18	218.5				

[a] Percent weight of the nonglycerol component in total mixture.
[b] Temperature at which the mixture becomes miscible.
Source: From Ref. 5.

average coefficient of thermal conductivity over a temperature range is equal to the true coefficient at the average temperature.

Eucken and Englert [58] found the thermal conductivity of liquid glycerol at 0°C to be 0.000691 cal/cm/deg/s and of vitreous glycerol at −78°C to be 0.000760×10^{-6} cal/cm/deg/s. The figures in Table 6.43 were obtained from curves plotted from experimentally determined values for coefficient of thermal conductivity and solution concentrations [59].

XXXV. HYGROSCOPICITY

Anhydrous glycerol has a high affinity for moisture and can, therefore, be used with advantage for the drying of gases, liquids or solids.

Table 6.41 Solubility of Some Gases in Aqueous Glycerol

Drucker and Mole				Hammel			
Glycerol (% wt)	Hydrogen λ 25°C	Glycerol (% wt)	Nitrogen λ 25°C	Glycerol (% wt)	Nitrogen λ 15°C	Glycerol (% wt)	CO_2 λ 15°C
4.0	0.0186	16.0	0.0103	0.0	0.01801	0.0	1.064
10.5	0.0178	29.7	0.0068	15.7	0.01478	26.11	0.829
22.0	0.0154	48.9	0.0051	29.9	0.01147	27.69	0.845
49.8	0.0099	74.5	0.0025	46.6	0.00886	43.72	0.675
50.5	0.0097	84.1	0.0024	57.6	0.00736	46.59	0.655
52.6	0.0090			67.1	0.00670	62.14	0.540
67.0	0.0067			72.8	0.00582	73.36	0.474
80.0	0.0051			74.7	0.00630	77.75	0.454
82.0	0.0051			77.0	0.00556	87.74	0.446
88.0	0.0044			85.1	0.00508	90.75	0.427
95.0	0.0034			87.3	0.00519	96.64	0.438
				88.5	0.00565	99.26	0.438
				99.25	0.00553		

Solubility, λ = $\dfrac{\text{volume of absorbed gas}}{\text{volume of liquid}}$

Source: From Refs. 56 and 57.

Table 6.42 Solubility of Various Compounds in Glycerol

Substance	Glycerol concentration (% wt)	Temp. (°C)	Soln. in parts per 100 parts of solvent	Reference[b]
Alum	a	15	40	6
Ammonium carbonate	a	15	20	8
	99.04	20	19.8	9
	87.27	20	13.7	9
Ammonium chloride	a	15	20.06	8
Anisic aldehyde	a	15	0.1	1
Arsenic acid	a	15	20	6
Arsenious acid	a	15	20	6
Atropine	a	15	3	5
Atropine sulfate	99.04	20	45.2	9
	87.27	20	45.8	9
Barium chloride	a	15	9.73	8
Benzoid acid	98.5	—	2	4
	95.1	23	2.01	—
	90	23	1.74	—
	86.5	—	1.18	4

Table 6.42   (*Continued*)

Substance	Glycerol concentration (% wt)	Temp. (°C)	Soln. in parts per 100 parts of solvent	Reference[b]
	75	23	1.02	—
	50	23	0.60	—
Benzyl acetate	a	15	0.1	1
Boric acid	98.5	20	24.80	4
	86.5	20	13.79	4
Brucine	a	15	2.25	5
Calcium hydroxide	35	25	1.3	2
Calcium hypophosphite	99.04	20	2.5	9
	87.27	20	3.2	9
Calcium oleate	45	15	1.18	6
Calcium sulfate	a	15	5.17	8
Calcium sulfide	a	15	5	6
Cinchonine	a	15	0.30	5
Cinchonine sulfate	a	15	6.70	5
Cinnamic aldehyde	a	15	0.1	1
Codeine hydrochloride	99.04	20	11.1	9
	87.27	20	4.7	9
Copper acetate	a	15	10	6
Copper sulfate	a	15	30	6
Ethyl acetate	99.04	20	1.9	9
	87.27	20	1.8	9
Ethyl ether	99.04	20	0.65	9
	87.27	20	0.38	9
Eugenol	a	15	0.1	1
Ferrous sulfate	a	15	25.0	6
Guaiacol	99.04	20	13.1	9
	87.27	20	9.05	9
Guaiacol carbonate	99.04	20	0.043	9
	87.27	20	0.039	9
Iodine	a	15	2	8
Iodoform	95	15	0.12	3
Iron and potassium tartarte	a	15	8	5
Iron lactate	a	15	16	5
Iron oleate	45	15	0.71	8
Lead acetate	a	15	10	8
	98.5	—	143	4
	86.5	—	129.3	4
Lead sulfate	a	15	30.3	8
Magnesium oleate	45	15	0.94	6
Mercuric chloride	a	15	8	8
Mercurous chloride	a	15	7.5	5
Mercurous cyanide	a	15	27	5
Morphine	a	15	0.45	5

Substance	Glycerol concentration (% wt)	Temp. (°C)	Soln. in parts per 100 parts of solvent	Reference[b]
Morphine acetate	a	15	20	5
Morphine hydrochloride	a	15	20	5
Novocaine	99.04	20	11.2	9
	87.27	20	7.8	9
Oxalic acid	a	15	15.1	6
Pentaerythritol	100	100	9.3	Anon
Phenacetin	99.04	20	0.47	9
	87.27	20	0.3	9
Phenol	99.04	20	276.4	9
	87.27	20	361.8	9
Phenylethyl alcohol	a	15	1.5	1
Phosphorous	a	15	ca. 0.25	8
Potassium arsenate	a	15	50.13	8
Potassium bromide	a	15	25	6
	98.5	—	17.15	4
	86.5	—	20.59	4
Potassium chlorate	a	15	3.54	8
	98.5	—	1.03	4
	86.5	—	1.32	4
Potassium chloride	a	15	3.72	8
Potassium cyanide	a	15	31.84	8
Potassium iodate	a	15	1.9	5
Potassium iodide	a	15	39.72	8
	98.5	—	50.70	4
	86.5	—	58.27	4
Quinine	a	15	0.47	8
Quinine sulfate	98.5	—	1.32	4
	86.5	—	0.72	4
Quinine tannate	a	15	0.25	5
	99.04	20	2.8	9
	87.27	20	2.45	9
Salicin	a	15	12.5	5
Salicylic acid	98.5	—	1.63	4
	86.4	—	0.985	4
Santonin	a	15	6	5
Sodium arsenate	a	15	50	8
	99.04	20	64	9
	87.27	20	44	9
Sodium biborate	98.5	—	111.15	4
	86.5	—	89.36	4
Sodium bicarbonate	a	15	8.06	8
Sodium tetraborate (borax)	a	15	60	8
Sodium carbonate (crystals)	a	15	98.3	8

Table 6.42   (*Continued*)

Substance	Glycerol concentration (% wt)	Temp. (°C)	Soln. in parts per 100 parts of solvent	Reference[b]
Sodium chlorate	[a]	15	20	5
Sodium hypophosphite	99.04	20	32.7	9
	87.27	20	42.2	9
Sodium pyrophosphate	87.27	20	9.6	9
Sodium sulfate $12H_2O$	100	25	8.1	10
Stearic acid	99.04	20	0.089	9
	87.27	20	0.066	9
Strychnine	[a]	15	0.25	5
Strychnine nitrate	[a]	15	4	5
Strychnine sulfate	[a]	15	22.5	5
Sulfur	[a]	15	ca. 0.1	8
Tannic acid	[a]	15	48.8	6
Tannin	[a]	15	48.83	8
Tartar emetic	[a]	15	5.5	5
Theobromine	99.04	20	0.028	9
	87.27	20	0.017	9
Urea	[a]	15	50	5
Zinc chloride	[a]	15	49.87	8
Zinc iodide	[a]	15	39.78	8
Zinc sulfate	[a]	15	35.18	8
Zinc valerate	99.04	20	0.336	9
	87.27	20	0.382	9

[a] Glycerol concentration not specified, probably 95–100%.

[b] References for Table:

1. Allen's Commercial Organic Analysis, 4th ed. P. Blakiston's Son & Co., Philadelphia, 1923, p. 461.
2. Cameron, F. K., and Patten, H. E., J. Phys. Chem., 15: 67–72, 1911.
3. Chiara, P., Giorn. farm. chim., 66: 94–96, 1917.
4. Holm, K., Pharm. Weekblad, 58: 860–862, 1921; ibid. 1033–1038, 1921.
5. Lawrie, J. W. Glycerol and the Glycols, The Chemical Catalog Co., Inc. (Reinhold Publishing Corp.), New York, 1928, p. 232.
6. Lewkowitsch, J., Chemical Technology and Analysis of Fats and Waxes, 6th ed., MacMillan & Co. Ltd., London, 1921, kp. 254.
7. Noble, M. V., and Garrett, A. B., J. Am. Chem. Soc., 66: 231–235, 1944.
8. Ossendovsky, A. M., J. Russ, Phys, Chem,. Soc., 37: 1071, 1906. Through Mac Ardle, D. W., The Use of Solvents in Organic Chemistry, D. van Nostrand Co., Inc. New York, 1925, p. 80.
9. Roborgh, J. A., Pharm. Weekblad, 64: 1205–1209, 1927.
10. Schnellbach, W. and Rosin, J., J. Am. Pharm. Assoc., 18: 1230–1235, 1929.

Table 6.43 True Coefficient of Thermal Conductivity of Glycerol-Water Solutions

Water glycerol (% wt)	Values for K (gcal, sec^{-1}, cm^{-2}, °C, cm^{-2})[a]								$\alpha20^{+}\%$, C^{o-1}	Equations for true COE of thermal conductivity
	10°C	20°C	30°C	40°C	50°C	60°C	70°C	80°C		
100 (Pure water)	0.00138	0.00141	0.00145	0.00149	0.00152	0.00156	0.00160	0.00163	0.26	Kt=0.00134+0.00000367(t)
95	0.00133	0.00137	0.00140	0.00144	0.00147	0.00151	0.00154	0.00158	0.25	Kt=0.00130+0.00000342(t)
90	0.00130	0.00133	0.00137	0.00140	0.00143	0.00146	0.00149	0.00152	0.24	Kt=0.00127+0.00000317(t)
85	0.00125	0.00128	0.00131	0.00134	0.00137	0.00140	0.00143	0.00146	0.23	Kt=0.00122+0.00000300(t)
80	0.00121	0.00124	0.00127	0.00129	0.00132	0.00135	0.00138	0.00141	0.23	Kt=0.00118+0.00000284(t)
75	0.00117	0.00119	0.00122	0.00125	0.00127	0.00130	0.00132	0.00135	0.22	Kt=0.00114+0.00000263(t)
70	0.00112	0.00115	0.00117	0.00120	0.00122	0.00124	0.00126	0.00129	0.20	Kt=0.00110+0.00000234(t)
65	0.00109	0.00111	0.00114	0.00116	0.00118	0.00120	0.00122	0.00124	0.20	Kt=0.00107+0.00000217(t)
60	0.00105	0.00107	0.00108	0.00110	0.00112	0.00114	0.00116	0.00118	0.17	Kt=0.00103+0.0000183(t)
55	0.00102	0.00103	0.00105	0.00106	0.00108	0.00110	0.00111	0.00113	0.15	Kt=0.00100+0.00000159(t)
50	0.00097	0.00099	0.00100	0.00101	0.00103	0.00104	0.00105	0.00107	0.13	Kt=0.00096+0.00000133(t)
45	0.00094	0.00095	0.00096	0.00098	0.00099	0.00100	0.00101	0.00102	0.12	Kt=0.00093+0.00000116(t)
40	0.00090	0.00091	0.00091	0.00092	0.00093	0.00094	0.00095	0.00096	0.10	Kt=0.00089+0.00000090(t)
35	0.00086	0.00087	0.00088	0.00089	0.00089	0.00090	0.00091	0.00091	0.08	Kt=0.00086+0.00000067(t)
30	0.00084	0.00084	0.00085	0.00085	0.00086	0.00086	0.00087	0.00087	0.06	Kt=0.00083+0.00000050(t)
25	0.00080	0.00081	0.00081	0.00081	0.00082	0.00082	0.00082	0.00082	0.04	Kt=0.00080+0.00000030(t)
20	0.00077	0.00078	0.00078	0.00078	0.00079	0.00079	0.00079	0.00079	0.04	Kt=0.00077+0.00000030(t)
15	0.00074	0.00074	0.00074	0.00074	0.00074	0.00074	0.00075	0.00075	0.01	Kt=0.00074+0.00000008(t)
10	0.00072	0.00072	0.00072	0.00072	0.00072	0.00072	0.00072	0.00073	0.01	Kt=0.00072
5	0.00070	0.00070	0.00070	0.00070	0.00070	0.00070	0.00070	0.00070	—	Kt=0.00070
100 (Pure glycerol)	0.00068	0.00068	0.00068	0.00068	0.00068	0.00068	0.00068	0.00068	—	Kt=0.00068

[a] Kt (cal, sec^{-1}, cm^{-2}, °C, cm^{-2}) 2900 = Kt (Btu, h^{-1}, ft^{-2}, °F^{-1} inch).

[b] $\alpha20$ as defined by Kt=K$_{20}$ [1+$\alpha20$(t−20)].

Cgs system English system

Source: From Ref. 5.

Table 6.44 Relative Humidity Over Glycerol-Water Mixtures at 25°C

Relative humidity	Glycerol (wt)	Specific gravity
%	%	
10	95	1.245
20	92	1.237
30	89	1.229
40	84	1.216
50	79	1.203
60	72	1.184
70	64	1.162
80	51	1.127
90	33	1.079

Source: From Ref. 5.

The behavior of aqueous solutions will depend on their strength. By choosing a solution which gives rise to a higher relative humidity than a substance with which it is brought into contact, the substance will take up a predetermined amount of moisture, thereby bringing both relative humidities into equilibrium. The relative humidities obtainable from various glycerol concentrations are shown in Table 6.44.

REFERENCES

1. Official Methods and Recommended Practices, 3rd. ed., American Oil Chemists' Champaign, IL, EA 7–50, 1973.
2. Mettler Instruments Corp., Density Meters, Mettler Technical Bulletin.
3. Bosart, L. W., and Snoddy, A. O., Ind. Eng. Chem., 20: 1377–1379, 1928.
4. Timmermans, J., and Mme. Hennaut-Roland, J. Chem. Phys., 32: 501–526, 589–16, 1935.
5. Physical Properties of Glycerine and Its Solutions, Glycerine Producer's Association, New York, 1969.
6. Miner, Jr., C. S., and Dalton, N. N., Glycerol, Reinhold Publ. Corp. 1953.
7. Int. Critical Tables, 3: 121, 1928.
8. Shell Chem. Corp., Tech. Report 115–55, p.21.
9. Green, E., and Parke, J. P., J. Soc. Chem. Ind., 58: 319–320, 1939.
10. Comey, A. M., and Backus, C. F., Ind. Eng. Chem., 2: 11–16, 1910.
11. Hoyt, L. F., Oil & Soap, 10: 43–47, 1933; Ind. Eng. Chem., 26: 329–332, 1934.
12. Newman, A. A., Glycerol, CRC Press, Cleveland, PA, 1968.
13. Von Rechenberg, C., Cinfache and Fraktionerte Distillation, Leipzig, 1923, p. 267.
14. Carr, A. R., Townsend, R. E., and Badger, W. L. Ind. Eng. Chem., 17: 643–646, 1925.

15. U.S. Bureau of Standards, Letter Circular 28.
16. Ross, H. K., Ind. Eng. Chem., 46: 601, 1954.
17. Sheeley, M. L., Ind. Eng. Chem., 24: 1060–1064, 1932.
18. Segur, J. B., and Oberstar, H., Ind. Eng. Chem., 43: 2117–2120, 1951.
19. Tammann, G., and Hesse, W., Z., Anorg, Allgem. Chem., 156: 245–257, 1926.
20. Vand, V., Research, 1: 44–45, 1947.
21. Bridgman, P. W., Proc. Am. Acad. Arts Sci., 61: 57–59, 1926.
22. Davis, P. B., Carnegie Inst. Wash. Pub., 260: 97–98, 1918. Davis, P. B., and Joes, H.C., Z. Physik. Chem., 81: 68–112, 1913.
23. Jones, H. C. et al., Carnegie Inst. Wash. Pub., 180: 153–178, 179–199, 1913.
24. Brisco, H. T., and Rinehart, W. T., J. Phys. Chem., 46: 387–394, 1942.
25. Pisarzhevikii, L., and Trachoniotoskii, P., Polytech. Kev J. Russ Phys. Chem. Soc., 42: 249–295, 1910.
26. Stedman, D. F., Trans Faraday Soc., 24: 289–298, 1928.
27. Perman, E. P., and Price, T. W., Cardiff Trans. Faraday Soc., 8: 68–81, 1912.
28. Campbell, F. H., Trans. Faraday Soc., 11: 91–103, 1915.
29. Olsen, J. C., Brunjes, A. S., and Olsen, J. W., Ind. Eng. Chem., 22: 1315–1317, 1930.
30. Gibson, G. E., and Giauque, W. F., J. Am. Chem. Soc., 45: 93–104, 1923.
31. Bennewitz, K., and Krats, L., Physick Z., 37: 496–511, 1936.
32. Gucker, F. T., Jr., and Marsh, G. L., Ind. Eng. Chem., 40: 908–915, 1948.
33. Fricke, R., Z. Elektrochem, 35: 631–640, 1929.
34. Kolossowsky, N., J. Chem. Phys., 22: 83–93, 1925; J. Russ Phys. Chem. Soc., Chem. Part 57: 17–21, 1925.
35. Lamm, O., and Sjostedt, G., Trans. Faraday. Soc., 34: 1158–1163, 1938.
36. Yasunori, N. and Oster, G., Bull. Chem. Soc. Japan, 39: 1649–1651, 1960.
37. Volmer, M., and Marder, M., Z. Physik. Chem., Abt. A: 154, 97–112, 1931.
38. Stedman, D. F., Trans. Faraday Soc., 29: 289–298, 1928.
39. Parks, G. S., West, T. J., Naylor, B. F., Fuju, P. S., and McClaine, L. A., J. Am. Chem. Soc., 68: 2524–2547, 1946.
40. Getman, F. H. and Daniel, F., Outlines of Physical Chemistry, J. W. Wiley and Sons, Inc., New York, 1943, p. 15.
41. Weast, R. C. Handbook of Chemistry and Physics, 64th ed., CRC Press, Boca Raton, FL, 1983–84, p. D289.
42. Segur, J. B., Heating Ventilating, 44: 86, 1947.
43. Masson, H. J., and Hamilton, W. F., Ind. Eng. Chems. 20: 813, 1928.
44. Michaelis, L., and Rona, P., Biochem. Z., 49: 232–248, 1913.
45. Smithsonian Physical Tables, Washington, D.C., 8th Rev. Ed., 475, 1933.
46. Scudder H., Electrical Conductivity and Ionization Constants of Organic Compounds, van Nostrand Co., Inc., New York, 1914, p.168.
47. Muller, E., Sitzber, Akad. Wiss. Wien. Math.-naturew. Klasse, Abt. 11a: 133, 133–147, 1924.
48. Bridgeman, P. W., Proc. Am. Acad. Arts Sci., 66: 185–233, 1931; 67: 127, 1932.
49. Pushin, N. A., and Grebenshchikov, E. V., J. Chem. Soc., 125: 2043–2048, 1924.
50. Comey, A. M., and Backus, C. F., Ind. Eng. Chem., 2: 11–16, 1910.
51. Gerlach, G. Th., Chem. Ind., 7: 227–287, 1884.

52. Freyer, E. B., Hubbard, J. C., and Andrews, D. H., J. Am. Chem. Soc., 51: 759–779, 1929.
53. Pushin, N. A., and Glagoleva, A. A., J. Chem. Soc., 121: 2813–2823, 1922.
54. Albright, G., J. Am. Chem. Soc., 59: 2098–2104, 1937.
55. Akerlof, G., J. Am. Chem. Soc., 54: 4125–5139, 1932.
56. Drucker, E., and Moles, E., Z. Physik. Chem., 75: 405–436, 1911.
57. Hammel, A., von, Z. Physik. Chem., 90: 121–125, 1915.
58. Euken, A., and Englert, H., Z. Ges, Kalte-Ind., 45: 109–118, 1938.
59. Bates, O. K., Ind. Eng. Chem., 28: 494–498, 1936.

7

Glycerine Analysis

David C. Underwood

Procter & Gamble, Cincinnati, Ohio

I. INTRODUCTION: SOURCES OF GLYCERINE METHODS

Glycerine is one of the most versatile chemicals because of its safety, humectant properties, and chemical properties. Therefore, several different organizations are involved in specifying methods for glycerine. It is helpful to classify these organizations into two groups. First, there are technical organizations that propose methods, but do not themselves specify limits for uses of glycerine. Methods proposed by the following organizations were consulted in preparing this chapter:

American Oil Chemists Society (AOCS), 1608 Broadmoor Drive, Champaign, IL 61821. *Official Methods and Recommended Practices,* 4th ed., 1989. Their methods are especially pertinent to analysis of crude and finished glycerine.

Association of Official Analytical Chemists, Inc. (AOAC), 1111 North Nineteenth Street, Suite 210, Arlington VA 22209. *Official Methods of Analysis of the AOAC,* 15th ed., 1990. Their methods are especially pertinent to analysis for glycerine in food products.

Another group of organizations set specifications and standards for glycerine. In all cases, they also define the analytical methods to be used in measuring the specifications. Methods from the following organizations were consulted in preparing this chapter:

The United States Pharmacopeial Convention Inc. (USP), 12601 Twinbrook
   Parkway, Rockville, MD 20852. *USP XXII NF XVII*, 1990. Standards per-
   tain to glycerine used in drug products.
British Pharmacopoeia (BP), Volume I, 1980, published in London by Her Maj-
   esty's Stationery Office. Standards apply to glycerine used in drug products.
The American Society for Testing Materials (ASTM), 1916 Race Street, Philadelphia PA 19103–1187. *1988 Annual Book of ASTM Standards* gives
   specifications for glycerine used in manufacture of plastics.
The Cosmetic, Toiletry and Fragrance Association, Inc. (CTFA), 1133 15th
   Street, N.W, Washington DC 20005. *TFA Standards-Methods*, 1974, describes standards pertaining to glycerine used in cosmetic and toiletries.
Food Chemicals Codex (FCC), 13th ed., 1981, Committee on Codex Specifications, National Academy Press, Washington DC, 1981. These standards
   apply to glycerine used in food products.
Reagent Chemicals, 7th ed., 1987, American Chemical Society (ACS), American Chemical Society Publications, Washington DC. Standards pertain to
   reagent-grade glycerol.

Because of the numerous uses of glycerine and the many sources of methods,
the analytical chemist in this area has an obligation to be knowledgeable in
two areas in addition to having a technical understanding of the methods.

First, the analytical chemist must understand the use of the glycerine being
analyzed, and make sure methodology from the appropriate standards-setting
organization is used to provide results. While the analyst can often substitute
methods (e.g., using the Food Chemicals Codex assay for glycerine rather
than the AOCS assay), the burden of proof is on the analyst to demonstrate
that both methods give equivalent and satisfactory results.

Second, for regulated products, the analytical chemist must use procedures
that are consistent with good laboratory practices. In general, this requires
trained personnel, use of methods that are known to work well, retrievable
written records, and clear identification of the raw data and results from each
sample that is analyzed. This is required for glycerine used in foods, cosmetics, and drugs. It is common sense for all analyses.

In recent years, many organizations have begun to work together toward
establish a common methodology for specifying glycerine ingredients. This is
a sensible and heartening trend. It makes the world of glycerine analysis much
less controversial and confusing than it was thirty years ago. But there still are
many unique methods, and the analytical chemist must continue to keep current on specifications and methods. It is certainly hoped that the trend toward
method uniformity across regulatory and trade organizations continues. It is
hoped further that precision information on results, such as those currently
being supplied by the AOCS become a growing trend. Many needless disputes
between customers and suppliers can be avoided by recognizing that all anal-

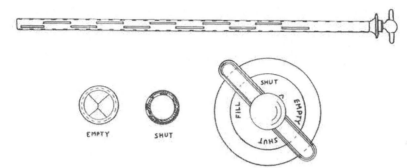

Figure 7.1 Drum sampler for glycerine. (Reprinted with permission of the American Oil Chemists Society.)

yses are subject to a predictable error (standard deviation, etc.) and that two laboratories will seldom get identical results on the same sample.

II. SAMPLING GLYCERINE

One of the most important—and difficult—requirements for obtaining accurate analyses on glycerine is the sampling procedure itself. Crude glycerine is normally shipped in tankcars or tank trucks. Since crude glycerine can contain unsaponified fats, precipitated salts, and particulate material, homogeneous samples are difficult to take. Separation and settling into different layers can easily occur during shipping. Therefore, extreme care and good equipment are requirements in obtaining representative samples of the glycerine.

AOCS method Ea 1-38 recommends several methods for uniform sampling. The drum sampler, pictured below can be used to obtain uniform samples of drums of glycerine. It is made from two brass tubes that fit inside each other tightly (Fig. 7.1). Each tube contains slots which permit uniform filling from the top, middle, and bottom of the drum. When the sampler is full, it is removed and the sample is placed in an appropriate storage container for analysis. The amount of sample to be withdrawn can be adjusted by selecting the inside diameter of the brass tubes.

For tank cars and tank trucks, a core sampler, (Fig. 7.2) can be used. This consists of a hollow tube with a valve at the bottom which is opened as the tube is immersed in the tank. The tube is lowered slowly enough so that glycerine from each level in the tank enters the tube. Before withdrawing the

Figure 7.2 Core sampler for glycerine. (Reprinted with permission of the American Oil Chemists Society.)

sample, the valve at the bottom is closed. This apparatus works well, but is rather heavy and requires special care to make sure the valve at the bottom opens fully and seals tightly without leaking.

Pulse or flow samplers are also very good ways of taking representative samples of a tankcar while loading or unloading. The filling hose is fitted with a sidestream with a valve that is opened, either manually or by means of a timer, at regular intervals during loading or unloading. The AOCS recommends that samples taken in this way represent at least 0.1% by weight of the total shipment, and that at least 25 evenly spaced withdrawals be made. However, sidestream valves are quite subject to plugging, so they must be monitored and maintained carefully. A more robust way of sampling is to use a dipper to take a number of samples from the discharge end of the pipe used to fill the tankcar.

Since salt in crude glycerine usually settles to the bottom of the tankcar during delivery, neither of these methods is entirely satisfactory for obtaining a good sample of glycerine as received. For this reason, the AOCS recommends that samples for analysis be made, by agreement between customer and supplier, at the time the tankcar or tank truck is loaded.

Homogeneity of the sample delivered to the laboratory is also essential for good analytical results. The sample obtained from the tankcar, drum, etc., must be thoroughly mixed before a portion of it is placed in the container to be sent to the laboratory for analysis.

III.   METHODS EMPLOYING PERIODATE

A.   Chemistry

The selective action of periodate as an oxidant for glycerine was first described by Malaprade in 1928 [1], and later developed as a quantitative method for glycerine analysis by Fleury and Lange [2]. The reaction is very convenient—and specific—for glycerine, compared with oxidants that are strong enough to oxidize almost any organic compound.

At room temperature, periodate does not oxidize alcohols with isolated hydroxy groups. However, materials containing adjacent hydroxyl groups are oxidized in a very specific and analytically useful way. Compounds containing two adjacent hydroxyl groups are oxidized to aldehydes, as shown below [3]:

$$CH_2OH\text{---}CH_2OH + IO_4^- \rightarrow IO_3^- + 2CH_2O + H_2O$$

A material such as 1,2–propanediol undergoes the following reaction:

$$CH_2OH\text{---}CHOH\text{---}CH_3 + IO_4^- \rightarrow IO_3^- + CH_2O + CH_3CHO + H_2O$$

With compounds containing three or more adjacent hydroxyl groups, the carbon atoms with the "internal" hydroxyl groups are oxidized to formic acid, as illustrated with glycerine:

$$CH_2OH-CHOH-CH_2OH + 2IO_4^- \rightarrow 2IO_3^- + 2CH_2O + HCOOH + H_2O$$

In general, the reaction can be expressed as:

$$CH_2OH-(CHOH)n-CH_2OH + (n + 1) IO_4^- \rightarrow (n + 1)IO_3^- + 2HCHO + nHCOOH + H_2O$$

Periodate is also useful as a structural identification technique for carbohydrates. This application has been the subject of an extensive review by Bobbitt [4]. Aldoses containing an aldehyde on the end carbon behave as though they had two "internal" hydroxyls; The aldehyde is oxidized to formic acid. So, the net reaction, as shown below, produces two moles of formic acid per mole of glucose, rather than the one mole that would be expected from glycerine [5]:

$$HCO-CHOH-CHROH + 2IO_4^- \rightarrow 2HCOOH + HCOR + 2IO_3^-$$

So, for glucose, the reaction becomes:

$$HCO(CHOH)_4CH_2OH + 5IO_4^- \rightarrow 5HCOOH + CH_2O + 5IO_3^-$$

In terms of the general equation given above, the stoichiometry of the reaction remains the same; there are four "internal" hydroxyls and 5 moles of periodate consumed in the reaction. But instead of 4 moles of formic acid, 5 moles are produced—with the extra mole arising from the aldehyde on the first carbon atom.

Ketoses react somewhat differently, hence structural information on sugars is obtained. Periodate cleaves these molecules on the carbon containing the carbonyl group, and produces a stable hydroxy acid:

$$CH_2OH-CO-CHROH + IO_4^- \rightarrow CH_2OH-COOH + HCOR + IO_3^-$$

Therefore, for fructose, the ketose equivalent of glucose, the reaction becomes:

$$CH_2OH-CO-(CHOH)_3-CH_2OH + 4IO_4^- \rightarrow CH_2OH-COOH + 3HCOOH + CH_2O + 4IO_3^-$$

As with glucose, only one formaldehyde molecule is formed, but the net use of periodate is decreased to 4 moles of periodate per mole of fructose, as opposed to the 5 moles used in oxidation of glucose.

As pointed out by Laitinen and Harris [6], the equilibrium between periodic acid and periodate is complex, and made even more so by a dehydration reaction. In acidic solution, periodic acid exists as a dihydrate, which dehydrates as it converts to iodate:

$$H_5IO_6 \rightarrow H^+ + H_4IO_6^- \qquad K_1 = 1 \times 10^{-3}$$
$$H_4IO_6^- \rightarrow H^+ + H_3IO_6 = \qquad K_2 = 3 \times 10^{-7}$$
$$H_4IO_6^- \rightarrow IO_4^- + 2H_2O \qquad K_{eq} = 43$$

Happily, in the aqueous environments commonly used for glycerine analysis, these equilibria do not play a significant role. While pH adjustment and acid-base titrations are commonly used in glycerine assay, periodate solutions do not act as buffers in the pH range between 5 and 10, and therefore do not interfere with the endpoint in formic acid titrations to determine glycerine levels.

B. Application of Periodate in Glycerine Analysis

1. Glycerine Assays

The most common method for establishing the purity of glycerine is the oxidation of glycerine with periodate, followed by acid-base titration of the resulting increase in acidity due to formic acid generation [7]. Small amounts of diols potentially arising from synthetic glycerine manufacture or from enzymic action on crude glycerine do not generate formic acid, and will not interfere. Other triols or carbohydrates are potential interferents, but are not commonly found in crude and refined glycerine. Over the years, assays similar to AOCS method number Ea 6-51 have become the standard criteria of purchasing crude glycerine and verifying the composition of finished glycerine.

This periodate method has several other advantages. It is very precise, as is common with volumetric methods, and normally has interlaboratory standard deviations of 0.5% or better. This is especially important for establishing the price of crude glycerine based on glycerine content. The analysis is performed with reagents and laboratory equipment that are within the economic range of small laboratories. More importantly, the analysis has stood the test of time. It has been studied extensively by several organizations and are quite well understood.

In general, the analysis is performed in the following way [8]:

1. A quantity of sample containing between 0.32 (3.47 mmol) and 0.5 (5.42 mmol) of glycerine is dissolved in 100–150 ml of distilled water. Accurate estimation of the expected glycerine level is important. The method becomes inaccurate if too much glycerine is present, since an excess of periodate is required for quantitative reaction. Similarly, if too little glycerine is present, the final titration of glycerine will take too little standardized base for accurate reading of the burret.
2. The solution is carefully neutralized to a pH of 8.1 with a pH meter or an indicator (e.g., bromthymol blue). This is required because it

corresponds to the neutralization point of the solution after the oxidation reaction.

3. A 50-ml aliquot of 0.28 M sodium metaperiodate (14 mmol) is added to the solution, and it is reacted at room temperature for about 30 minutes. Since each millimole of glycerine requires 2 mmol of periodate for reaction, these conditions are sufficient to assure an excess of periodate, even when 0.5 gram (5.42 mmol) is present.

4. After this 30 minute reaction period, excess ethylene glycol is added to react with the periodate. The ethylene glycol is oxidized to formaldehyde.

5. The formic acid produced is determined by titration with standardized sodium hydroxide to a pH of 6.5. This lower pH endpoint is used because it corresponds to the equivalence point of the formic acid/formate titration.

A similar blank analysis is determined, but the final titration is done to pH 8.1, since no formic acid is present to buffer the solution.

2. Determination of Glycerine in Soap

Glycerine is commonly added to soaps as a humectant and emollient. Since glycerine levels in soaps are below 10%, and significant amounts of soap are present that would interfere with the titration of formic acid, a straightforward redox reaction is used for glycerine analysis. AOCS method Da 23-56 is a good example of this analysis. It consists of the following steps:

1. About 10 g of soap, 91 ml of chloroform and 25 ml of glacial acetic acid are mixed in the presence of water and the solution is diluted to 1 liter. This converts the soap into its conjugate acid and permits its extraction into the chloroform layer. Glycerine remains in the aqueous layer and is largely free of organic matter that would otherwise interfere with the periodic acid oxidation of glycerine.

2. Next, 100 ml of the aqueous soap layer is pipetted into a beaker and reacted with a 50 ml solution of approximately 0.012 M periodic acid. This solution is allowed to react for 30 minutes.

3. Excess potassium iodide solution is added to dispel unreacted periodate by oxidation to iodine.

4. The amount of iodine formed is determined by titration with 0.1 N sodium thiosulfate, using a starch-iodine endpoint.

This titration is compared with a blank titration. The difference in periodic acid used indicates the amount of glycerine present. As is pointed out in the method, most accurate results are obtained when the titration of the sample is at least 80% of the blank titration. The method is quite precise: at the 0.5% level of glycerine in soap, the absolute standard deviation is about 0.01%.

IV. METHODS EMPLOYING DICHROMATE

Dichromate in acid solutions is a good oxidant for many organic compounds—including trace organics in water and wastewater, where it is used to determine chemical oxygen demand [9]. It is no surprise, then, that dichromate is also used for glycerine analysis, especially for very low levels of glycerine.

Glycerine analysis is typically carried out in the presence of strong sulfuric acid at temperatures of 80–90°C. Glycerine is oxidized completely to carbon dioxide [10]:

$$3CH_2OHCHOHCH_2OH + 7Cr_2O_7{-} + 56H^+ \rightarrow 9CO_2 + 40H_2O + 14\,Cr^{+3}$$

Official AOAC methods for the following products utilize dichromate oxidation followed by determination of excess dichromate:

Sample	AOAC Method
glycerine in beer	10.032
glycerine in vinegar	30.079
glycerine in lemon extract	19.064
glycerine in vanilla extract	19.005
glycerine in wines	11.012

Because dichromate is a very nonselective oxidant, the success of these methods depends largely on the prior separation schemes to remove essentially all other organics. In general, isolation schemes utilize the following steps:

1. The product is evaporated to a smaller volume to remove some of the organics. Vinegar is evaporated several times, with intermediate addition of water, to expel all traces of acetic acid. The sample should not be allowed to go dry, or some glycerine may be lost.
2. Milk of lime (calcium oxide) solution is added to precipitate additional carboxylic acids. The solution is then filtered and evaporated almost to dryness.
3. The filtrate is diluted with a mixture of absolute alcohol and anhydrous ether (2:3, vol:vol) and filtered again, presumably to remove sugars by precipitation.
4. The alcohol/ether solution is evaporated almost to dryness several more times, with intermediate additions of water to remove all traces of alcohol and ether.
5. Proteins and other remaining organics are then precipitated from the aqueous solution by addition of silver carbonate and lead acetate solutions.
6. An excess of dichromate solution is added and the solution is reacted at elevated temperature (80–90°C) for about 30 minutes.

An interesting and rather unusual aspect of these analyses is the method by which excess dichromate is determined. Rather than titrating the dichromate with a reducing agent, an aliquot of the reaction product, containing excess dichromate, is added to a burette. This is then used as the titrant for titrating a known weight of ferrous ammonium sulfate which has been standardized against a standard solution of potassium dichromate.

V. ENZYMATIC METHODS OF ANALYSIS

While enzyme reactions are not commonly used by many analytical chemists, they are certainly quite specific. Several enzymatic methods work very well for glycerine analysis, and are quite popular in the area of the life sciences. They are fairly specific for glycerine in serum samples and are sensitive enough to determine glycerine at low levels in blood. Most enzymes needed for glycerine analysis are now commercially available. Earlier workers were often obliged to isolate and prepare their own enzymes—an exacting task that undoubtedly called for laboratory techniques not known to the average analytical chemist!

In 1954, Bublitz and Kennedy [11] described a two-step glycerine assay consisting of the following steps:

$$\text{glycerokinase}$$
1. Glycerine + ATP $\xrightarrow{\hspace{2cm}}$ L-glycerol-3-phosphate + ADP

$$\text{glycerophosphate dehydrogenase}$$
2. L-Glycerol-3-phosphate + NAD$^+$ $\xrightarrow{\hspace{3cm}}$ dihydroxyacetone + NADH

Boltralik and Noll [12] studied the kinetics and mechanism of these reactions and suggested several changes to make it a one-step analysis. They replaced the original glycerokinase obtained from rat liver with a purified extract from a *Mycobacterium tuberculosis* strain grown in glycerine.

The authors cited above found several criteria had to be met to obtain optimum quantitation. First, reaction 2, given above, is the rate-limiting step. In order to keep it rapid, they used a relatively large excess of glycerophosphate dehydrogenase (GDH) and added albumin to the reaction mix to prevent the deactivation of the GDH from the hydrazine that was added as a trapping agent. A relatively large amount of NAD+ is also required to react rapidly with the L-glycerol-3-phosphate as it is formed. They stabilized the NADH formed by adding ethanol at the end of the reaction. At the end of the reaction (60 min at 37°C), the amount of NADH formed is determined spectrometrically by its absorbance at 340 nm. Under these conditions, good results are obtained in the range of 0.2-1.0 mmol liter, with recoveries of spiked samples of 97 + 2%.

A rapid method for glycerine analysis in serum has been described using phosphorylation of glycerine coupled with reduction of NADH in the following way [13]:

glycerokinase
1. Glycerine + ATP————————→L-glycerol-3-phosphate + ADP

pyruvate kinase
2. ADP + phosphoenolpyruvate————————→ATP + pyruvate

lactate dehydrogenase
3. Pyruvate + NADH————————————→lactate + NAD⁺

Ultraviolet spectrometry is used to monitor the disappearance of NADH at 340 nm. At a pH of 7.6, all the forward reactions are favored. Blood plasma samples are prepared for analysis by deproteinating with trichloroacetic acid and extracting the glycerine with ether. The reaction is essentially complete within 3–5 minutes in blood plasma sample. Recoveries of spiked plasma samples are in the range of 95%, with a relative standard deviation about 2%.

Hagen [14] proposed a method using fewer enzyme catalysts, by the use of glycerol dehydrogenase to catalyze the reduction NAD^+:

glycerol dehydrogenase (GDH)
glycerine + NAD⁺————————————————→dihydroxyacetone + NADH

Glycerol dehydrogenase (GDH) was prepared from *Aerobacter aerogenes*. Protein-free plasma samples were added to the appropriate reagents in a pH 9.5 glycine buffer and reacted for 30 minutes at room temperature. Glycerine levels were quantitated with ultraviolet spectrometry by monitoring absorbance increase at 340 nm. Hagen found that glucose, lactate, malate, beta-hydroxybutyrate, succinate, glutamate, alpha-glycerophosphate, and alpha-oxoglutarate did not interfere, but relatively large amounts of ethanol, glyceraldehyde-3-phosphate produce interferences. This same chemistry has been used by several other workers, as part of instrumental analysis systems.

Pardue and Frings [15] added a blue dye (2,6-dichlorophenolindophenol) to the solution. As NADH is generated, it reduces the blue dye to a colorless form. A spectrometer monitors this process. Reaction rates are then monitored and the glycerine levels are calculated based on the reaction rate during about the first minute of the reaction. They used spiked blood plasma samples and found good accuracy (within 2% relative of expected results) in the 50–200 part per million (ppm) range, with relative standard deviations of 1–2%.

Masoom and Worsfold [16] adapted the GDH method for flow injection analysis, with good results. They used an automate merging-zones manifold with a spectrophotometer to monitor absorbance at 340 nm. In their procedure, 20 µl of analyte is injected into a stream of Tris buffer (pH 8.0). Twenty microliters of reagent (GDH, ammonium chloride, and NAD+ in buffer) is injected in a second stream of Tris buffer, and the two streams are merged. After mixing with a length of tubing containing glass beads, the absorbance of the NADH produced by oxidation of glycerine is measured at 340 nm.

If lipase is added to the reagent, it hydrolyzes triglycerides to glycerine. The glycerine then reacts with NAD+ and can be analyzed as well. The method appears to work well in the 5×10^{-6} to 10^{-2} M region, where the reaction rate after mixing (stopped flow for 120 seconds) can be correlated to glycerine concentration. With such equipment, about 25 samples per hour can be analyzed automatically.

A fairly complex but interesting application of this same reaction chemistry is described by Morishita and Co-workers [17]. They use similar flow-injection equipment but immobilize the enzymes on the wall of a glass reactor. The glycerine sample is injected into a stream containing buffer and NAD^+. After mixing, the stream is split into three tubes containing immobilized lactate dehydrogenase, GDH, or glucose dehydrogenase. With appropriate tubing lengths and timing, lactate, glycerine, and glucose can all be determined individually by passing the reacted effluents through a fluorescence detector (340 nm excitation, 470 nm emission) to monitor the NADH produced.

Recently, Kiba and co-workers [18] combined an immobilized GDH enzyme reactor with high-performance liquid chromatography (HPLC) equipment to overcome the lack of specificity of the glycerine–GDH reaction. They point out that GDH catalyzes the reduction of NAD+ with materials other than glycerine, such as 1,2,3-propanetriol, 1,2-ethanediol, 1,2-propanediol, and 1,2-butanediol. Since propanediol is used in the pharmaceutical industry as a solvent for drugs, it might constitute an important interference in analysis of blood samples from patients undergoing chemical therapy. Serum samples are separated on a reversed-phase liquid chromatography column using water as the mobile phase. GDH is covalently bonded to alkanolamines on polystyrene beads and packed into a postcolumn reaction tube. The column effluent is then mixed with an NAD^+ solution buffered at pH 10 and passed through the GDH column. NADH produced by the glycerine oxidation is monitored with a flowthrough fluorescence detector (348 nm excitation, 465 nm emission).

The method was validated or certified serum samples. In the range of 0.01–2.0 mmol liter of glycerine and 1,2-propanediol, results were within 2% relative of the expected results, with day-to-day relative standard deviations of 4% or less.

The activity of the enzyme postcolumn reactor varies with time. However, when proper calibrations are made to compensate for the slow change in column activity, the useful lifetime of an enzyme column was found to be three months or more.

VI. LIQUID CHROMATOGRAPIC TECHNIQUES

High-pressure liquid chromatography (HPLC) has become widely used in recent years for a number of nonvolatile compounds. It normally requires little, if any sample derivitization and tends to be especially useful with samples where high temperatures could produce deleterious changes in the analyte or in other components associated with the sample.

Early work by Schwartzenbach [19] and Scobell et al. [20] demonstrated the utility of silica columns containing bonded aminoalkyl groups for separations of polyhydric alcohols. Nagel et al. [21] expanded on this work and demonstrated its utility in an especially difficult matrix: glycerine in meat.

Meats that have been prepared to contain water levels of 10–40% can be stored almost indefinitely at room temperature, and are known as intermediate moisture (IM) meats [22]. Humectants such as sucrose, sorbitol, glycerine, and propylene glycol are used to help reduce water activity in the meats and to provide a plastic mouthfeel. Meats are prepared by soaking in infusion solutions containing about 50% water and 50% humectant. During this time, water diffuses from the meat to the infusion solution because of its lower water activity, and some of the humectants from the infusion solution diffuse into the meat. In order to obtain material balance of the humectants, it was necessary to develop a procedure that worked on both the humectant solution and the meat.

Samples of meat, 100 g, were prepared in a Waring Blender by blending with 300 ml of water. After dilution to 500 ml, the aqueous meat slurry was clarified by centrifugation and aliquots containing 10–100 mg of each sample were taken for cleanup. This aliquots were passed through a strong acid cation-exchange column (Dowex 50 H$^+$) and a strongly basic anion-exchange resin (Amberlite IR 45 CP OH$^-$). This removed most of the sodium chloride, amino acids, and protein from the solution.

The purified sample was then analyzed on an Aminex HPX-87 column (7.8 mm i.d. $\times$ 300 mm) using 0.013N aqueous sulfuric acid as the mobile phase. Peaks were detected on a refractive index detector and analyzed by comparing area of the sample peaks against reference curves prepared from pure standards.

To validate the analysis, meats were prepared from infusing solutions containing fixed ratios of sucrose, sorbitol, glycerine, and propylene glycol, but with varying amounts of water (50–75%). Recovery of each humectant was then calculated from the amount of humectant found in the infusing solution and the meat. Results are shown in Table 7.1. The authors note that high glycerine recoveries in these samples may reasonably be explained by partial hydrolysis of meat triglycerides during the infusion process. This was borne out by analysis of a meat control, where the meat was warmed with only distilled water. The analysis showed small amounts of glycerine and glucose in the meat.

Table 7.1   % Recovery of Humectants Recovered from Meat

	Sucrose	Sorbitol	Glycerine	Propylene glycol
Solution 1				
In solution	75.0	74.6	73.7	67.5
In meat	22.2	24.5	28.1	27.2
Total	97.2	99.1	101.8	94.7
Solution 2				
In solution	74.9	78.0	81.1	74.1
In meat	24.7	26.8	31.0	30.4
Total	99.6	104.8	112.1	104.5
Solution 3				
In solution	73.3	79.5	86.2	67.4
In meat	23.1	24.6	28.6	27.7
Total	96.4	104.1	114.8	95.1

In 1984, Pecina et al. [23] made an extensive study of the effect of temperature and flow rate on resolution of a number of materials, including polyols and carbohydrates. They also used columns (300 × 7.8 mm i.d. Aminex HPX-87-H) and a mobile phase (0.01 N sulfuric acid) similar to Nagel's. While the article does not demonstrate the separation and quantitative analysis of glycerine from "real-world" sample matrices, it is a useful reference source for glycerine analysis, since it shows the retention behavior of glycerine and 62 other analytes in great detail. Interestingly, The capacity factor for glycerine (retention time of glycerine/retention time of an unretained material) is essentially independent of temperature between 40 and 80°C.

Glycerine levels in soap have been successfully analyzed by George and Acquaro using a reversed-phase column (Waters Carbohydrate Analysis column, 3.9 mm × 30 cm) and an acetonitrile:water mobile phase [24]. As the authors point out, traditional methods of glycerine (AOCS official method Da 23-56, etc.) are tedious and time-consuming. It is also fair to say that they could be subject to interference if other polyols are used as humectants in the bars of soap. Samples were prepared for analysis by blending them for 10 minutes with mobile phase. The resulting slurry was filtered to remove insoluble soaps, diluted and injected into the column. An internal standard of 1,2,4-butanetrol was used for quantitation. This particular mobile phase is beneficial for glycerine analysis, since soap is not very soluble in acetonitrile, and water-rich samples containing both soap and glycerine produce poor resolution, presumably due to mutual solubility problems.

For confirmation, known amounts of glycerine were added to soap in the 0.66–9.66% concentration range. Recoveries were quite good, ranging from a low of 94.3% to a high of 104.8%. When compared with the standard AOCS method, results in the 1–10% range were similar and generally within 5% relative of each other. The method also separates glycerine cleanly from mixtures of ethylene glycol, propylene glycol, and diethylene glycol, which are used as humectants in the soap and cosmetic industry.

VII. GAS CHROMATOGRAPHIC METHODS

Gas chromatography [GC] is an increasingly popular method for separating glycerine from complex mixtures. As early as 1962, Ghanayem and Swann [25] reported successful separation of glycerine from other glycols. They used a 4 ft × ¼ in. column packed with polyphenyl ether and carbowax on Fluoropak 80. They found the combination of polyphenyl ether and carbowax to be more applicable than others available at that time (Ucon greases, silicone oils, and Igepol) at separating glycerine from triethylene glycol. Methanolic solutions containing about 5% glycerine and ethylene glycol, diethylene glycol, and triethylene glycol were used for GC analysis. Electronic integrators were not common at that time, so concentrations were calculated from peak heights in the chromatograms. Multiple injections of the same sample gave a relative standard deviation of about 2%.

Later, Gross and Jones [26] and Gross [27] developed a GC method for glycerine in cosmetic products. The column consisted of a sucrose acetate-isobutyrate stationary phase on Chromosorb G. These stationary phases are no longer in common use. But, at that time, it was the best of several candidates in reducing tailing of the glycerine peak. The sample was dissolved in methanol–water and extracted several times with chloroform to remove oils and triglycerides. The methanol–water solution was then evaporated almost to dryness several times after addition of 2-methoxyethanol. This removed methanol and water from the solution, which otherwise caused excessive tailing. Evaporation to dryness was found to be detrimental, because it caused low glycerine recoveries, presumably from evaporation.

In some cases, it was noted that large amounts of alkanolamines—generally 2 to 3 times as much as the glycerine in the sample—tended to cause high glycerine recoveries. So, samples known to contain such materials were dissolved in water and made slightly acidic to convert amines to their ammonium salts. The ammonium salts were then removed by passing the aqueous solution through an Amberlite CG 120 ion-exchange resin in the acid form. This exchanged the ammonium salts in solution with hydrogen ion from the ion-exchange resin. While this made the eluate more acidic, it did not interfere in the analysis.

The authors noted the glycerine response curve was quite nonlinear, presumably because of irreversible adsorption the column at low glycerine levels. To obtain satisfactory results, they kept glycerine levels in the samples in the

Table 7.2

Sample type	Sample weight	Glycerine added (mg.)	Percent recovery
Water	20 ml	39.9	100
Glyceryl monostearate	150 mg	50.3	95
Vanishing cream	500 mg	39.9	100
Vanishing cream	500 mg	34.0	95
Vanishing cream	500 mg	17.0	100
Toothpaste	500 mg	34.0	103
Toothpaste	2.0 g	34.0	105

same range as in the standards. Modifications of this procedure were used on a variety of known formulations with good success, as indicated in Table 7.2.

This study was supplemented by a collaborative study among six laboratories using vanishing cream to which two different levels of glycerine were added. Recovery was reasonable, ranging from 82% to 102%, but some of the laboratories experienced difficulties with column bleed.

An early gas chromatographic analysis of humectants in tobacco by Cundiff, Greene, and Laurence [28] was modified by Giles [29] and studied collaboratively. The method used columns packed with Carbowax 20M–terephthalic acid on an inert support. The method was designed to measure three common humectants in tobacco: propylene glycol, glycerine, and triethylene glycol. Tobacco samples were prepared by extracting about 10 g of tobacco with 100 ml of methanol which contained anethole as a gas chromatographic internal standard. After settling, the supernatant solution was injected directly into the gas chromatograph. Standard curves were constructed from glycerine/propylene glycol/triethylene glycol solutions and used for quantitation of the tobacco samples.

Later, Williams [30] used a very similar GC method for tobacco, but replaced anethole with 1,3-butylene glycol as the internal standard. In the original method developed by Giles, the anethole internal standard was not baseline resolved from triethylene glycol. This may have contributed to the lack of precision and accuracy in the results. The 1,3-butylene glycol was more cleanly separated from the other analytes, and so made a more appropriate internal standard.

A collaborative study under the auspices of the AOAC showed improved results, and the method was adopted as an official first action method by the AOAC (Table 7.3).

Glycerol and other diols are used as flavor bases in flavored wines, and a gas chromatographic method has been developed for its determination [31]. Samples are prepared by diluting with ethanol and adding 1,4-butanediol as the internal standard.

Table 7.3

Component analyzed	Approximate analyte range in tobacco (%)	Standard deviation, Williams study (%)	Standard deviation, Giles study (%)
Propylene glycol	0.5	0.05	0.06
	1.0	0.05	0.08
	1.5	0.06	0.1
Glycerol	1	0.1	0.2
	2	0.1	0.2
	3	0.2	0.4
Triethylene glycol	0.5	0.1	0.1
	1.0	0.1	0.1
	1.5	0.1	0.1

Three different kinds of chromatographic columns were studied: Chromosorb 101, a porous polymer; Chromosorb 101 coated with SP-1000; And SP-1000 coated on Chromosorb W, a diatomaceous earth. Standard deviations on the SP-1000/Chromosorb W column were best and further efforts were focused on that system. Their results are shown in Table 7.4.

It is likely that future analytical developments for glycerine will move toward capillary gas chromatography. Capillary column technology utilizes stable stationary phases coated on very inert capillary columns, typically glass or fused silica, resulting in less tailing of peaks, superior resolution, and less tendency for degradation at high temperatures. It is also common to react hydroxide-containing materials with silylating reagents to decrease their tendency to bond to hydrogen. This reduces tailing and results in better chromatographic resolution. It also reduces irreversible adsorption and results in more linear calibration curves.

Table 7.4

Component	% added	Relative ret. time[a]	% found	SD
1,2-Propanediol	0.80	0.32[b]	0.80	0.01
2,3-Butanediol	0.80	0.32[b]	0.80	0.01
1,3-Propanediol	0.80	0.63	0.81	0.01
1,3-Butanediol	0.80	0.54	0.77	0.01
Glycerine	2.00	4.08	1.98	0.03

[a] Relative to 1,4-butanediol, the internal standard.
[b] Separated at a lower temperature.

A report by Molever [32] illustrates these trends for the analysis of glycerine in soap. Samples are prepared by blending about 10 g of soap with 200 ml of N.N-dimethyl formamide. After filtration to remove insoluble soap, a portion of the solution is reacted with BSTFA (bis-trimethylsilyltrifuroacetamide) to produce the silylated glycerine derivative. The column used is a 12 m × 0.2 mm i.d. fused silica capillary coated with methyl silicone fluid. Glycerol was analyzed isothermally at 100°C. Quantitation was done by comparing peak areas of the sample with peak areas of external standards.

Gas chromatographic results were checked against AOCS method #Da 23-56 (periodic acid titration). Bars of soap ranging from about 1% to about 3% glycerine gave satisfactory agreement, with an average difference between methods of about 0.03% absolute. Bars of soap were also spiked with 0.05% and 0.10% glycerine, and gave recoveries of 97.1% and 100.3%, respectively.

VIII. COLOR AND SPECIFIC GRAVITY

Several organizations have limits on color and specific gravity (Table 7.5). Details of the methods follow.

A. Color Measurement Methods

Commercially traded refined glycerine is generally light in color, and simple visual tests are normally satisfactory to demonstrate suitable color. In general, Nessler tubes are used for these inspections. These are glass cylinders with optical-glass tops that contain 50–100 ml of solution. The glycerine sample is placed in a Nessler tube and put in a container illuminated from the bottom with a daylight-type fluorescent tube. The analyst judges the darkness (intensity of light passing through the Nessler tube) of the glycerine sample compared with a series of color standards which are placed in identical Nessler

Table 7.5

Organization:	CTFA	USP	BP	FCC	ACS Reagent glycerol	ASTM High-gravity glycerin
Description:	Glycerin 95%	Glycerin	Glycerol	Glycerin		
Used for:	Cosmetics	Drugs	Drugs	Food	Reagents	Plastics
Color:						
color scale	APHA	$FeCl_3$	BP	$FeCl_3$	APHA	APHA
maximum	30	0.4 ml	see text	0.4ml	10	20
Specific gravity min, 25/25°C	1.2491	1.249	1.258–1.263	1.249	1.2570	1.2587

tubes. These are straightforward tests, and require little in the way of training or equipment for good results.

1. APHA Color Comparisons Using Platinum–Cobalt Solutions

Glycerine colors are compared with dilute platinum–cobalt color standards. This is commonly referred to as APHA color. The technique was first described in the 1890s [33] and later adopted by the American Public Health Association for measuring the color of water [34]. Since that time, the method has become widely used for many lightly colored solutions. As is correctly pointed out in ASTM method D-1209-84, it is probably more correct to describe this as a platinum–cobalt color test, since the designation ASTM refers primarily to water.

The platinum–cobalt stock is prepared by dissolving 1.245 of potassium chloroplatinate and 1 g of cobalt chloride hexahydrate in water, adding 100 ml of concentrated hydrochloric acid and diluting to 1 liter with water. This is referred to as the 500 color standard (it contains 500 ppm of platinum) and is used to prepare other color standards. Simplified, the 250 color standard is made by a 1:1 dilution of the 500 color standard, the 50 color standard is made by a 10:1 dilution, etc.

Standards are stable for at least a year without color change. The color test sets spectrometric absorbance limits for the stock solution at several wavelengths from 430 to 510 nanometers. Glycerine colors are compared with the appropriate platinum-cobalt solutions (typically ranging from 5 APHA units upwards in increments of 5 units) in matched 100 ml Nessler tubes having a liquid depth between 275 and 295 mm.

2. $FeCl_3$: Ferric Chloride Color Comparisons

FCC and USP standards call for comparison of glycerine colors with aqueous ferric chloride solution. Typically, a concentrated solution of ferric chloride is made by dissolving excess ferric chloride hexahydrate in 25 volumes of hydrochloric acid and 975 volumes of water. The ferric chloride concentration is calculated iodometrically after appropriate dilution, addition of iodide, and titration of the iodide with thiosulfate. The solution is then diluted with 25 volumes of hydrochloric acid and 975 volumes of water to correspond exactly to 45.0 g/L of ferric chloride.

Color comparisons are then made with reference to the ferric chloride color solution, using 50 ml Nessler tubes to compare samples. To pass FCC and USP limit tests, glycerine must be lighter than a solution made from 0.4 ml of the ferric chloride color solution diluted to 50 ml with water.

3. BP Color Scale

British Pharmacopoeia (BP) color limits specify that 2 ml of glycerine, when placed in glass tubes in 12 mm outside diameter and viewed in diffuse daylight against a white background, should have the same color as distilled water in an identical tube. This is classified as colorless. To judge the glycerine color-less against a solution having a very faint (and probably visually undetectable) color, the BP methods give directions for preparing a very light brown color standard, designated as B_9, which is made in the following way:

1. Prepare a yellow primary solution by making a 45 mg/ml solution of fer-ric chloride hexahydrate in a liter of a solution containing 25 ml of con-centrated hydrochloric acid and 975 ml of distilled water. This is prepared the same as the ferric chloride solutions used by the USP.
2. Prepare a red primary solution by making a solution containing exactly 59.5 mg of cobalt(II) chloride hexahydrate per ml of solution (25 parts con-centrated hydrochloric acid, 975 parts water). Directions for standardizing the solution by redox titration using thiosulfate as the titrant are given.
3. Prepare a blue primary solution containing exactly 62.4 mg of cuprous sulfate pentahydrate per ml of solution (25 parts concentrated hydrochlo-ric acid, 975 parts water). Directions for standardizing the solution by redox titration using thiosulfate as the titrant are given.
4. Make a brown standard solution by mixing 30 ml of yellow primary solu-tion, 30 ml of red primary solution and 24 ml of blue primary solution with 16 ml of 1% (w/v) hydrochloric acid.
5. Dilute 1 ml of the brown standard solution with 99 ml of 1% (w/v) hy-drochloric acid. This is labelled reference solution B_9, and is the most colorless of the reference solutions.

It might be noted that the British Pharmacopoeia, by specifying these three primary color solutions, has done an admirable job of creating a flexible sys-tem of color designations for visual inspections. By proper mixture of each of the primary solutions, they can produce brown, brownish-yellow, yellow, greenish-yellow, and red solutions of different intensities. Very handy!

B. Specific Gravity

The specific gravity of glycerine is the weight ratio of a given volume of glyc-erine to the same volume of water at a specified temperature. Specific gravity measurements are normally taken at 25°C for both the glycerine and the water, and reported as "apparent specific gravity at 25/25°C." This is a common method for spot-checking glycerine concentrations in crude glycerine as well, although the relationship between specific gravity and concentration only holds for pure binary mixtures of glycerine and water.

United States tables used to relate specific gravity to concentration are al-most universally based on the very careful work of Bosart and Snoddy in 1927 [35]. Until their work, several different tables were in use. The tables used several different temperatures, and varied greatly in accuracy.

Figure 7.3 Specific gravity bottle used in AOCS and CTFA methods. (Reprinted with permission of Thomas Scientific Company.)

The work of Bosart and Snoddy remains interesting reading today because of the care that they used in obtaining their results. They started with 98.5% pure commercial glycerine and bleached it with activated charcoal to remove color. Distillation apparatus was constructed to permit sparging with dry nitrogen that had been passed through a sulfuric acid and phosphorous pentoxide train. After the first distillation at 2 mm of pressure, the middle third of the distillate was collected and redistilled. The middle third of the second distillate was then stored in the collection receiver without exposure to damp air. Water used to make the solutions was distilled from potassium permanganate and boiled before use. Solutions were carefully mixed several days before use to permit complete separation of air bubbles.

As a result of the care taken in experimentation, Bosart and Snoddy produced results accurate to the fourth decimal place—probably for the first time—with fairly small divergence of replicates in the fifth decimal place. Results were also calculated as "apparent," and "true" specific gravity, although only the apparent specific gravity tables are in common use today.

Several styles of pycnometer are in use, and are quite satisfactory if they are calibrated carefully and used with a water bath capable of good temperature control. One such pycnometer is pictured in Figure 7.3. The bottle is filled completely with glycerine at room temperature, being careful to avoid air bubbles. It is then equilibrated in a constant-temperature bath at 25°C + 0.1°C, with the sidearm cap removed. This permits glycerine to expand out of the bottle as it warms. The bottle is removed from the bath, and the sidearm cap is placed on the bottle. The outside of the bottle is dried carefully and the bottle is weighed. The weight is compared with the weight of water prepared the same way. AOCS method Ea 7-50 reports that duplicate determinations made in two different laboratories should be within 0.00025 units of each other.

In recent years, equipment for simply and reproducibly measuring the specific gravity of glycerine has become available. It is usually referred to as a density meter [36] and consists of a glass U-tube inside a temperature-controlled chamber, as diagrammed in Figure 7.4. The U-tube is vibrated at a known and measurable frequency. When the sample is introduced into the U-tube, its mass changes the vibrational frequency. This can be related to the density of the sample. Since shifts in frequency can be measured very accurately by electronic devices the instrument measures densities to 4 and even 5 decimal places; comparable to pycnometer accuracies. The density meter is calibrated with glycerine–water solutions of known specific gravities and produces specific gravity results equivalent to pycnometer methods. Compared with pycnometer methods, it is faster and requires less skill of the part of the analyst. Samples can normally be introduced into the instrument by syringe and results are obtained after a 5-min temperature equilibration period. Water baths and analytical balances are not required so total system training and up-keep are very satisfactory.

C. Minor Component Analysis

The principal distinctions among different kinds of finished glycerine are the percentage of glycerine, the physical properties (color, odor, etc.) and minor chemical components.

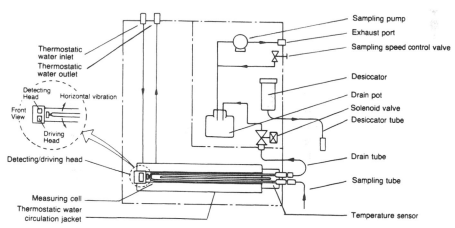

Figure 7.4 Density meter suitable for glycerine measurements. (Reprinted with permission of Mettler Ins. Co., Hightstown, NJ and Kyoto Electronics, Ltd., Kyoto, Japan.)

Table 7.6

Organization:	CTFA	USP	BP	FCC	ACS	ASTM
Description:	Glycerin 95%		Glycerol	Glycerin	Reagent glycerol	High-gravity glycerine
Used for:	Cosmetics	Glycerin Drugs	Glycerol Drugs	Glycerin Food	glycerol Reagents	glycerine Plastics
Free acids/bases	yes		yes		yes	yes
Ash	yes	yes	yes	yes	yes	yes
Acids and esters	yes	yes	yes	yes		
Sulfates	yes	yes			yes	
Chloride	yes	yes	yes			
Arsenic	yes	yes		yes		
Chlorinated compounds	yes	yes	yes	yes		
Heavy metals		yes	yes	yes		
Periodate assay		yes		yes	yes	

Comparing specifications among different organizations is no simple task, because each organization tends to specify its own analytical method. The following discussion is intended to give an overview of the similarities and differences among many of the common minor component methods. It is important to recognize that specifications are reviewed periodically by the issuing organizations, so they can change from year to year. So, the prudent analytical chemist will keep up to date on his/her customer's finished product usage and make sure that the analytical methods are current.

Table 7.6 indicates the limit tests required by different organizations. As expected, glycerine for food and drug use requires more testing than glycerine used as a chemical reagent or chemical intermediate.

D.   Methods for Measuring Acids/Bases in Glycerine

During glycerine purification, fatty acids are commonly converted to their soaps by base neutralization to prevent codistillation with the glycerine. Residual fatty acid esters are removed by fractional condensation of the distilled glycerine. The amount of acids and esters remaining in the finished glycerine measures the quality of the refining process. It is common to set specifications for glycerine to have upper limits for the presence of acids and esters. In some cases, there are separate limits for free acids or bases, and in other cases total amount of acid and ester is established by saponification and titration.

1.   Combined Acids and Esters

Table 7.7 illustrates the major differences among the various methods for determining total acids and esters. These methods specify a certain amount of glycerine (25, 40 or 50 grams) to be added to water 25 or 50 mml). Sodium

Table 7.7 Overview—Total Acid and Ester Methods

	Method title	Glycerine used (g)	Water used (ml)	NaOH amt/conc	Reflux time (min)	Titrate with	Indicator used	Upper limit
BP	Ester	25	25	10 ml/0.1 N	5	0.1 N HCl	Phenolphthalein	0.8 mEq/25 g
FCC CTFA USP	Fatty acids and esters	50	50	5 ml/0.5 N	5	0.5 N HCl	Phenolphthalein	0.5 mEq/50 g
ACS	Fatty acids and esters	40	50	10 ml/0.1 N	45	0.1 N HCl	Bromthymol blue	0.23 mEq/40g

hydroxide (5 ml of 0.5 N. or 10 ml of 0.1 N) is added and refluxed (5 or 45 minutes) and titrated to neutrality with hydrochloric acid. The difference between the titration and a blank titration indicates the number of milliequivalents of saponifiable material in the sample. It should be recognized that these tests are designed to specify upper limits of saponifiable material. They do not distinguish among organic acids and any partial glycerides which may be present in the samples. Any material that reacts with hydroxide will count as an ester or fatty acid and ester. These materials can include the following:

Fatty acids: $NaOH + RCOOH \rightarrow RCOONa + H_2O$
Monoglycerides: $NaOH + H_2COOR(HCOH)H_2COH \rightarrow RCOONa + H_2O$
    $+ H_2COH(CHOH)CH_2OH$
Diglycerides: $2NaOH + H_2COOR(HCOR)H_2COH \rightarrow 2RCOONa + 2H_2O$
    $+ 2H_2COH(CHOH)CH_2OH$
Inorganic acids: $2NaOH + H_2SO_4 \rightarrow Na_2SO_4 + 2H_2O$

2. Free Acids and Bases

These methods are usually quite simple, as seen in Table 7.8. They consist of measuring the pH of a glycerine solution, or of titrating the sample with base. The American Oil Chemists' Society methods for acids and bases are more comprehensive, due largely to their use in determining appropriate conditions for neutralizing and refining glycerine. These methods are titled "Ash, Alkalinity or Acidity and Sodium Chloride" and described in method Ea 2-38. The method consists of the following steps:

For glycerine samples that are basic to phenolphthalein in 10% aqueous solution:

1. Two to 5 g of glycerine is ashed at 500°C. The ash, consisting of sodium oxide from sodium hydroxide, sodium carbonate, and sodium soaps is titrated with sulfuric acid and reported as "total alkalinity," in units of % sodium oxide.

Table 7.8 Overview: Free Acid/Base Methods

	Title	Sample amount	Indicator	Titrant	Upper limit
BP	Acidity	25 grams	Phenolphthalein	0.10 N NaOH	.02 MEq
ASTM	Acid value	50 ml (about 63 g)	Phenolphthalein	0.05 N NaOH	0.3 as mg KOH/g (about 0.34 mEq)
CTFA	Free acids and alkalies	10% in water	Litmus		Neutral
ACS	Neutrality	10% in water	Litmus		Neutral

2. Barium chloride is added to a fresh sample to precipitate carbonates and soaps, and an aliquot of the clear supernatant solution is titrated with hydrochloric acid. This gives the amount of sodium hydroxide in the sample, reported as "free caustic alkalinity," in units of % sodium oxide.
3. Another sample is neutralized with sulfuric acid. This converts carbonate to carbon dioxide, which is then removed by boiling. The sample is then titrated to a phenolphthalein endpoint to give the amount of soap present, expressed as "alkalinity combined with organic acids," in units of % sodium oxide.
4. Carbonate content is calculated by deducting hydroxide (2, above) and soap (from 3, above) from total alkalinity (1, above)

For glycerine samples that are acidic to phenolphthalein in 10% aqueous solution:

1. The sample is titrated with sodium hydroxide.
2. Results are reported as "acidity as sodium oxide equivalents" in units of % sodium oxide.
3. Another sample is refluxed with sodium hydroxide to saponify free acids and esters present, and the amount of sodium hydroxide used is compared to a blank titration.
4. Again results are reported as "alkalinity equivalent to acids and esters," in units of % sodium oxide.

E. Chloride Methods

Because almost all commercial glycerine is distilled, inorganic chlorides are only present because of entrainment during distillation. As with sulfate, only traces of chlorides are likely to be present in refined glycerine. Simple visual turbidimetry matching methods are normally used for specifying the upper limit of chloride in glycerine. While these methods do not give absolute amounts of chloride, they are simple to perform and sensitive. Essentially, the glycerine is diluted with water, the solution is made slightly acidic with nitric acid and silver nitrate is added to it. The presence of chlorides is indicated by precipitation of silver chloride, which will make a slightly turbid solution. The turbidity of this solution is then matched with a standard aqueous solution made from sodium chloride and silver nitrate to match the upper limit. Limits and amounts of glycerine used for the various tests are listed in Table 7.9. The American Oil Chemists Society, which often provides methods that are suitable for crude glycerine where levels of inorganics are large and of interest in determining processing conditions. AOCS method Ea2-38 uses a titration with silver nitrate. Two to five grams of sample is ashed, dissolved in water, and made acid to phenolphthalein. Potassium chromate (Mohr titration method) is

Table 7.9

Organization	Glycerine used (g)	Solution volume (ml)	Upper limit
USP, CTFA	7	50	10 ppm as chloride
ACS	2	20	5 ppm as chloride
B.P.	0.5	15	10 ppm as chloride

added as an indicator. When all the chloride has precipitated, a reddish precipitate of silver chromate is visible. Results are reported as sodium chloride. While this method is a precise determination of chloride content as opposed to a go/no go limit test, it is not so suitable for determining very small amounts of chloride, unless larger amounts of glycerine are ashed.

F. Determination of Arsenic

Surprisingly, oxides of arsenic are fairly soluble in glycerine. This is apparently due to the formation of glyceryl arsenite, $C_3H_5AsO_3$ [37]. In the past, arsenic in glycerine may have come from neutralization of crude glycerine with low grades of hydrochloric acid containing arsenic. During glycerine distillation volatile arsenic (III) halides could then be transferred into the distilled glycerine. In any event, even though today's finished glycerine is unlikely to contain any significant amounts, upper allowable limits for arsenic are common specifications.

1. Limits

The Food Chemicals Codex and CTFA specify limits equal to or less than 3 parts per million of arsenic. USP/NF limits are 1.5 parts per million.

2. Test Methods

The most common method for measuring arsenic content uses silver diethyldithiocarbamate as a colorimetric reagent. Food Chemical Codex and USP require wet-ashing the sample before the colorimetric analysis is performed. CTFA specifies diluting the sample without wet-ashing and proceeding directly to the color-generating step. Wet-ashing destroys organic matter and converts arsenic to the arsenic (V) oxidation state without loss of volatile arsenic compounds, which would occur if the sample were ashed at muffle oven temperatures. About 1 g of glycerine is digested with sulfuric acid until charring begins. Thirty percent hydrogen peroxide is added, very slowly and carefully at first, to prevent too violent a reaction. Oxidizing conditions are maintained by adding more peroxide whenever the solution becomes dark. When oxidation is complete, the temperature is increased until copious fumes of SO_3 evolve. Then, the sample is cooled, water is added and the solution is fumed again to remove traces of hydrogen peroxide.

To generate the color, the sample is treated with stannous chloride to reduce arsenic (V) to arsenic (III). Zinc is added to the solution to evolve hydrogen and arsine (AsH_3). The Arsine and hydrogen gases pass through a lead acetate scrubber and the arsine is trapped in a solution of silver diethyldithiocarbamate in pyridine. This produces a red complex with an absorbance maximum around 535 nm. The absorbance of the sample is compared with arsenic tri-oxide standards prepared in the same way.

3. Glassware

Several versions of glassware are used for this analysis, but all share common features. A glass flask is used to digest the sample and generate the arsine. An absorber tube packed loosely with lead acetate-impregnated cotton wool con-nects the flask to an absorber tube that contains the silver diethyldithiocarbam-ate solution. As the arsine bubbles through the absorber tube, the red color is formed. A common type of glassware is pictured in Figure 7.5.

4. Atomic Absorption Methods

While the silver diethyldithiocarbamate method is the most common official method for arsenic analysis, it requires care and significant effort in preparing fresh reagents and wet-ashing the sample. For research purposes, a simpler technique described by Ullman [38] is very suitable. A glycerine sample is

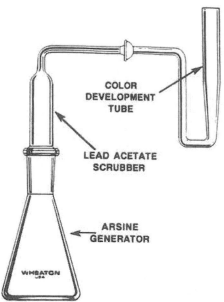

COLOR
DEVELOPMENT
TUBE

LEAD ACETATE
SCRUBBER

ARSINE
GENERATOR

WHEATON
USA

Figure 7.5   Arsenic test apparatus—USP, Food Chemicals codex. (Re-printed with permission of Thomas Scientific Co.)

Table 7.10

Arsenic added (ppm)	Atomic absorption results (ppm)	Average USP results (ppm)
0	0.03	0.04
0.1	0.08	0.11
0.5	0.46	0.52
1.0	1.1	0.9
7.0	7.4	6.1

placed in a sealed reaction flask and reacted with sodium borohydryde to release arsine. The arsine is swept into a heated quartz tube placed in the light path of an atomic absorption spectrometer. As the arsine passes through the heated tube, it produces elemental arsenic which absorbs light from an arsenic lamp. The absorbance is compared with standards, and is proportional to the original arsenic concentration of the glycerine. The relative standard deviation of the method at the 1 ppm is about 4.4%, and somewhat better than can be expected from the colorimetric method.

Finished glycerine samples were spiked with varying amounts of standard arsenic solutions and analyzed by both USP and atomic absorption methods, with good agreements (Table 7.10). Tioh and co-workers [39] extended this work by using flow-injection equipment to automate mixing the glycerine with the sodium borohydride generating solution.

G. Heavy Metals

The USP, BP, and FCC all have upper limits of 5 ppm heavy metals as lead. The ACS has an upper limit of 2 ppm. All three organizations use similar techniques. Sulfides in an aqueous solution of the glycerine are precipitated at pH 3.5 with a fresh solution of hydrogen sulfide (USP and FCC) or thioacetamide (BP). This produces a brown suspension of sulfides that can be visually matched in Nessler tubes against a suitably prepared standard. Other metals precipitate with sulfide in this test, including silver, arsenic, bismuth, cadmium, copper, mercury, antimony, and silver—the old "Group 1" metals from qualitative analysis days.

The USP and FCC both specify that a glycerine sample known as a "monitor preparation" be tested to make sure there are no interferences with the test. A separate glycerine sample is prepared and to it is added the equivalent of 5 ppm lead. This solution is then tested along with the original glycerine solution. If there is no interference, the "monitor preparation" should be at least as dark as the 5 ppm reference heavy metal solution. If it is lighter, the glycerine sample must be ashed to remove organics and retested.

H. Sulfated Ash/Residue on Ignition

The USP, CTFA, FCC, and BP all have limits of 0.01%. The ASTM has limits of 0.1%. These are straightforward tests, usually calling for the ignition of about 50 g of glycerine over a flame. When the sample has been completely consumed, the residue is moistened with about 0.5 ml of concentrated sulfuric acid and the sample is heated, either over a flame or in a muffle oven at 800°C, to a constant weight.

I. Sulfates

The USP and CTFA share methodology and upper limits of 20 ppm. The ACS calls for a limit of 10 ppm. The test is performed by placing 10 g of sample in a viewing tube and precipitating sulfates with barium chloride solution. After mixing, the solutions are allowed to stand for 10–20 minutes. Then, turbidity is compared with a fresh barium sulfate solution containing the appropriate amount (10–20 ppm) of barium sulfate.

J. Chlorinated Compounds/Halogenated Compounds

The CTFA, USP, and FCC have equivalent limits of 30 ppm, expressed as chloride. Their limit tests are also essentially the same. About 5 g of glycerine is added to a flask with 15 ml of morpholine and refluxed for 3 h. After refluxing the solution is acidified with nitric acid. It is then transferred to a Nessler tube and silver nitrate is added to precipitate silver chloride. The turbidity is compared with a fresh silver chloride solution containing 30 ppm of silver.

The BP "halogenated compound" test is similar in comparing turbidity of silver chloride with a reference solution. Their limits are also about 30 ppm. But the BP uses a different procedure for releasing chlorides from the sample. To 5 ml of glycerine, 1 ml of 21 M sodium hydroxide, 10 ml of water, and 50 mg of Raney nickel catalyst is added. This mixture is heated for 10 minutes on a water bath and filtered. This is washed with water until 215 ml of aqueous solution is obtained. A 5 ml aliquot of this is mixed with 4 ml of ethanol, 2.5 ml of water, 0.5 ml of nitric acid to acidify, and silver nitrate to precipitate the silver chloride.

REFERENCES

1. Malaprade L., Compt. rend., 186:382, 1928.
2. Fleury, P. F., and Lange, J., Compt. rend. 195:1395, 1932.
3. Cheronis N. D., and Ma, T. S., Organic Functional Group Analysis, Interscience, New York, 1964, p. 193.

4. Bobbitt, J. M., in Advances in Carbohydrate Chemistry, Vol., 11, Academic Press, New York, 1956, p. 1.

5. Cheronis, N. D., and Ma, T. S., Organic Functional Group Analysis, Interscience, New York, 1964, p. 158.

6. Laitinen, H. A., and Harris, W. E., Chemical Analysis, 2nd ed., McGraw Hill, New York, 1975, p. 370.

7. Pohle, W. D., and Mehlenbacher, V. C., J. Am. Oil Chem. Soc. 24:155, 1947.

8. Miner, Carl S., and Dalton, N. N., Glycerol, Reinhold Publishing Corp., New York, 1953, p. 190.

9. Standard Methods for the Examination of Water and Wastewater, 14th ed., American Public Health Assn., New York, 1975.

10. Miner, C. S., and Dalton, N. N., Glycerol, Reinhold Publishing Corp., New York, 1953, p. 204.

11. Bublitz, C., and Kennedy, E. P., J. Biol. Chem., 122:951, 1954.

12. Boltralik, J. J., and Noll, H., Anal. Biochem., 1:269, 1960.

13. Garland, P. B., and Randle, P. J., Nature, 196:987, 1962.

14. Hagen, H., Biochem. J., 82:23, 1962.

15. Pardue H. L. and Frings, C. S., Anal. Chim. Acta, 34:228, 1966.

16. Masoom, M., and Worsfold, P. J., Anal. Chim. Acta, 188:281, 1988.

17. Morishita, F., Nishikawa, Y., and Kojima, T., Analyt., Sci., 2:411, 1986.

18. Kiba, N., Goto, K., and Furusawa, M., Anal. Chim. Acta, 185:287, 1986.

19. Schwartzenbach, R., J. Chromatogr., 140:304, 1977.

20. Scobell, H. D., Brobst, K. M., and Steele, E. M., Cereal Chem., 54:905, 1977.

21. Nagel, C. W., Brekke, C. J., and Leung, H. K., J. Food Sci., 47:342, 1981.

22. Kaplow, M., Food Technol., 24:889, 1970.

23. Pecina, R., Bonn, G., Burtscher, E., and Bobleter, O., J. Chromatogr., 287: 245, 1984.

24. George, E. D., and Acquaro, J. A., J. Liq. Chrom., 5:927, 1982.

25. Ghanayem, I., and Swann, W. B., Anal. Chem., 34:1847, 1962.

26. Gross, F. C., and Jones, J. H., J. Assoc. Off. Anal. Chem., 50:1287, 1967.

27. Gross, F. C., J. Assoc. Off. Anal. Chem., 50:1292, 1967.

28. Cundiff, R. H., Greene, G. H., and Laurene, A. H., Tobacco Sci. 8:163, 1964.

29. Giles, J. A., J. Assoc. Off. Anal. Chem., 53:655, 1970.

30. Williams, J. F., J. Assoc. Off. Anal. Chem., 54:560, 1971.

31. Martin, G. E., Dyer, R. H., and Figert, D. M., J. Assoc. Off. Anal. Chem., 58:1147, 1975.

32. Molever, K., J. Am. Oil Chem. Soc., 64:1356, 1987.

33. Hazen, A., American Chemical Journal, XIV:300, 1892.

34. Standard Methods for the Examination of Water, 1st ed., American Public Health Assn., New York, 1905.

35. Bosart, L. W., and Snoddy, A. O., Ind. Eng. Chem., 19:506, 1927.

36. Mettler Instrument Co., Density/Specific Gravity Meters, Highstown, N.J.

37. Martin, G., The Manufacture of Glycerol, 2nd ed., vol. III, p. IV–11, The Technical Press, Ltd., London, 1956.

38. Ullman, A. H., J. Am. Oil Chem. Soc. 60:614, 1983.

39. Tioh, M., Israel, Y., and Barnes, R. M., Anal. Chim. Acta 184:205, 1986.

8

Handling, Safety, and Environmental Aspects

Norman O. V. Sonntag

Ovilla (Red Oak), Texas

I. HANDLING

A. Thermal Constraints in the Production (Distillation) and Industrial Use of Glycerol for Derivatives

The handling of glycerol industrially is dependent upon the optimum temperature at which it is thermally stable, and the specific conditions of thermal exposure, such as the neutrality, alkalinity, or acidity of the medium. Glycerol suffers two major changes when subjected to heat (1) dehydrative polymerization through etherification (written as the first in a series of polymerizations with α, α', etherification):

$$2 \text{ HOCH}_2\text{CHOHCH}_2\text{OH} \rightarrow \text{HOCH}_2\text{CHOHCH}_2\text{OCH}_2\text{CHOHCH}_2\text{OH} + \text{H}_2\text{O} \tag{1}$$

and, (2) dehydration to acrolein:

$$\text{HOCH}_2\text{CHOHCH}_2\text{OH} \rightarrow \text{CH}_2 = \text{CHCHO} + 2 \text{ H}_2\text{O} \tag{2}$$

Glycerol appears to be especially prone to dehydration to acrolein under acidic conditions. A 33–48% yield of acrolein results when glycerol is heated with potassium acid sulfate in an oil bath to 190–200°C for about one hour, then at 215–230°C after the initial reaction has abated [1]. Glycerol, distilled with 5% infusorial silica, yields acrolein. There is considerable decomposition

of glycerol when vapors are passed through a pumice-filled tube at 430–450°C
[2]. Acrolein and aldehyde condensation products result when glycerol vapors
are passed over alumina at 360°C. The dehydration reaction is also possible
under non-acidic conditions, apparently to a somewhat lesser degree. At 330°C
finely divided copper causes glycerol to decompose to acrolein, allyl alcohol,
ethyl alcohol, hydrogen and small amounts of carbon monoxide and methane.
Glycerol vapor on uranium dioxide at 350°C gives 10% acrolein, 35% ethyl
alcohol, and some water and allyl alcohol [3]. Alkali phosphates have also
been employed for the vapor phase production of acrolein from glycerol at
temperatures in the 300–600°C range [4].

On the other hand, traces of polyglycerols are detected in glycerol which
has been heated to temperatures as low as 200°C, and polyglycerols are always
found in the still residues from glycerol distillations. Glycerol has a boiling
point at atmospheric pressure of 290.0°C with slight decomposition. Accordingly, its distillation must be accomplished under substantially reduced pressure considerably below temperatures of 200°C. The distillation of anhydrous
glycerol can be carried out at reduced pressures from 30 mm Hg (194°C,
381.2°F) preferably at 10 mm Hg (170°C, 338°F). By distilling with steam at
45 mm Hg with the partial pressure of steam at 30 mm Hg and glycerol at 15
mm Hg, the glycerol vaporizes at 347°F, thus lowering the glycerol vaporization temperature from 400° to 347°F.

Note that when glycerol is heated in the presence of strong alkalies like
sodium or potassium hydroxides at 350°C there is extensive pyrolytic decomposition. Fry and Schulze [5] proposed that the reactions:

$$C_3H_5(OH)_3 + 6NaOH \rightarrow 3Na_2CO_3 + 7H_2 \tag{3}$$
$$C_3H_5(OH)_3 + 4NaOH \rightarrow 2Na_2CO_3 + 3H_2 + CH_4 + H_2O \tag{4}$$

occurred in about a ratio of (3):(4) = 1:3.6 at 350° and at a ratio of (3):(4) =
3:7.1 at 450°C.

Based upon numerous studies of the use of excess glycerol in the glycerolysis of triglycerides for the manufacture of mono- and diglyceride mixtures
ordinarily carried out under moderate or strong alkaline catalysis [6], the optimum temperature for the production of edible products appears to be 255°C
when hydrated lime [Ca(OH)_2] at 0.06–0.1% concentration is the catalyst,
and perhaps 245°C when stronger alkalies, NaOH or KOH, are used at from
0.05–0.20%. Products subjected to prolonged periods of time at these temperatures, or for shorter exposure at higher temperatures begin to be characterized
by sophisticated taste panels variously as "burnt," "acrid," or "acid." Possibly, this is the threshold detection level for the taste response for traces of
toxic lachrymatory acrolein derived by glycerol dehydration.

B. Materials of Construction and Storage

Use of copper in any of its alloyed forms should be avoided in tankage or lines and equipment. Brass or bronze valves are especially vulnerable sources of copper contamination. There are two major reasons for preventing copper contamination in glycerol. The first is that copper is a universal toxicant and a powerful prooxidant. Glycerol is a raw material for mono-, di-, and triglyceride production as well as a host of ester derivatives used in the food, cosmetic and pharmaceutical industries. Many producers of these products have stringent specifications for minimum copper content. Furthermore, glycerol in many of its uses, will be in contact with many organic materials in its final use form in cosmetics, pharmaceuticals and foods, where autoxidation is a primary concern. Thus, copper traces are detrimental both for toxicity and for prooxidant reasons. Second, copper traces are catalysts for the dehydration of glycerol to acrolein with heat. This is a toxic, lachrymatory material of relatively high hazardous nature (LD_{50} orally in rats is 26 mg/kg body weight [7]).

Equally objectionable is the contamination of glycerol with soluble iron down to the possible level of 1–5 ppm or higher. This applies to both glycerol and its derivatives. Recently, the presence of particulate iron in highly divided form possibly derived from friction of moving parts such as agitator shafts, etc, has been recognized as especially detrimental. Corrosion of cast iron equipment, or even low-grade steels are the usual sources of soluble iron contamination. The use of stainless steels of at least 304-grade, preferably 316-grade, is recommended for tanks, lines, and equipment for glycerol refineries and for all operations. Materials of construction for glycerol tankage include 304 or preferable 316 stainless steels, nickel-clad steel or aluminum, although use of the latter appears to be diminishing.

Certain resin linings such as Lithcote have also been used. Glycerol does not seriously corrode ordinary steel at ambient temperatures but absorbed moisture has an effect on older steel tanks, and iron content should be intermittently monitored in products stored in steel tanks more than 5 years old.

The recognized hydroscopic nature of glycerol, especially in its higher concentrations, requires that during tank storage the entrance of moist air be denied in the normal "breathing" of tanks used for storage. Glycerol in its finished forms is an excellent humectant, and its absorption of water is increased dramatically by thin-film exposure to ordinary air or by pumping into an empty tank. All tankage should be equipped with an air breather trap or similar protective device. Obviously, stirring glycerol in a tank by blowing with air is always inadvisable.

Although earlier experience indicates that glycerol could be processed in plants which contained cast iron filter presses (covered with canvas), mild steel, wrought iron or cast iron tanks and piping, modern international practice tends to eliminate these materials to a large extent. Many older stills were made of mild steel; today it is considered better long-range economical practice to construct entirely of high-grade 316-type stainless-steel or equivalent.

C. Shipping and Transfer

Optimum transfer temperatures for pumping glycerol is in the 37–45°C range. Pumps are constructed of stainless steel, but those made of bronze and cast iron, while less satisfactory, still prevail in use in older plants. CP or USP grades of glycerol are shipped mainly in bulk in tank cars or tank wagons. Today these are largely stainless, steel, aluminum, or lacquer-lined. Received in standard 8000 gal (30.3 m^3, 36.3 t) tank cars, this will be contained in a storage tank of 38–45 m^3 (10–12 × 10^3) gal capacity. About 4 h would be required to unload a tankcar of warm glycerol using a 3.15 L/s (50 gal/min) pump.

Unlined standard steel tank cars, provided they are kept clean and in rust-free condition, are satisfactory for glycerol shipments. Refined glycerol is also available in 4.5 kg (3.8 L or 1 gal) tinned cans, or more frequently, in 250–259 kg (208 L or 55 gal) drums of a nonreturnable type (ICC-17E) containing a phenolic-resin lining.

II. SAFETY

Undoubtedly, glycerol is one of the safest materials encountered in the food, cosmetic, and pharmaceutical industries. A comment in the *Journal of the American Medical Association* of May 20, 1933 is illustrative:

> If it is assumed that the average daily intake of fat in the present day-diet approximates 100 gm, this foodstuff will liberate about one tenth its weight, or 10 gm of glycerine (chemically designated glycerol) in the alimentary tract as a consequence of the lipolytic digestive changes. Glycerine also finds its way into the gastro-enteric canal from other sources, it is mixed with certain commercially processed foods, is present in some pharmaceutical preparations, and is sometimes fed as such, offering through its sweetness a substitute for sugar. It seems almost gratuitous to question the physiologic wholesomeness of a normal digestive product of one of the familiar nutrients, yet this seems to have been done from time to time.

The Glycerine Producers Association (New York, NY) in its *Glycerine, Preferred for Product Conditioning* brochure (about 1955) states relative to nontoxicity of glycerine

Both by experiment and experience, glycerine has been proven nontoxic as a medicinal or food ingredient for human or animal consumption. Glycerine in amounts as high as 30 percent of the total diet (of rats) was fed for many weeks with no reduction of growth rate. Dosages of 100 grams of glycerine per day have been fed to humans without gastrointestinal disturbances.

As in the case with many food ingredients, such as table salt, massive doses of highly concentrated glycerine are to be avoided, but these are well above limits normally employed. Intravenous and intraperitonal dosage beyond well-defined limits may have serious effects.

In applications on the surface of the skin, glycerine in most concentrations has been shown to have emollient properties, and to be less irritating than some other polyols. Preparations containing glycerine have been recommended for topical application on wounds, as well as on normal skin.

Many antiseptic, analgesic, vaginal, dermal, nasal and burn ointment and jellies are made in water soluble bases compounded with glycerine. Cough syrups and elixirs, both National Formulary preparations and proprietaries, containing substantial quantities of glycerine, have had generations of accepted use.

Glycerol has a generally recognized as safe (GRAS) status [8] as a miscellaneous or general-purpose food additive and is permitted in certain food packaging materials.

The half lethal dose (LD_{50}) for glycerol orally for mice is determined as 31.5 (7.86 g/kg) [9], with the lethal dosage as 25 g by mouth, 10 g subcutaneously, and 6 g intravenously, with 4 to 5 ml stated as "may be fatal to rats," and 8 ml/kg the lethal dose for rats [10]. A mutagenic evaluation of glycerine (Litton Bionetics, Inc. Kensington, MD for FDA 1975) stated "glycerine did not exhibit genetic activity in any of the in vitro microbial assays employed in this evaluation."

Oral LD_{50} levels have been determined in the mouse at 470 mg/L kg [9] and the guinea pig at 7750 mg/kg [11]. A number of other studies [12–14] have shown that large quantities of both synthetic and natural glycerol can be administered orally to experimental animals and man without the appearance of adverse effects. Intravenous administration of solutions containing 5% glycerol to animals and humans has been found to cause no toxic or otherwise undesirable effects [15].

M. E. Hanke in Miner and Dalton's *Glycerol* [16], adequately treated the physiological action of glycerol as was evident at that time. He discussed glycerol as a food, the harmful effects of unphysiological amounts of glycerol and other physiological effects of glycerol. He summarizes:

Glycerol is a common energy yielding food, essentially equivalent to glucose in nutrition and metabolism. No harmful effects have been noted following the ingestion of glycerol by dogs and rats in amounts of 35 to 50 per cent of the diet, or 10 to 15 g/kg body weight per day, and by man in daily quantities of 50 g by children or 150 g by adults. As the amount of the glycerol in the diet is increased, the efficiency of glycerol as a food is decreased, because of the excretion of an increasing proportion of the glycerol in the urine. In man, the threshold dose at which glycerol begins to appear in the urine is about 25 g.

The lethal dose of glycerol in mice and rats in g per kg body weight is 25 orally, 10 subcutaneously, and 6 intraperitoneally or intravenously. Death is due to failing circulation and respiration. The parenteral administration of glycerol is followed by fall in blood pressure, hemoglobinuria, alburminuria, anemia, and local inflammation. These effects are not observed when glycerol is taken by mouth.

In doses of 50 g or more in man glycerol is a diuretic, and stimulates uric acid excretion. Several clinicians have reported favorably on its use in the treatment of urinary calculi. Glycerol has a mild germicidal action and has been recommended for use in surgical dressings especially for topical application to suppurative wounds.

III. ENVIRONMENTAL ASPECTS

The complete biodegradability of glycerol in nature and its overall safety as a foodstuff testify to the innocuous nature of this material in the environment. There is another aspect to this situation that is well-appreciated among commercial producers of glycerol. Among industrial fatty acid producers, for example, where glycerol today is priced at about 80¢/lb, this amounts to roughly four times the value of one of the industry's principal products, stearic acid. Consequently, there is a very strong desire to waste (or to transfer to the environment) as little as technologically possible. This also infers that there is ample economic justification for the recovery of that little amount which is unintentionally lost.

The aquatic toxicity (TLm96) for glycerol is > 1000 mg/L [17], which is defined by NIOSH as an insignificant hazard.

G. N. McDermott (Procter & Gamble Co.), in *Bailey's Industrial Oil and Fat Products* [18] gives the BOD of lost glycerol from wastewaters in soapmaking during the two operations of glycerol concentration and glycerol distillation as 15 and 5 mass BOD units, respectively, per 1000 units of glycerol. This data was obtained from mass quantities of pollutant in wastewaters following removal of floatable oil by passing through gravity settling tanks. Further, the suspended solids in these wastewaters from these materials were 2 and 2 units, respectively, per 1000 units of glycerol. Based upon average costs (1982) of 8¢/lb for the removal of one pound of BOD and of 6¢/lb for the removal of one pound of suspended solids, the processing costs alone could be approximated as $0.00135/lb glycerol produced. An additional cost of about 15¢/lb per 1000 gallons of wastewater processed must be imposed upon this cost plus the cost of the water itself. The total cost for the loss of one pound of glycerol through wastewater can be approximated as 50.2¢/lb, assuming that glycerol is valued at 50¢/lb.

REFERENCES

1. Adkins, H. and Hartung, W. H., In Organic Syntheses Coll., Vol. I, 2nd Ed., A. H. Blatt, ed., John Wiley & Sons, New York, 1944, pp. 15–18.
2. Hurd, C. D. The Pyrolysis of Carbon Compounds, The Chemical Catalog Co., New York, 1929, p. 184.
3. Sabatier, P. and Gaudion, G., Compt. Rend., 166: 1033–1039, 1918.
4. U.S. Patent 1,916,743, 7/4/33, Schwenk, E., Gehrke, M. and Aichner F. (to Schering Kahlbaum A.–G).
5. Fry, H. S., and Schulze, E. L. JACS, 50: 1131–1138, 1928.
6. Sonntag, N. O. V. JAOCS, 59 (10): 795A–802A, 1982.
7. Christensen, H. E. (ed.), The Toxic Substances List, NIOSH, US Dept. of Health, Education & Welfare, Rockville, MD, 1973, p. 35.
8. CFR, Title 21, Section 182.1320.
9. Smyth, H. F., Seaton, J., and Fisher, L., J. Ind. Hyg. Toxicol., 23: 259, 1941.
10. Deichmann, W., Ind Med., Ind. Hyg. Sect., 1: 60, 1940; Latven, A.R., and Molitor, H. J. Pharm., 65:89, 1939; Schubel, K., Sitzungsher der physikalisch-medizerichen Sozietat zu Erlanger, 68: 263, 1936; Arch exp. Path. Pharmakol., 181: 132, 1936; Pfeifer, C., and Arnove, I. Proc. Soc. Expt'l. Biol Med., 37: 467, 1937.
11. Anderson, R. C., Harris, P. N., and Chen, K. K., J. Am. Pharm. Assoc. Sci. Ed., 39: 583, 1950.
12. Johnson, V., Carlson, A. J., and Johnson, A., Am. J. Physiol., 103: 517, 1933.
13. Hine, C. H., Anderson, H. H., Moon, H. D., Dunlap, M. K., and Morse, M. S., Arch Ind. Hyg. Occup. Med., 7: 282, 1953.
14. Deichmann, W., Ind. Med. Ind. Hyg. 9(4): 60, 1940.
15. Sloviter, H. A., J. Clin. Inv., 37: 619, 1958.
16. Hanke, M. E., In Glycerol (C. S. Miner and N. N. Dalton, eds.), Reinhold Publishing Corp., New York, 1953, pp. 402–422.
17. Hann, W., and Jensen, P. A., Water Quality Characteristics of Hazardous Materials, Texas A & M University, College Station, TX, 1974, p. 4.
18. McDermott, D. N., In Bailey's Industrial Oil and Fat Products, 4th ed., Vol. II (D. Swern, ed.), Wiley, New York, N.Y., 1982, pp. 527–586.

9

Economics

Norman O. V. Sonntag

Ovilla (Red Oak), Texas

I. INTRODUCTION

Quoting C. E. Gentry, Procter & Gamble Co. glycerol product manager in 1984 [1], "glycerol may be the most versatile chemical known to man. It is used to make glue to stick things together, and in dynamite to blow things apart. It is used in cough suppressants and suppositories. Glycerol is used in hair sprays and house paint. It is an ingredient in expensive liqueurs and cheap pet foods."

Gentry's statement just begins to illustrate the ultimate versatility of this unique and interesting material. In food applications, where its uses will undoubtedly increase over the next two decades, glycerol is used in products as commonplace as chewing gum as well as such sophisticated products as caviar. There are current applications for glycerol that can be expected to decrease substantially or even disappear entirely within the next two decades, such as its use in some alkyd resins, and possibly tobacco; and there are new applications for it that were not even conceived three years ago. Leffingwell and Lesser [2] tabulated the distinct and separate applications and listed 1583 uses.

But glycerol has other unusual features. Few chemicals, if any, can be manufactured from as great a variety of raw materials, both synthetic and natural. Synthetic glycerol is and has been produced from petrochemical feedstocks derived from propylene such as allyl chloride, epichlorhydrin, glycerol chlorohydrins, and acrolein (see Chap. 4). It may be produced from a wide variety of carbohydrate or carbohydrate-containing materials such as sucrose, molasses, glucose, amylose, starch, cellulose, wood chips, and straw. Its synthesis from sucrose or glucose provides an example in which glycerol has been manufactured from naturally occurring raw materials by two entirely different technologies, fermentation, and hydrogenolysis. Glycerol is also produced from a large variety of natural fats and oils of animal, marine, and vegetable origin. Potentially, it could even be produced from certain fat/wax natural raw materials, jojoba included, were this not so economically prohibitive. Glycerol is a byproduct from the manufacture of soap, fatty acids, methyl esters, and fatty alcohols, where it is usually more valuable on a ¢/lb basis than the major product from each operation. It can be manufactured from vegetable oil soapstocks derived from the chemical refining of soybean, cottonseed, corn, canola, peanut, sunflower, safflower, and other oils if the volume processed is sufficiently large. In fatty acid manufacture it has been, over the last half century, a significant contributory factor to keeping a number of small and medium-sized fatty acid producers from bankruptcy on those occasions where depressed prices for stearic acid have prevailed over extended periods of time.

In this chapter we attempt a brief treatment of several of the economic factors of glycerol marketing and manufacture. One manifestation of this is the sporodic fluctuation in glycerol prices on a local and international scale (Table 9.1). Such factors as the scarcity of glycerol during and immediately after times of war, price increases during periods of raw material unavailability, higher costs for manufacturing by unit processing that become progressively obsolete, and the relative costs of separative processing from competitive technologies (as in the separation of water from glycerol in fractional distillation as a function of column design, etc.) can be easily pinpointed. Other factors are not so easily characterized. The effect of a natural disaster such as a typhoon in the Philippines upon the local and international pricing for glycerol derived from coconut oil is much more difficult to explain than is the partial diversion of epichlorhydrin, a petrochemical glycerol raw material, to epoxy resin production rather than to glycerol manufacture. If, in fact, all of the pertinent factors were thoroughly understood and appreciated the prediction of glycerol price trends would be a simple matter. Such is never the case. In fact, glycerol economic prediction is still among the most difficult among all of the world's chemicals. One can well appreciate that there is much more to the problem than the realization that the "raw material costs are too high" or "glycerol is too expensive produced by this process."

II. ECONOMICS OF GLYCEROL PRODUCTION FROM FATTY ACID MANUFACTURE

In the Colgate-Emery fat-splitting process [3] glycerol is a byproduct from the manufacture of fatty acids like those from beef tallow, cottonseed, palm, palm

Table 9.1 Quoted U.S. List Prices for Various Glycerines from 1920 to 1980
(Prices are those published last week of June for each year)

Year	¢/lb drums or bulk, del'd	Crude Saponif. 88%	Soap lye 88%	Dyna-mite, drums, cl	High-grav. drums, cl	refined BP, 98% drums, cl	USP 95% drums cl	Yellow dist'd drum, cl	Synthetic
1920	20.50								
1921	15.25—bulk; 17.50—low vol.								
1922	14.25								
1923	14—dynamite; 16.25—c.p.; 15.50—bulk								
1924	16.50								
1925	16.50–16.75								
1926	19–19.50 -c.p. 29–32								
1927	25–25.50								
1928	15–15.50—drums; 16.50–17—cans								
1929	14–14.50								
1930	13.50–14								
1931	11.50–12								
1932	10.75–11								
1933	10.25–10.50								
1934	13.50								
1935	14								
1936	16.50–17.50								
1937	23.50–24								
1938	16.25–16.75								
1939	14.50–15								

Table 9.1 (*Continued*)

Year	¢/lb drums or bulk, del'd	Crude Saponif. 88%	Soap lye 88%	Dyna-mite, drums, cl	High-grav. drums, cl	refined BP, 98% drums, cl	USP 95% drums cl	Yellow dist'd drum, cl	Synthetic
1940	14.50–15								
1941	16.50–17								
1942	18.75; 18 tanks								
1943	18.75								
1944	18.75								
1945	12.13—crude saponif; 17.25—c.p. 18.25—USP								
1946									
1947	—								
1948		28–30	27–31	39.5–40.25	39.25–40–50	39.75 41.13	39.50–41.25 39.25–40 tks	39.83–40.13 39.25–40 tks	
1949		16.75–18.75	15–18.25 tks	24 24.5 lcl	24.25	25.50 26 lcl 25 tks del	24.75 25.25 lcl 24.25 tks del	23.63	
1950		16.75–18.75	15.25–16.50	24	24.25	25.50 26 lcl del	24.34 25.25 lcl del 24.25 tks del	24.3 24.2 tks 23.6 tks del	
1951		44–46	40–42	54 54.5 lcl del	54.25 54.75 lcl	55.5 56 lcl del	54.75 55.25 lcl del	54.5 54.2 lcl	

Year				53.75 tks		54.25 tks	53.2 tks	
1952	24	22	33.75 / 34.50 / lcl / 33.25 tks	34 / 34.50 / lcl / 33.50 tks	35.63 / 36.13 / lcl / 35.13 tks	34.50 / 35 lcl / 34 tks	33.88 / 34.88 lcl / 33.13 tks	
1953	33.50	30–50–31	43.75 / 44.25 / lcl / 43.25 tks	44 / 44.50 / lcl / 43.25 tks	45.88 / 46.348 / lcl / 45.38 tks	44.50 / 45 lcl / 44 tks	43.88 / 44.38 / lcl / 43.13 tks	
1954	23	21	30.50 / 31 lcl / 30 tks	30.50 / 31 lcl / 30 tks	29.50 / 30 lcl / 29 tks	30 / 30.50 / lcl / 30 tks	30.50 / 31 lcl / 30 tks	30.50 / 31 lcl / del / 30 tks
1955	23– 23.50	21	30.50 / 31 lcl / 30 tks	30.50 / 31 lcl / 30 tks	30.50 / 31 lcl / 30 tks	29.50 / 30 lcl / 29 tks	30 / 30.50 / lcl	30.50 / 31 lcl / 30 tks
1956	18.50	16.50	31.25 / 31.75 / lcl / 29.75 tks	31.25 / 31.75 / lcl / 29.75 tks	31.50 / 32 lcl / 30 tks	30.50 / 31 lcl / 29 tks	30.50 / 31 lcl	31.50 / 32 lcl / 30 tks
1957	17.60– 18	15.50– 16.25	29.25 / 29.75 / lcl / 27.75 tks	29.25 / 29.75 / lcl / 27.75 tks	29.38 / 29.88 / lcl / 27.88 tks	(96%) 28.50 / (96%) 29 lcl / (96%) 27 tks	28.50 / 29 lcl	29.50 / 30 lcl / 28 tks

Table 9.1 (*Continued*)

Year ¢/lb drums or bulk, del'd	Crude Saponif. 88%	Soap lye 88%	Dyna-mite, drums, cl	High-grav. drums, cl	refined BP, 98% drums, cl	USP 95% drums cl	Yellow dist'd drum, cl	Synthetic
1958				29.25 29.75 lcl 29.64 tks		29.36 (CP 99%) 27.88 27.63 (98%) 28.50 (98%) 29 lcl (98%J) 26.75 tks		29.50 30 lcl 27.75 tks
1959	20.50	19.50 (80%) 17.00		29.25 29.75 lcl 27.50 tks		(96%) 28.50 (96%) 29 lcl (96%) 28.75 (96%) tks (99%) 29.36 (99%) 29.88 lcl (99%) 27.63 tks		29.50 30 lcl 27.75 tks
1960	21.50– 21.75	(80%)19– 19.75		30.75 31.50 lcl 29 tks		(99%) 30.88 (99%) 31.63 lcl (96%) 29.13 (96%) 30 30.75 lcl 29.25 tks		31 31.75 lcl 28.25 tks
1961	13.50– 14	12.50– 12.75		(nat) 26.50 27.75 lcl 24.75 tks	25.75 26.50 lcl 24 tks	26.38 27.38 lcl 24.88 tks (98%) 25.75		(99.5%) 26.75 27.50 lcl 25 tks

Year				26.50 lcl 24 tks	27.50 lcl 25 tks
1962	13	11–12	21.50	(99%) 21.63 (99%) 29.75 lcl (99%) 21.50 tks	21.75
1963	13.25–13.50	(80%) 11–12	18.25	(99%) 18.63 (96%)17.75	18.50
1964	15.75–16	14.25–14.50	22.25	(99%) 22.63 (96%) 21.75	22.50
1965			22.25	(CP) 22.38 (96%) 21.75	22.50
1966			23.50	(99%) 23.63 (96%) 23	23.75
1967			24.75	(99%) 24.88 (96%) 24.50	25
1968			25.50	(99%) 25.63 (96%) 25	(99.5%) 25.75
1969			24	(99%) 24.13 (96%) 23.50	(99.5%) 24.25
1970			29.50	(99%) 20.63 (96%) 20	(99.5%) 20.25–22
1971			21.50	(99%) 21.63 (96%) 21	(99.5%) 21.75–22

Table 9.1 (*Continued*)

Year	¢/lb drums or bulk, del'd	Crude Saponif. 88%	Soap lye 88%	Dyna-mite, drums, cl	High-grav. drums, cl	refined BP, 98% drums, cl	USP 95+% drums cl	Yellow dist'd drum, cl	Synthetic
1972					22.50		(99%) 22.63 (96%) 22–22.25		(99.5%) 22.75–23
1973					22.50		(99%) 22.63 (96%) 22–23.25		(99.5%) 22.75–24
1974							(99.5%) 49–51 (96%) 48–49.50		(96%) 42.50–49.50tks (99.5%) 44–50
1975							(cp 99.5%) 45.75–49.75 (96%) 44–48		(96%) 48.25 (99.5%) 50
1976							(99.5%) 45.75–49.75 (96%) 44–49		(96%) 48.25 (99.5%) 50
1977							(99.5%) 49–52.25		(96%) 48.25–50.75

1978	(99.5%) 50–52.50	(96%) 47–50.25
	(96%) 48.25–50.75	(99.5%) 49–52.25
	(99.5%) 52.50	(96%) 48–50.25
1979	(96%) 53.25	(99.5%) 49–51.50
	(99.5%) 55	(96%) 47.25–52.38
1980	(96%) 62.25	(99.5%) 59.50–65.50
	(99.5%) 68	(96%) 57.75–63.38
1981	(96%) 74.25	(99.5%) 76
	(99.5%) 76	(96%) 74.50
1982	(96%) 78.75–79.25	(99.5%) 80.50–81
	(99.5%) 80.50–81	(96%) 78.75–79.25
1983	(96%) 71.25	(99.5%) 66.50–72.50
	(99.5%) 73	(96%) 66.75–70.75

Table 9.1 (*Continued*)

Year	¢/lb drums or bulk, del'd	Crude Saponif. 88%	Soap lye 88%	Dyna-mite, drums, cl	High-grav. drums, cl	refined BP, 98% drums, cl	USP 95+% drums cl	Yellow dist'd drum, cl	Synthetic
1984							(99.5%) 78.50–79.50 (96%) 76.75–77.75		(96%) 79.25 (99.5%) 81
1985							(99.5%) 89.50 (96%) 87.75		(96%) 89.25 (99.5%) 91
1986							(99.5%) 89.50 (96%) 87.75		(96%) 89.25 (99.5%) 91
1987							(99.5%) 89.50 (96%) 87.75		(96%) 89.25 (99.5%) 91
1988							(99.5%) 66 (K) 66.50 (96%) 64.25–76.25		(96%) 66.75 (99.5%) 68.50
1989							(99.7%) 82—tks (K) 82.50 (96%) 80.25		(96%) 82.75 (99.7%) 84.50

Source: *Oil, Paint & Drug Reporter*, Schnell Publ. Co., 100 William St. and later 80 Broad Street, New York, NY; 1920–1971—*Chemical Marketing Reporter*, Schnell Publ. Co., 80 Broad Street, New York, NY 10004; 1972–1989.
Abbreviations: tks, tanks; lcl del, less than carload delivered; del, delivered; cl, carload.

kernel, coconut, fish, rapeseed, castor oil, and other fats and oils. The fatty acids are processed by separative and other upgrading technology to the various grades of stearic, palmitic, oleic, lauric, myristic, erucic, behenic, arachidic, hydroxystearic, ricinoleic and other fatty acids.

The heat that is recovered from the hydrolysis of fats and oils, generally carried out at 260°C and 750 psig, is more than sufficient to evaporate the sweetwater obtained to from 83% to 88% glycerol concentration. Usually, but not always, existing processing requires that a separate still be employed for distillation to the USP grade (95% min), dynamite-grade (99%), or the CP grade (96, 99, or 99.7%). Unless the fatty acid producer is in the refined glycerol business and carries out the final operation at his own plant, the evaporated glycerols are sold, preferably to a nearby glycerol refiner. In Table 9.2 we tabulate the glycerol byproduct credit and the data that determine it for eight representative fats and oils in the production of 83% glycerol, ultimately used for distillation to the various grades greater than 95%. Except perhaps in the case of fish oil byproduct glycerol, no isolation of glycerol by source is practiced and it is customary to combine separate glycerols from any fat and oils source at the 83–88% range of concentration prior to final distillation, if they have not, in fact, already been combined earlier in the processing sequence.

Traditionally, the fatty acid manufacturer has usually considered his glycerol production as a means to lower, through byproduct credit, the cost of manufacture of his fatty acid line. In the past two decades, however, with the inflationary spiral of glycerol prices, especially during the 1980s, this thinking has changed markedly. As total glycerol production increased from fatty acid manufacture, most large or medium-sized producers entered into USP glycerol production themselves, as soon as it became economically evident that this was relatively more profitable than was the manufacture of the basic fatty acid line. A few producers of fatty acids find their own production sufficient to sustain a USP or CP glycerol program. Others augmented the available crude glycerol stocks through use of captive ester glycerol derived from concurrent methyl ester production, and others compete with existing glycerol refiners for crude soap glycerol, and ester crude glycerol supplies on the open market. Suffice to say, the overall natural glycerol economic situation is somewhat complex. It occasionally has resulted in the bizarre situation of crude glycerol selling on the open market for a higher price than refined glycerol.

One of the largest producers of natural glycerol in the United States is in the advantageous position of having captive crude glycerol available from soap manufacture, from fatty acid hydrolysis, from methyl ester as ester glycerol as intermediate in a large fatty alcohol program, while at the same time offering to purchase crude glycerol on the market if quantities are at least 500,000 lbs/year 100% glycerol basis. This is necessary to the operation of a multirefinery operation located strategically in the major U.S. markets.

Table 9.2 The Economic Importance of By-Product "Evaporated" 83% Glycerol From Colgate-Emery Fat Splitting

Raw material fat or oil	Fat or Oil					% Prodn yield 83% glycerol[b]	%Theo. Yield crude fatty acid[c]	Glycerol credit[d]	
	Mol. wt. fat or oil[a]	S V	A V	Unsaponifi- ables %	Mol. wt. fatty acid[a]			¢/lb Fat or oil used	¢/lb Crude fatty acid produced
Tallow, beef BFT grade IV = 40.5	850.0	198.0	3.6	0.2	270.7	12.13	97.8	5.89	6.30
Soybean oil, selectively hydrogenated, IV = 69	873.2	192.7	0.4	0.5	278.4	11.96	97.5	5.81	6.22
Cottonseed oil IV = 105.2	860.5	195.0	0.2	0.52	274.2	12.15	97.2	5.90	6.34
Coconut oil IV = 8.5	684.5	245.8	0.3	0.1	215.5	15.34	97.8	7.45	8.05
Palm oil, IV = 53.5	846.8	195.0	2.5	0.6	269.6	12.19	97.4	5.92	6.36
Palm kernel oil, IV = 17.8	702.5	250.0	0.4	0.4	221.5	14.89	97.6	7.23	7.83
Rapeseed oil, high erucic type IV = 100.6	984.2	171.0	0.2	0.4	315.4	10.63	97.6	5.16	5.50
Japanese sardine oil, hydrogenated IV = 2.5	907.9	194.3	2.0	0.8	290.0	11.38	97.0	5.77	5.93

[a]Calculated from fatty acid composition.

[b]Practical production yield as 83% "evaporated" glycerol corrected for A. V. and non-saponifiable content of fat or oil. Yield corresponds to about 5% loss of total available glycerol from respective fat or oil, which is near optimum by modern separative technology.

[c]Corresponds to a fatty acid yield loss of about 2%.

[d]Based on assumption that 83% glycerol credit price is 48.56¢/lb delivered to nearest glycerol refinery USP—Glycerol price Sept. 1988 at 84¢/lb; credit payment for 100% glycerol is 69.5% (at 82.5¢/lb) to 70.5% (at 92.5¢/lb) of the prevailing USP price.

III. ECONOMICS OF GLYCEROL PRODUCTION FROM METHYL ESTER MANUFACTURE

In batch fat methanolysis processing, after 1 hour of reaction, the lower glycerol layer generally contains not less than 90%, usually 92% of the theoretically available glycerol. The upper layer contains another 2% as free glycerol, with the balance, based upon an overall 98% degree of completion, as mono- and diglycerides. Centrifugal layer separation after a 3-h reaction increases the glycerol yield from the lower layer to nearly 94%; with the inclusion of free glycerol from the upper layer, this can bring the overall yield of nearly anhydrous glycerol to 95.5% of the theoretical. Incremental addition of methanol and other process innovations in the methanolysis result in an optimum glycerol recovery of 96% of theoretical, but very few existing plants can operate at this efficient recovery level. The methyl ester producer has some limitations in refining the crude glycerol obtained from his lower layer. Methanol can be recovered for reuse from the crude glycerol by careful evaporation, but soap, excess alkali catalyst, and color contaminants pose some separative problems, especially when it is realized that treatments with aqueous solutions or with water dilutes the "anhydrous" glycerol crude and this increases evaporation and distillation costs in refining and upgrading.

Table 9.3 details the glycerol byproduct credit and the data that determine it for eight representative fats and oils in the production of "near-anhydrous" 94% glycerol, ultimately used for distillation and refining to the 95+% grades.

IV. ECONOMICS OF GLYCEROL PRODUCTION FROM SOAP MANUFACTURE

Godfrey, in Miner and Dalton's *Glycerol* [4], cites 25 points of loss of glycerol during the course of manufacture of soap:

1. Loss by spills and leaks in unloading fats and failure to clean containers
2. Loss by washing away fat and glycerol in water from steaming out
3. Loss of fat by leakage in storage and handling
4. Loss resulting from the hydrolysis of fats during storage
5. Loss of fat during acid washing of low-grade stock
6. Loss of fat in earth or chemical bleaching
7. Losses in the foots from refining fats
8. Reduction of yield as a result of glycerol left in the soap
9. Loss of glycerol left in unsplit fat from hydrolyzers
10. Loss of glycerol in lyes run to sewer
11. Losses in separating sweetwaters from fatty acid stocks
12. Loss of glycerol by fermentation in lyes

Table 9.3 The Economic Importance of By-Product Glycerol From Methyl Ester Production

Raw material fat or oil	Fat or oil				Mol. wt. methyl ester[b]	% Prodn yield 94% glycerol[b]	% Prodn yield crude methyl esters[c]	Glycerol credit[d]	
	Mol. wt. fat or oil[a]	S V	A V	Unsaponifiables %				¢/lb Fat or oil used	¢/lb Crude methyl ester produced
Tallow, beef BFT-grade IV = 52	850.0	198.0	0.07	0.2	284.7	11.02	98.0	6.06	6.16
Soybean oil, selectively hydrogenated, IV = 69	873.2	192.7	0.05	0.5	292.4	10.69	97.7	5.87	6.00
Cottonseed oil, IV = 105.2	860.5	195.0	0.05	0.52	288.2	10.70	97.6	5.88	5.99
Coconut oil, IV = 8.5	684.5	245.8	0.05	0.1	229.5	13.58	98.1	7.46	7.57
Palm oil, IV = 53.3	846.8	195.0	0.06	0.6	283.6	11.02	97.6	6.06	6.18
Palm kernel oil, IV = 17.8	702.5	250.0	0.05	0.4	235.5	13.30	97.8	7.30	7.44
Rapeseed oil, high erucic type IV = 111	984.2	171.0	0.07	0.4	329.4	9.50	97.8	5.22	5.32
Japanese sardine oil, hydrogenated IV = 2.5	907.9	194.3	0.10	0.8	304.0	10.25	97.4	5.63	5.76

[a]Calculated from fatty acid composition.
[b]Practical production yield as 94% glycerol corrected for A.V. and non-saponifiable content of fat or oil. Yield loss of glycerol corresponds to about 4% loss of total available glycerol from respective fat or oil.
[c]Corresponds to a methyl ester yield loss of 1.8%.
[d]Based on assumption that 94% glycerol credit price is 55.00¢/lb, delivered to nearest glycerol refinery—USP Glycerol price Sept. 1988 at 84¢/lb; credit payment for 100% glycerol is 69.5% (at 82.5¢/lb) to 70.5% (at 92.5¢/lb) of the prevailing USP price.

13. Loss in salt discarded from system
14. Loss by lye treatment through leakage, spills, careless handling of skimmings, etc.
15. Loss in filter press cake from lye treatment
16. Loss by entrainment, foaming, etc., in the evaporator
17. Loss through inadequate removal of glycerol from salt
18. Loss through washings from drums, tanks, floors, etc.
19. Loss through entrainment during distillation
20. Loss by incomplete condensation during distillation
21. Loss by decomposition during distillation
22. Loss occasioned by handling sweetwaters
23. Loss in foots from still
24. Loss in washing still or other equipment
25. Loss in filter press cake from bleaching

Note, that items 1–6 are applicable to glycerol production from fat hydrolysis, and 1–7 apply to glycerol production from fat methanolysis, whereas items 18–25 apply to any glycerol production in which dilute glycerols or sweetwaters are evaporated and distilled for upgrading.

While the purification of glycerol liquors resulting from saponification of fats can be accomplished practically with losses of glycerol from lye to crude glycerol in the 1–3% range, it should be emphasized that some glycerol is purposely left in the soap, and this cannot be considered as a true loss. Usually about 2% yield loss is experienced in the evaporation step to 83–90% glycerol, and about 1% in the glycerol distillation. Generally, the evaporation of high salt-content liquors from soapmaking yield lower glycerol recoveries than does evaporation of liquors from the hydrolysis or methanolysis routes of production. Overall, yields of 90–95% of theoretical glycerol from soapmaking result from the relatively large number of processes available.

Table 9.4 gives some data for the production of a "high-gravity" glycerol (99.5%) in the manufacture of a toilet soap, based on an anhydrous organic soap component, assuming the soap manufacturer has the equipment for 95+% glycerol production at the soap plant site. The increase in credit of 7.46 ¢/lb fat/oil blend over that from a tallow/coconut oil blend calculated from Table 9.2 at 6.28 ¢/lb, reflects the on-site production of 99.5% glycerol by the soap manufacturer.

Table 9.5 gives data for the production of a "high-gravity" glycerol (99.5%) in the manufacture of a toilet soap, based on an anhydrous organic soap component, assuming the soap manufacturer has the equipment at the soap plant site, and manufactures soap by the countercurrent washing system in order to reduce to a minimum the amount of water to be evaporated in the recovery of the glycerol.

Table 9.4   Economics of 99.5% Glycerol Production from Toilet Soap Manufacture

			Glycerol credit	
Fat/oil blend[a]	Prodn yield, % practical 99.5% glycerol[b]	Prodn yield, % practical dry, organic soap[c]	c/lb of Fat/ oil blend used[d]	c/lb of dry organic soap produced[d]
75% Soap-grade tallow				
	10.45	102.4	7.46	7.29
25% Soap-grade coconut oil				

[a]Generally, by blending soaps from separate tallow and coconut oil saponifications.
[b]Glycerol overall loss assumed at 8% of theoretical available.
[c]Anhydrous soap yield is 99.2% of theoretical from fat/oil blend.
[d]Based upon assumption that 99.5% glycerol has a value of 85% of the price of USP glycerol, which in September 1988 was 84¢/lb.

Table 9.5   Economics of 99.5% Glycerol Production from Toilet Soap Manufacture with Concurrent Washing System

			Glycerol credit	
Fat/oil blend[a]	Prodn yield, % practical 99.5% glycerol[b]	Prodn yield, % practical dry, dry, organic soap[c]	c/lb of Fat/ oil blend used[d]	c/lb of Dry organic soap produced[d]
75% Soap-grade tallow				
	10.99	102.4	7.84	7.66
25% Soap-grade coconut oil				

[a]Generally, by blending soaps from separate tallow and coconut oil saponifications.
[b]Glycerol overall loss claimed to be 5% of theoretical available.
[c]Anhydrous soap yield is 99.2% of theoretical from fat/oil blend.
[d]Based upon assumption that 99.5% glycerol has a value of 85% of the price of USP glycerol, which in September 1988 was 84¢/lb.

V.   ECONOMICS OF MISCELLANEOUS FAT AND OIL PROCESSING FOR GLYCEROL PRODUCTION

Soap, methyl ester, and fatty acid production by the Colgate-Emery fat-splitting process account for well over 95% of all the natural fat and oil-derived glycerol produced in the United States. Older technology like Twitchell fat splitting is practiced in other parts of the world. European producers still employ this method to a greater extent than does the United States. Occasionally, fats and oils, especially less common or specialty oils, are split by various modifications of the batch autoclave method. Typical oils treated this way are dehydrated castor oil and babassu.

In Twitchell fat splitting successive water batches are boiled with the fat stock, allowed to settle, and withdrawn for glycerol recovery until the glycerol concentration is too low to justify the cost of evaporation and distillation, or, until the hydrolysis is sufficiently complete. Operation is usually a compromise. The first boil is carried out by adding sufficient water and sulfuric acid so that after about 9–10 hours of boiling, the glycerol concentration from typical fats and oils is around 8–12%, depending upon the fat or oil. If the glycerol content is lower than this, insufficient degree of completion of hydrolysis will accrue during the first boil. If lower-grade stock is split, such as greases, soapstock or other materials, the glycerol concentration can be much lower.

Twitchell fat splitting suffers from several disadvantages compared with Colgate-Emery or autoclave methods: acidity of stock is higher and neutralization results in more soluble salts, color of glycerol and fatty acids are usually darker as a consequence of traces of sulfonated products formed during the hydrolysis, and glycerol and mono- and diglyceride residues in the fatty acids can cause reesterification during fatty acid distillation, and, thus, an even higher loss of residue. Like washing spent lye cakes from soap manufacture, washing filter cakes from glycerol refining can recover up to 3% of the available glycerol. The washings will not contain enough glycerol to be evaporated directly, but glycerol can be conserved by using the washes as water charges in subsequent Twitchell splits. For these reasons, although Colgate-Emery fat-splitting technology may afford a glycerol recovery of 96%, Twitchell process recovery is limited to about 94% of theoretical (presumably, some mono- and diglycerides are lost to the fatty acids).

Recovery of glycerol from autoclave processing sweetwaters poses no special problems, and, when zinc oxide or other catalysts are employed there are no yield losses because the fatty soaps are reconverted to fatty acids. There is no apparent effect upon glycerol yields.

In several large installations, continuous autoclaves are sometimes used, and the sweetwater is discharged from the autoclaves directly to the evaporators without previous treatment, thus saving pre-evaporator treatment as well as predistillation treatments.

The sodium reduction of fatty triglycerides, which yields glycerol as a byproduct, was developed at Dupont de Nemours & Co. [5] during the period 1933–1947. Its obvious advantage over catalytic high-pressure hydrogenolysis technology was that unsaturated fatty alcohols, like oleoyl alcohol, could be prepared without saturation of double bonds. Another advantage was that the byproduct glycerol was not contaminated with propylene glycol. For saturated fatty alcohol production, sodium reduction was subject to several process objections: sodium handling was hazardous, air or water had to be scrupulously eliminated from the zones of reaction, solvent and solvent recovery were involved, and overall yields were no greater than 90%. It quickly became apparent that the use of sodium metal for the replacement of hydrogen at high pressures was entirely too expensive. The other advantage, namely that less sodium was consumed for longer chain fatty alcohols with C-16 and C-18 carbon atoms than for shorter chains, was dissipated when it became obvious that the C-12 and C-14 fatty alcohols were better detergent materials than their longer-chain homologs, and were in far greater market demand. Finally, when technology was developed for unsaturated fatty alcohol manufacture employing copper chromite-catalyzed hydrogenolysis, sodium reduction was discarded. Archer-Daniels-Midland, Chemical Products Division closed down its sodium reduction plant at Ashtabula, OH in the mid-1950s, and Procter & Gamble Co., which had started its first sodium reduction plant at Ivorydale, OH, in 1942, and its second one there in 1947, shifted to catalytic hydrogenation at three other plants. Today, there is virtually no glycerol produced in the United States by sodium reduction technology.

One essentially untapped source of natural glycerol is soapstock derived from the chemical refining of edible vegetable oils like soybean, cottonseed, corn, canola, sunflower, safflower, peanut, and olive oils. In the United States it is estimated that about 300 million pounds per year of acidulated soapstocks result from soybean oil processing alone. These are already used as a source of inexpensive, low-grade fatty acids by the U.S. oleochemical industry. This developing use is slowly replacing the present use of vegetable oil soapstocks in animal feed, which during the last two years has shown a trend of consistently depressed prices.

Depending upon the refining process, acidulated soapstocks contain from 15 to 25% of free triglyceride oils, and, thus, are a potential source of glycerol. Table 9.6 affords economic data on the recovery of glycerol from a typical acidulated soybean soapstock obtained from centrifugal caustic refining of soybean oil and acidulation with sulfuric acid.

Small volume glycerol recovery from vegetable oil soapstock appears to be an economically marginal technique. Unless the vegetable oil refiner or the oleochemical manufacturer conserves optimum heat from a series of batch autoclaves or can assign a Colgate-Emery splitter to the operation, the glycerol credit hardly pays for the cost of evaporation to the 88% concentration. Further, splitting operation is inefficient with 82.5% of the charge to the splitter already as the fatty acid. Small-quantity operation is limited because the glycerol refinery usually will only purchase 88% crude glycerol if the minimum quantity is 500,000 pounds, 100% glycerol basis. Consequently, it is likely that a vegetable soapstock program could only be sustained if multiple acquisitions of vegetable oil acidulated soapstocks from several refineries were col-

Table 9.6 Economics of Glycerol Production from Soybean Soapstock-Recovered Triglyceride Oil

Raw material dry acidulated soybean soapstocks[a]	Mol. wt.[b]	% Prodn yield practical 88% glycerol[c]	Glycerol credits[d]
82.5% Soybean fatty acids	275.5	1.92% From acidulated soapstocks	0.99¢/lb Acidulated soapstock
16.9% Soybean oil	870.4		
0.6% inerts, color bodies, nonsaponifiables	—	11.41% From soybean oil in soapstocks	5.87 ¢/lb Soybean oil in soapstocks

[a]Price, November 1989, fob Chicago, IL 10.5 ¢/lb.
[b]Calculated from fatty acid distribution in soybean oil.
[c]Generally, soybean soapstocks are split by the batch autoclave method catalyzed by zinc oxide using steam below 500 PSIG. This unit operation costs between 1 and 3¢/lb, largely dependent on the cost of steam generation at the plant and kind of fuel used in the steam plant (gas, liquid fuel, coal). The alternate method for splitting still residue after distillation of acidulated soapstocks yields a lower grade glycerol frequently difficult to bleach, and restricted to dynamite grade glycerol manufacture.
[d]Based upon assumption that 88% glycerol credit price is 51.49¢/lb delivered to nearest glycerol refinery—USP Glycerol price Sept. 1988 at 84¢/lb; credit payment for 100% glycerol is 69.5% (at 82.5¢/lb) to 70.5% (at 92.5¢/lb) of the prevailing USP price.

lectively processed at a common site, preferably at an existing glycerol refinery, in which the soapstock volume amounted to 50 million pounds per year acidulated soapstock basis, equivalent to about 1 million pounds per year 100% glycerol.

Although sodium reduction has failed economically to provide byproduct glycerol from a primary product fatty alcohol program by triglyceride hydrogeneolysis, research and development has continued on this approach because it represents potentially a one-step synthesis. In 1968 Hydrocarbon Research [6] developed one-step hydrogenolysis processing for tallow with the use of alumina or bauxite-type catalysts promoted with Group VII and VIII metals to afford a higher yield of desired lower chain fatty alcohols. Despite the elimination of propylene glycol formation, 12–24% hydrocarbons in the reaction products proved undesirable.

Henkel K.-G.a.A. patented triglyceride hydrogenolysis technology in 1988 [7, 8] in which propylene glycol is the only byproduct. In Table 9.7 we contrast the economics for coconut oil catalytic hydrogenolysis in which glycerol (84¢/lb) propylene glycol (69¢/lb) or mixtures of the two are generated as byproducts.

Table 9.7 Economics of Glycerol/Propylene Glycol Production by Coconut Oil Hydrogenolysis

Raw material	Relative % of polyols formed[a]	% Prodn yield 96% glycerol[b]	% Prodn yield 95% propylene glycol[b]	Glycerol and propylene glycol credits			
				¢/lb of coconut oil used[c]		¢/lb of coconut alcohol produced[c]	
				(I)	(II)	(I)	(II)
Coconut oil	100% Glycerol	13.26	0	7.29		8.26	
Mol. wt. = 684.5	75% Glycerol	9.47	3.14	6.46	6.76	7.33	7.67
SV = 245.8	25% Propylene Glycol						
AV = 0.3	50% Glycerol	6.03	6.03	5.45	6.09	6.17	6.81
Unnsap. = 0.1%	50% Propylene Glycol						
	25% Glycerol	2.89	8.65	4.70	5.49	5.32	6.23
	75% Propylene Glycol						
	100% Propylene Glycol	0	11.06	4.34	5.52	4.93	6.16

[a]Refer to European Patent 254,189; 1/27/88 and German Patent 3,624,812; 1/28/88, both assigned to Henkel K.-G.a.A.
[b]The all-glycerol-producing reaction tends to give 96% glycerol product; the all-propylene glycol reaction tends to give 95% propylene glycol (presumably, water is produced during propylene glycol formation).
[c]96% Glycerol is credited at 55.58¢/lb, delivered to the nearest glycerol refinery—USP Glycerol price, Sept. 1988 at 84¢/lb; credit payment for 100% glycerol is 69.5% (at 82.5¢/lb) to 70.5% (at 92.5¢/lb) of the prevailing USP price. In Case (I) 95% propylene glycol is credited at 39.23¢/lb, or 70% of the present (Nov. 1989) USP price; Arco Chemical Co., 11/6/89, 59¢/lb, fob Bayport, TX, tankcars. Case (II) is for distilled propylene glycol credited at 49.9¢/lb, or 85% of the prevailing USP price, with a 1% loss in fractional distillation separation of propylene glycol and glycerol and a 1.5% loss during propylene glycol distillation for Case (II) only.

VI. ECONOMICS FOR GLYCEROL PRODUCTION FROM CARBOHYDRATES BY HYDROGENOLYSIS AND FERMENTATION

In 1947, common sugar sold for about 7.5¢/lb. For a short time during 1974, the price had soared to 56.5¢/lb. However, with the advent of the International Sugar Agreement, negotiated in 1977 the world price of sugar has been kept at $0.286–0.505/kg [12.9–22.9¢/lb] through export quotas, buffer stocks, and various other curious mechanisms. In the United States, when sugar prices are low, import duties, fees, and quotas are imposed to keep the raw sugar price above $0.37/kg [16.8¢/lb], thus helping domestic cane and beet growers remain in business. The 1947–1974 inflationary price spiral for sugar not only defeated the intended use for sugar esters as biodegradable detergent basestocks, but it made sugar hydrogenolysis uneconomical for glycerol production.

Table 9.8 outlines the effect of raw material price increase for sugar on three typical sugar hydrogenolysis processes developed during the period 1945–1974. The intended processes exhibited characteristic features of conventional carbohydrate to glycerol technology: too high a proportion of the product was unaccounted for, was of dubious value, or was detrimental as a contaminant in the separated glycerol, propylene glycol, or in both, separative techniques were difficult and expensive, lower-quality and cheaper raw materials inevitably afforded poor quality and additional separation or purification problems for the major products; and catalysts were relatively expensive and reuse performance was economically prohibitive.

The processes in Table 9.8 had high unit processing costs that were prohibitive. The necessity to separate and upgrade the various products formed could not be accomplished in an economically feasible way. Atlas Chemical Industries Inc. found it desirable in 1969 to close down its glycerol operation in order to optimize production of more lucrative sorbitol at the Atlas Point plant (New Castle, DE). It was a relatively easy move, requiring only a small change in reaction conditions and a shift to a less severe splitting catalyst.

VII. SOME ECONOMIC FACTORS INVOLVED IN PETROCHEMICAL SYNTHESIS OF GLYCEROL

Table 9.9 shows the raw-material costs for synthetic glycerol produced by the so-called "epichlorohydrin" chlorination route calculated on prices for 1948, 1975 and estimated produced in 1990, with all figures corrected for credits due to the coproduction of 32% muriatic acid, 1, 3-dichloropropene and 1,2-dichloropropane byproducts. The 1975 raw material costs amount to 45% of the average list price that year for glycerol. When unit processing costs and other fixed costs were added to the raw material costs the total manufacturing cost for synthetic glycerol was about 38¢/lb, corresponding to a return on investment

Table 9.8 Economics for Sugar Hydrogenolysis Processes

Year	Price sugar ¢/lb[a]	Process[b]	Yields, %, production				Raw material Cost ¢/lb[c]	Adjusted Raw material Cost ¢/lb[d]	Price, USP glycerol ¢/lb
			Glycerol	Propylene glycol	Ethylene glycol	Others			
1945–6	2.5	I	28.8	43.8	—	8.8–"glycerol-like"	8.56	−9.36	19.9
1967–8	glucose, 5.0	II	39.0	PG + EG = 38.0		Sorbitol–22.1	12.6	0.37	25
1970–1	3.4	III	31.3	17.9	15.6	[e]	10.86	−8.49	32
1975–6	A 56.5 B 15.3	IV	19.4	13.5	2.5	Sorbitol–64.6	A $2.21 B 60	A $1.73 B 12.15	45

[a]Process:

I—Assoc. of Am. Soap & Glycerine Producers process, British Patent 499,417; 1/24/39.

II—Atlas Chemical Industries, Inc. process: U.S. Patent 3,396,199; 8/6/68.

III—Von Ling, et al., Ind. Eng. Chem., Prod. Res. Dev., 9(2), 210–212 (1970).

IV—Atlas Chemicals Industries, Inc. process: U.S. Patent 3,691,100; 9/12/72.

[b]Prices during the 1974–1975 period were erratic, ranging from 15.3 to 56.5¢/lb.

[c]Raw material costs are based on yields of glycerol obtained with no credit for any byproducts.

[d]Adjusted raw material costs are corrected for optimum credits due to byproducts. Propylene glycol and ethylene glycol are credited at 75% of the prevailing market prices, sorbitol is credited at 50% of its market price.

[e]Hexitols are assumed all sorbitol, pentitiols and methyl-D-fructofuranoside credited at only 2¢/lb, no credit for dehydrated hexitols and unknowns.

Table 9.9 Raw Material and Adjusted Raw Material Costs For Synthetic Glycerol
Produced By "Epichlorhydrin" Route Calculated For 1948, 1975, and 1990

Year	Bulk prices, raw materials Texas Gulf Coast, ¢/lb				Raw material cost[a] ¢/lb	Adjusted raw material cost[b] ¢/lb	List price USP glycerol ¢/lb
	Propylene	Chlorine	Caustic soda	Slaked lime			
1948	3.0	4.5	3.5	2.5	13.71	7.78	38
1975	8.5	7.25	5.5	5.1	20.20	10.84	45
1990	13.5	9.75	15.5	5.5	31.14	23.44	84

[a]Raw material costs assume theoretical yields of glycerol from propylene of 72.3%
(1948), 78% (1975) and 80% (1990) and no credit for any byproducts.
[b]Adjusted raw material costs are corrected for optimum credits due to byproducts:
37% hydrocholoric (muriatic) acid and 15% combined yield of 1,3-dichloropropene
plus 1,2-dichloropropane. 1,3-dichloropropene is marketed (1990) by Dow Chemical
Co. as *Telone II*, 94% active, 6% inerts, a soil fumigant and "restricted-use" pesti-
cide, price $0.98/lb, and also as *Telone C-17*, 74% active, containing 16.5% chloro-
picrin and 9.5% inerts as a soil fumigant and nematicide, price $1.47/lb 1,3-
Dichloropropene also in *Vorlex*, 40% active with 20% methyl isothiocyanate, NOR-AM
Chemical Co., price $1.84/lb. 1,3-Dichloropropene was credited at 75% of its yearly
market price, 1,2-dichloropropane at 45% of its yearly market price, FOB Illinois. 37%
hydrochloric (muriatic) acid was priced at 2.25¢/lb bulk in 1990 and is credited at 50%
its market price.

of only 18%. Shell decided to shut down its synthetic glycerol operation
and to divert its epichlorohydrin raw material to sales in the epoxy resin market, which at that time, showed good growth potential. On the other hand,
Dow Chemical built additional epichlorhydrin production capacity and optimized byproduct production and marketing from its synthetic glycerol operation in order to remain in the U.S. market. Today, Dow is the only
organization capable of producing synthetic glycerol.

FMC's 40 million pounds per year synthetic glycerol operation commenced
in 1969 at Bayport, Texas at a most inopportune moment. The technology
involved use of a process that began with an unusual catalytic conversion of
propylene oxide into allyl alcohol, which was treated with peracetic acid to
yield glycerol. The plant also was scheduled to produce about 45 million
pounds per year of byproduct acetic acid. Unfortunately, glycerol demand was
off, production of natural glycerol from fatty acid operation was high, and the
major soap-based glycerol makers cut prices 3.5¢ per pound toward the end of

the year. Obviously, synthetic glycerol producers, including FMC, had to follow suit. FMC was finally forced to shut down this operation in 1982.

VIII. SOME ECONOMIC FACTORS INVOLVED IN INTERNATIONAL PRODUCTION AND MARKETING OF GLYCEROL

A. General Factors

The development of new use applications for glycerol would be welcomed enthusiastically by the entire glycerol industry. Although glycerol continues to be a very versatile product, there are a few application areas that are showing decreasing annual consumption such as alkyd resin manufacture and use in tobacco products. Food uses tend to be increasing gradually. Competition for glycerol markets from substitutes or alternates like penetarythritol, propylene glycol, sorbitol, sugar polyols, trimethylolpropane, trimethylolethane, and others is a constantly changing factor.

The availability of large quantities of glycerol from a new source of supply is likely to be a price-depressing influence, at least for a short time. For example, the advent of Procter & Gamble's *Olestra*, a low-calorie fat substitute manufactured from soybean oil and sugar, and prepromoted for a target volume of 100 million pounds per year, could increase U.S. glycerol supplies by 7.5 million pounds per year.

Aside from periods of warfare, shortages of glycerol are not often experienced. They frequently are the result of raw material diversion to other uses. Even Shell Chemical's decision to concentrate on epichlorhydrin manufacture instead of synthetic glycerol in 1975, resulted in a temporary shortage.

The United States is not dependent upon the availability of coconut oil- or palm oil-derived glycerol; this country has cheap and abundant beef tallow and petrochemical propylene as alternate raw materials.

International glycerol prices are, of course, dependent upon efficient marine transport and changes, usually upward, in marine freight rates. At times, even a strike at the ship unloading or loading facility can create unanticipated variations in price.

Tariffs, embargos, quotas and the like play an insidious role and their overall influence is sometimes difficult to pinpoint.

B. Factors With Natural Glycerol

Fats and oils are diverse and in plentiful supply. They are *replenishable*. Glycerol results from a wide choice of fats and oils, and with the possible exception of marine and fish fats and oils, is available in satisfactory quality from all these raw materials.

Animal fat such as beef tallow is a dependable U.S. raw material for natural glycerol production, but for fat-bearing animals, disease is occasionally a problem. In some parts of the world, lamb tallow is available, for example, in New Zealand, where a potential of about 30 million lbs/yr exists. While lamb tallow offers a potential for glycerol production, the lower amount of oleic acid homolog than that found in beef tallow does not encourage use as a replacement for beef tallow that is much used for oleic acid production.

Adverse weather conditions such as typhoons, floods, drought, and other natural disasters limit the growth of many vegetable crops, thus determining availability and price. Bacterial and fungus plant infections can take their toll. Animal (e.g., rodents) and insect crop damage can be controlled somewhat more easily but are never eliminated entirely from most vegetable crops. Crop damage due to fire, civil strife, sabotage, and other causes are important in those areas where government stability has not yet been established.

Despite the significant advantages of natural raw materials, fats and oils over the past 70 years have not provided a stable and predictable supply of glycerol for all parts of the world. There still appears to be a moderating and stabilizing effect for synthetic glycerol, where and when it can be economically produced, on the international market.

C. Factors With Synthetic Glycerol

Ultimately, the availability of propylene and its price is determinative for synthetic glycerol manufacture. The critical factor becomes the relative demand for propylene as a raw material for other products. If this occurs, perhaps it is best if glycerol is manufactured from replenishable fats and oils, and propylene is used for material that cannot be made from anything else or cannot be produced so economically from natural raw materials. Still, synthetic glycerol from propylene and its derivatives provides a safety gap for those instances in which natural products fail to provide, for one reason or another, the glycerol required. Shrewd and intelligent petrochemical management has provided us synthetic glycerol that competes successfully with the natural product of today.

Table 9.10 outlines the trends in glycerol production and stock disappearance in the United States from 1920 to 1989.

Table 9.10 Glycerol Production and Disappearance in the United States (in 1000 lb units)[a]

Year	Stocks at beginning	Production of crude	Imports	Exports	Stocks at end	Dis-appearance	Excess of disappear-ance over production
1920	11,239.8	43,750.6	17,893.6	1,707.8	18,260.1	52,916.1	9,165.5
1921	18,260.1	51,157.4	2,216.3	2,346.8	21,328.8	47,958.2	−3,199.2
1922	21,328.8	68,269.6	2,909.7	2,813.1	20,714.8	68,980.2	710.6
1923	20,714.8	79,663.0	12,213.0	1,732.1	20,764.3	90,094.5	10,431.4
1924	20,764.3	76,123.0	13,012.3	1,287.6	21,035.2	87,476.9	11,353.8
1925	21,035.2	82,725.6	17,414.5	1,339.8	12,370.9	107,464.5	24,738.9
1926	12,370.9	93,095.4	32,749.2	752.3	20,971.3	116,491.8	23,396.4
1927	20,971.3	102,566.9	20,083.7	679.3	36,518.5	106,424.2	3,857.3
1928	36,518.5	104,399.6	7,789.3	2,010.9	31,646.6	115,049.9	10,650.3
1929	31,646.6	112,063.7	17,065.0	1,346.1	26,664.7	132,754.4	20,700.8
1930	26,664.7	110,940.2	11,989.4	595.5	25,879.9	123,118.8	12,178.6
1931	25,879.9	112,001.3	10,032.6	321.6	34,427.2	113,165.0	1,163.7
1932	34,427.3	107,135.1	6,606.4	255.1	44,501.9	103,411.6	−3,723.4
1933	44,501.9	95,849.3	7,898.6	N.S.	24,196.9	124,052.9	28,203.6
1934	24,196.9	122,402.4	14,090.2	N.S.	32,982.4	127,797.1	5,304.8
1935	32,082.4	112,947.9	6,643.9	3,286.6	34,402.7	114,885.0	1,937.2
1936	34,402.7	123,276.8	12,297.8	1,123.1	31,183.8	137,670.2	14,393.5
1937	31,183.8	135,231.0	18,109.3	1,347.5	56,079.7	127,096.9	−8,134.1
1938	56,079.7	129,696.1	12,994.1	3,671.3	77,669.1	117,429.4	−12,266.6
1939	77,669.1	147,581.1	9,113.7	7,250.7	70,888.9	156,224.3	8,643.1
1940	70,888.9	157,856.0	7,569.3	12,130.9	72,120.0	152,063.3	−5,702.7
1941	72,120.0	195,283.0	8,304.8	9,449.8	60,349.0	205,909.0	10,626.0
1942	60,349.0	177,435.0	4,883.8	32,638.5	56,659.0	153,370.3	−24,064.7
1943	56,659.0	171,933.0	8,220.2	24,499.6	80,914.0	131,398.6	−40,534.4
1944	80,914.0	199,834.0	4,866.5	9,142.7	87,681.0	188,790.8	−11,043.2
1945	87,681.0	172,450.0	7,676.5	7,598.0	45,720.0	214,489.6	42,039.6
1946	45,720.0	156,823.0	20,230.6	653.6	42,695.0	179,425.0	22,602.0
1947	42,695.0	207,768.0	3,062.4	4,268.2	48,582.0	200,675.2	−7,092.8
1948	48,582	195,837.0	6,394.3	3,834.5	41,851.0	205,127.8	9,290.8
1949	41,851.0	193,849.0	18,519.2	11,036.6	48,551.0	194,631.6	782.6
1950	48,551.0	225,512.0	23,781.8	7,254.7	50,300.0	240,290.1	−14,778.1
1951	50,300	211,348	14,606	4,856	55,727	215,671	−4,333
1952	55,727	187,902	15,203	8,971	37,716	212,245	−24,343
1953	37,716	214,997	35,378	4,262	59,989	223,840	−8,843
1954	59,989	207,092	14,433	16,860	42,359	222,294	−15,202
1955	42,359	227,999	26,959	9,680	51,977	235,660	−7,661
1956	51,977	244,177	18,043	10,233	67,247	236,718	7,459
1957	67,247	239,743	26,862	9,509	81,033	243,310	−3,567
1958	81,000	213,300	17,900	17,300	58,300	236,600	−23,300

Table 9.10   (*Continued*)

Year	Stocks at beginning	Production of crude	Imports	Exports	Stocks at end	Dis-appearance	Excess of disappear-ance over production
1959	41,200	267,700	10,400	21,400	38,100	259,900	7,800
1960	38,100	301,800	14,700	20,000	56,200	278,300	23,500
1961	56,200	279,200	14,800	15,600	69,300	265,500	13,700
1962	69,300	249,200	9,300	13,400	57,500	256,900	−7,600
1963	57,500	302,100	2,300	30,700	40,700	290,600	11,500
1964	40,700	328,100	9,400	28,100	61,200	288,900	39,200
1965	61,200	346,500	4,200	52,000	47,500	312,400	34,100
1966	47,500	354,800	4,500	43,400	39,000	324,700	30,100
1967	39,000	366,200	3,500	43,800	51,000	313,900	52,300
1968	51,000	362,600	800	55,800	53,200	305,300	57,300
1969	53,200	348,200	1,600	52,800	62,400	287,800	60,400
1970	62,400	339,300	200	73,600	50,400	277,900	61,400
1971	50,400	335,300	300	68,000	49,500	286,200	64,600
1972	48,100	350,800	1,100	64,300	49,500	286,100	64,600
1973	49,500	360,000	200	60,900	43,900	304,900	55,200
1974	43,900	358,500	6,300	65,100	45,500	298,000	60,500
1975	45,500	272,100	400	43,900	55,800	218,100	53,900
1976	55,800	324,600	4,300	56,600	49,500	278,600	45,600
1977	49,500	310,700	12,800	31,600	57,100	285,000	25,700
1978	57,100	302,600	7,700	41,500	42,800	283,100	19,500
1979	42,800	345,700	600	53,700	37,400	298,000	47,700
1980	37,400	301,100	19,300	57,100	33,700	266,900	34,200
1981	33,700	280,000	40,400	30,100	29,400	294,600	−14,600
1982	29,400	232,000	32,300	14,100	38,300	241,400	−9,400
1983	38,300	241,700	28,900	11,800	33,300	263,800	−22,100
1984	33,300	281,600	40,900	35,400	20,900	299,400	−17,800
1985	20,900	310,500	36,700	28,700	24,200	315,200	−4,700
1986	24,200	320,800	63,200	15,200	49,900	343,100	−22,400
1987	49,900	307,200	49,600	16,500	56,400	333,800	−26,600
1988	56,400	296,800	48,700	33,900	31,600	336,300	−39,500
1989	31,600	293,700	61,700	32,600	46,600	307,700	−14,100

[a]Soap & Detergent Association, Glycerine & Oleochemicals Div., New York, N.Y.

REFERENCES

1. Gentry, C. E., Glycerine, Short Course on Fatty Acids, AOCS, King's Island, OH, September 26, 1984.
2. Leffingwall, G., and Lesser, M., Glycerin, Brooklyn, 1945.
3. Sonntag, N. O. V., in Fatty Acids in Industry, (R. W. Johnson and E. Fritz, eds.), Marcel Dekker, Inc., New York, 1989.
4. Miner, C. S., and Dalton, N. N. (eds.), Glycerol, Reinhold Publishing Corp., New York, 1953.
5. Hansley, V. L., Ind. Eng. Chem., 39: 55–62, 1947.
6. Schuman, R. C., and Wolk, R. H., U.S. Patent 3,363,009, 1968.
7. German Patent 3,624,812, Carduck, F. J., Folbe, J., Fleckenstien, T., and Pohl, J., (to Henkel K.G.a.A.) 1988.
8. European Patent 254,189, Carduck, F. J., Folbe, J., Fleckenstien, T. and Pohl, J. (to Henkel K.G.a.A.) 1988.

10

Functions of Glycerine in Cosmetics

Rolf Mast

Neutrogena Corporation, Los Angeles, California

I. OVERVIEW

A. Introduction

The balance dichotomy of the glycerine molecule is its signature and the key to its decade of use as a major constituent in the cosmetic industry. Inside a hydrogen perimeter are three atoms of carbon, the structure of the organic world, balanced with three atoms of moisture-seeking oxygen. Moisture is the key to skin flexibility and health [1]. Driven by this molecular anisotropy, glycerine exerts its benign influence on every facet of the cosmetic product.

Glycerine helps in the creation of different product forms, including sticks, gels, microemulsions, and creams. The maintenance of product integrity in the package, including water retention and sometimes even microbiological stability is facilitated by its use. The positive organoleptic experience of cosmetic application, so essential to that feeling of well-being engendered by cosmetics, is often enhanced. Its major role, however, is as a moisturizing humectant in skin and hair. The preservation of skin structure and function, with concomitant mildness and metabolic inactivity, truly makes glycerine the quintessential cosmetic ingredient.

Appropriate to its widespread use in cosmetics and foodstuffs is the high degree of purity to which this colorless and odorless material is made. Use and purity are discussed in manufacturers' technical bulletins [i.e., 2,3], which also provide detailed background information on the chemical [4]. The typical specifications given will not be relisted here, except to make the point that the overall purity of commercially available material is as high as 99.7% when obtained from synthetic petrochemical sources, and 99.8% from "natural" oleochemical production. About 350 million pounds of glycerine are consumed in the United States annually. Of this, about 25% is used in the cosmetic and pharmaceutical industries, of which about 5%, or 17.5 million pounds are used in cosmetics [5]. To make the point that nearly 40 million pounds a year of glycerine is used as a direct food additive in the United States, emphasizes both its harmless nature and its utility. As with cosmetics, the humectancy, viscosity, antimicrobiological characteristics, and nonvolatility, all work to preserve product freshness and texture. With foods, an added dimension is its flavor-enhancing sweetness.

Among the general reviews on the uses of glycerine in cosmetics, toiletries, and pharmaceuticals, are those by Moxey [6] which gives a number of detailed formulations, the Glycerine Producers Association [7], and Newman [8]. Included in the wide range of products which benefit from glycerine's incorporation are the different varieties of skin creams/lotions, cleansing lotions, makeup creams/cakes, barrier/protective creams and gels, clear soaps, shaving products, clear gel cosmetics of different types, perfumes, etc., not to forget its large volume use in toothpaste. The use of glycerine in various contexts will be described, mainly from a functional point of view, as this chapter progresses.

First, however, it will be instructive to consider some of its physical properties and the relationship of those properties to its use in formulation work.

B. Solubility Characteristics

Hildebrand defined the solubility parameter, δ, as

$$\delta = \sqrt{(-E_v/V)}$$

where V = molecular weight/density, and $-E_v$ = heat of vaporization. The greater the similarity in δ, the more readily will the materials mix. This can be readily understood when it is considered that there is no hydrophobic-hydrophilic repulsion as such. However, there is a large loss of energy when molecules that are strongly attracted through oscillating dipoles, permanent dipoles, hydrogen bonds, or ionic attractions (hydrophilic molecules) are interdispersed by hydrophobic molecules without these intermolecular forces, resulting in the appearance of such a repulsion. Vaughan, in an interesting and cogent article, has listed the solubility parameters of numerous cosmetic materials [9]. In a solubility parameter list which ranges from methane at 4.70 to water at 23.4, Vaughan reports the solubility parameter of glycerine at 16.26. This value makes it a solvent for lower alcohols, and glycols as well as phenol, a partial solvent for acetone, low molecular weight esters and ethers, and insoluble in oils and higher alcohols.

Not surprisingly glycerine is used as a barrier for oily materials (see section below). On the other hand, with lower alcohols, water and soaps, as well as inorganic salts, it is used in the formulation of clear gels and sticks. For example, in a recent patent [10] glycerine is an essential ingredient in forming a microemulsion in an aqueous soap/hydrophilic polymer mixture. Similarly, glycerine is compatible with inorganic salt mixtures such as an aluminum chlorate-based styptic gel [11], and an aluminum acetate nail-hardening formula [12]. Further it is a solvent for boric acid used in different cosmetic formulations. Another example is its use to blend indomethacin (1,1-O-octyl-3-O-methyl-2-O-2',3'-dihydroxypropylglycerol) into an aqueous eye lotion [13]. There are many more examples of the usefulness of glycerine's hydrophilic solubility parameter, but probably the most famous is its proclivity to make clear toilet soap bars such as the famous "Pears" soap, that originated in the nineteenth century.

Other dimensions in the consideration of the solubility parameter as the physical basis for glycerine's utility in formulation work are cosolubilization, oil-based microemulsions, and pigment dispersion. Here one can utilize the effect that glycerine has on materials with which it is neither a solvent, nor directly miscible. For instance it is used as a carrier, both for perfumes in shaving lotions, and flavors in toothpastes, as well as a cosolvent for different drugs [2,7]. It can play a key role in the formation of clear mixtures with surfactants and oil. One patent, for instance describes the formation of transparent cleansing compounds containing a complex nonionic + anionic surfactant mixture (about 12%), with glycerine (about 15%), and an oil mixture (about 60%) [14]. In makeups and toothpastes alike, glycerine aids in the dispersion of the coloring or abrasive pigments. For example, it is used at 5% in an eyeshadow which also contains pigments, hydrolyzed gelatin and urea in 69% water [15].

C. Humectancy Characteristics

Glycerine is added to many products because of its ability to attract and hold water. Various aspects of this behavior, especially quantitative relationships, will be dealt with below. Remarks here will be limited to an overview of why and how glycerine's humectancy contributes to its wide use. First, the formulator must consider how humectancy contributes to the stability of the product itself, and secondly the effect of humectancy upon skin integrity after application of the product. Further, when interpreting the water absorption isotherms in relation to function it must be remembered that the effective concentration of an active material on the skin is much higher than the concentration in the product due to the evaporation of water. In the case of glycerine this concentrating effect on the skin will be further enhanced by its insolubility in many of the oils present in the formulations. Thus even small amounts of glycerine in a formula, once dried on the skin, can have the same effective hygroscopicity as 100% glycerine directly applied. Details of its resultant beneficial effects on the skin, measured both clinically and instrumentally will be given elsewhere in this report.

The effect of glycerine in the product per se, as well as after application to the skin, is relevent. Thus, in a review on polyols in creams and lotions, Courtney, discusses their use in maintaining the moisture content of a finished formulation [16]. He quotes Griffin et al. [17], where even 20% glycerine in an oil in water emulsion will lose water at ambient relative humidity of anything less than 95%. Any beneficial effect of humectants in water retention of aqueous systems, therefore is related to the rate of loss, not the equilibrium levels of water, in air-exposed product. In addition to Griffin, Courtney quotes several other studies: Strianse [18], Segur and Miner [19], Henny et al. [20], on the rate at which moisture loss is retarded by different humectants from various water based formulations. Under nonequilibrium conditions in a soap-based cream, neither glycerine nor propylene glycol were useful in retarding the rate of water loss, even at 70% relative humidity. Surprisingly, sorbitol, which is a much weaker humectant under equilibrium conditions, does in fact significantly retard the rate of water loss in the simple soap-based creams studied by Griffin even at 2% sorbitol and 30% relative humidity (RH) where water loss is retarded by 10–50% over a 30 minute to 48 hour period. More complex anionic and nonionic formulations gave different types of behavior depending upon the specific formula being studied, as well as the humectant type, humectant concentration, and external relative humidity. In some of the formulations, glycerine had an insignificant effect in reducing the rate of water loss from an air-exposed cream even at 70% RH and 15% glycerine concentration. In other formulations (different workers), the rate of water loss was nearly halved under the same conditions. Further, under some conditions glycerine is better than sorbitol and vice-versa. The surprising thing is that sorbitol, a much weaker humectant, is not always inferior to glycerine. In summary, one can conclude that the effect of glycerine in reducing water loss from an air-exposed emulsion is marginal, and not a reason to use it in formulation work, even though there is a benefit under high glycerine and RH conditions.

As a footnote to this section it should be added that glycerine has a much more significant effect in maintaining the water content of an exposed solid, where the glycerine is not significantly water diluted. To this end it is used with great effect by the foodstuffs industry for maintaining the consistency of sweets, dried fruits, bakery goods, etc. [2]. A quantitative example of the use of glycerine in foods is in the work of Lo et al. [21]. Glycerine and propylene glycol were much more effective than glucose or sucrose in lowering water activity. In general it was noted that humectants are used to confer stability to nonrefrigerated foods due to their water-binding ability, and that in the egg system studied, glycerine was easily miscible with the fat and water content of the egg. Similarly, in low water content formulations such as face powders, it can also be effective in preserving the integrity of the product itself, due in direct measure to its humectant effect. Other uses in this context are possible. In part, this must be the reason for using glycerine in a cleaning cloth that also contains polyethylene glycol and a finely divided solid on a fabric or paper substrate [22].

D. Freezing Point and Volatility Characteristics

Pure glycerine freezes at approximately 19°C (depending upon who is reporting the data). Placed in an aqueous environment it causes the overall freezing point of the system to be lowered, a fact which can be helpful in enhancing the low temperature stability of a product in much the same way that ethylene glycol is used to prevent the freezing of car radiators. In simple aqueous systems the effects are not large as shown in the following figures [23]:

Percent of glycerine in water = 5   10   20   30   40
Freezing point depression °C = 1   2.3   5.5   9.7   15.5

In formulated systems, the inhibition of freezing can still be useful. For instance, 4–5% in a shampoo can act both as freezing point depressant and as a thickener [8].

Because it has very low volatility, another major benefit is in its use in preventing the drying of cosmetics into hard films and residues. This can prevent problems like the plugging up of valves and spouts as product residue dries.

E. Miscellaneous General Effects of Glycerine in Formulations

Some properties of glycerine do not lend themselves to ready classification as in the above sections, but should be mentioned here for completeness. Glycerine can act as a fixative for perfumes, ensuring consistency of fragrance note during product use-up. It is said that at sufficient concentration it can inhibit electrolyte corrosion of tinplate and aluminum containers, and certainly as documented below, it can act as a preservative.

During product application to the skin, especially of potentially high friction products like stearic acid-based vanishing creams, glycerine can play a significant role in the smoothness of application. In essence, in this context it acts as a lubricating agent, preventing balling and rolling of the product.

Depending upon the formulation, it has been described as having plasticizing, binding, glossiness, and smoothness properties. These are, in essence, more general descriptions of the effects imparted by the solubility, humectancy, and physical characteristics already discussed.

II. THE QUANTITATIVE HUMECTANCY OF GLYCERINE

Humectancy, as a general characteristic in formulation work, has been discussed above. Further, the role of glycerine and water in specific skin interactions will be considered in detail below. The objective of this section is to describe the more quantitative aspects of this phenomenon.

Zanker has presented a nomograph which relates the refractive index of an aqueous glycerine mixture (calculated in terms of the refractive index at 25°C), to the equilibrium relative humidity that the mixture will maintain in small vessels, as a function of the temperature [24]. The results are represented here (Fig. 10.1) in terms of the concentrations that those refractive indexes (measured at 25°C) represent. The relative humidity of a particular glycerine solution at a particular temperature is obtained by drawing a line from the concentration to the temperature and extending it to the relative humidity scale.

Table 10.1 lists some equilibrium relative humidities of different water/ glycerine mixtures at 25°C as taken from the nomograph. The supplemental data needed to construct this table was taken from the *CRC Handbook of Chemistry and Physics* [23]. It should be noted that the differences in refractive index in going from 20°C to 25°C were ignored when relating refractive index to concentration in the *CRC Handbook*.

Figure 10.2 lists this same data graphically, along with similar data from another source [25], that includes propylene glycol and sorbitol. Data taken from reference 16 alone shows that glycerine has somewhat stronger water

Table 10.1 Equilibrium RH at Different Glycerine Concentrations, at 25°C

Glycerine concentration in water (% by wt)	Refractive index	Equilibrium relative humidity
14	1.350	97.5
22	1.360	95.5
30	1.370	93.0
37	1.380	89.0
44	1.390	85.0
52	1.400	81.0
58	1.410	75.0
65	1.420	69.0
72	1.430	63.0
78	1.440	55.0
84	1.450	43.0
91	1.460	27.0
98	1.470	7.0

Source: Adapted from Rep. 24.

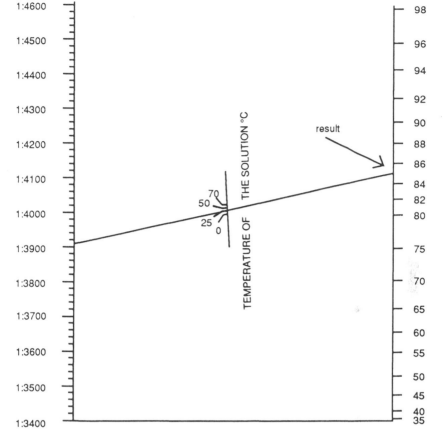

Figure 10.1   Relative humidity of aqueous glycerine solutions (from Ref. 24).

absorption than either propylene glycol or sorbitol, which agrees well with the results from Figure 4. I do not know why there is such a large disparity between these values and those adapted from reference 23. Nevertheless, the general trend of equilibrium relative humidity versus humectant concentration is quite clear.

Measuring the relative humidity of aqueous glycerine, as a function of temperature in a closed environment, is one way to quantify its humectant power. Another way to look at this is to consider how glycerine affects the chemical activity of water in mixtures. The chemical activity is a thermodynamic quantity which is related to the Gibb's free energy, G, of the water present. This relationship is derived through the chemical potential, μ, as follows:

$$\mu_i = (\delta G/\delta n_i)_{T,P,n_i}$$

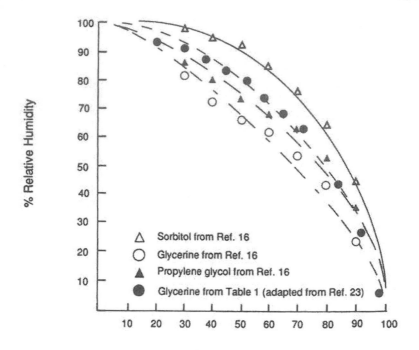

% Relative Humidity

% Humectant in water

△   Sorbitol from Ref. 16
○   Glycerine from Ref. 16
▲   Propylene glycol from Ref. 16
●   Glycerine from Table 1 (adapted from Ref. 23)

Figure 10.2   Humectant concentration vs. equilibrium RH at 25°C.

where T = temperature
      P = pressure
      n_i = molar concentration of species i

Thus, the chemical potential, which defines the propensity that a material has to participate in a chemical or physical change, is related to the activity as follows:

$$\mu_i - \mu_i^* = RT\ln a_i$$

where μ_i^* is the chemical potential of the pure reference state, and a_i is the chemical activity.

In short, the chemical activity, a_i, defines reactivity more precisely than concentration. It is related to concentration, X, through the activity coefficient, y:

$$a_i = y_i X_i$$

Sloan and Labuza [26] have listed the water activity, a_w, of several water humectant mixtures as measured by the vapor pressure manometer method at 23°C. Table 10.2 lists some of their results, recalculated in terms of percent humectant mixed with the water.

Table 10.2   Water Activities (a_w) with Different Humectants at 23°C

a_w	Glycerine	Percent humectant in water			
		Propylene glycol	1,3 Butylene glycol	PEG 400	Sorbitol
.95	14	13	14	24	17
.90	32	27	33	44	43
.85	41	37	45	55	53
.80	48	47	56	62	60
.75	55	54	64	68	65
.70	61	61	68	72	68
.60	71	71	77	79	77
.50	79	80	85	85	97
.40	85	86	89	91	98
.30	89	91	91	94	98
.20	91	94	93	96	98
.10	95	96	95	98	98

Source: From Ref. 26.

In essence, a reduction of the a_w below 1.0, as measured by vapor pressure at any particular water concentration, is a measure of the chemical interaction between the humectant and the water molecule and the lowering of the water free energy. It can be seen that glycerine and the two glycols are very similar in this respect, and better than the other humectants listed. Further, in keeping with the above discussion on the utility of humectants in retaining water in formulated products, the humectants have to be at very high concentrations to exert an appreciable effect. It must be remembered, however, that on the skin glycerine is indeed present at high concentration. This is because not being volatile, it remains, when the water has already evaporated.

In passing, it might be mentioned that the measurement of vapor pressure is only one way of measuring a_w. In another method, Lo et al. [21] used the relative isothermal equilibrium absorption onto a dry powder (microcrystaline cellulose) to determine the a_w of humectant-egg mixtures. Indeed, the methods are probably limited only by the imagination, and need only measure the potential for certain concentrations of water to react or change its physical state relative to a standard condition without humectant. For example, Sakamoto et al. used nuclear magnetic resonance (NMR) to correlate the hygroscopicity of various humectants, including glycerine, Na pyrrolidone carboxylate, L-lysine, and Na lactate, with the amount of unfrozen water at −20°C [27].

For the interested reader a physical chemical study on the hydration energies and shell volumes of conformationally flexible solute molecules has been published [28].

III. TOXICITY AND IRRITATION: GLYCERINE AND ALTERNATE HUMECTANTS

A. Glycerine: A Very Safe Ingredient

The general presumption that glycerine is a very safe ingredient is derived from its overwhelming widespread use in a multitude of products which can be ingested or topically applied. Individual daily usages in the United States alone would extend into the millions, so one can safely assume that any untoward interactions would become apparent.

Fisher, who is well known for his experience in the dermatological effects of cosmetics, calls glycerine nonirritating and a very rare sensitizer [29]. Fisher quotes Hannuksela as having tested several thousand patients with 50% glycerine [30], where only one patient apparently had an allergic response. Other, extremely rare, instances have been reported, although Fisher himself has not witnessed any such reactions.

Guillot et al. describe 26 cosmetic ingredients that were evaluated, in rabbits, for ocular and skin irritation [31]. Included in this list were 15 glycols and glycol derivatives, glucose derivatives, sorbitol, sodium lactate, lactic acid, and urea as well as glycerine. Ocular irritation, primary cutaneous irritation tests under occlusion, as well as six-week cumulative cutaneous irritation tests (sensitization) were done. Concentrated lactic acid was poorly tolerated in all tests, presumably due to its acidity. Interestingly the only other bad reaction was the ocular irritation of hexylene glycol. Glycerine was tested at both 100% and 20% and showed very little, if any, irritating effects.

In work from a recent patent study, aqueous solutions containing about 40% glycerine were safety tested [32]. Aqueous glycerine was used as a carrier for topically applied human interferon. These very interesting preparations are used to treat keratosic disorders such as lichen planus and leukoplakia. No abnormalities were observed when the preparations were occlusively patch tested on rabbits (primary skin irritation), intradermally injected into guinea pigs (skin sensitization), human occlusive patch tested, and irritation tested on human oral mucosa. This is a stringent selection of tests, and shows once again what an excellent carrier glycerine is for drug and cosmetic items. The intradermal injection into 23 guinea pigs, however, should be compared with the human intradermal testing described in the next section.

B. Glycerine in Subcutaneous and Wound-Healing Situations

Although the humectant effect of glycerine makes it a wonderful cosmetic ingredient for reasonably intact skin, there are situations in which its osmotic pressure effects can be detrimental. One example of this is given in a study of the pain produced by mafenide acetate preparations in burns [33]. Thus when glycerine is used as a carrier for 11.2% solutions of mafenide acetate, and the mixture is applied to burns, considerable pain is caused by the hypertonicity which derives equally from the carrier (1,080 mOsm/kg) and the drug (1,100 mOsm/kg). Reducing the osmotic pressure by eliminating the glycerine greatly reduced the degree of pain experienced. Pain reduction was not affected by the pH variations observed. Similarly, other work has indicated that increasing osmolality overrides the effect of pH when studying the irritant action of various buffers on mouse skin [34]. This complicated report studies the effect of "water-related" phenomena, pH, osmolality, and buffer capacity on mouse skin irritation. Among its findings, is the direct relationship between osmolality and the intensity of edematous reaction with increasing sorbitol and glycerine levels. Again, as with the other negative reactions of glycerine alluded to in this section, this is essentially a subdermal interaction, as interactions in open wheals are being measured. Thus it can be concluded that the hypertonicity, osmotic pressure, or humectancy (in this context call it what you will) is a detriment when the material is applied intradermally or to an open wound. However, as amply demonstrated elsewhere in this text, this physical chemistry of glycerine is of great benefit in the cosmetic sense both in formulation, and on the skin.

Glycerine can also provoke reactions when injected intradermally. Thus workers have claimed that 2% glycerine is a frequent cause of skin reactions when it is introduced intradermally in the testing of allergenic reactions in patients suffering from allergies [35]. It might be noted that the 2% glycerine is present because 50% of the material is used in preservation of the concentrated allergens prior to injection, which is yet another example of its ubiquitous use in products. It is not clear whether phenol is also present in these tests, although the authors definitely ascribe the numerous whealing reactions observed with the allergen-free controls to the glycerine. In similar work with intradermal allergenicity tests, 5% glycerine was shown to produce a small flaring reaction in 75% of the patients tested [36]. A nonspecific histamine-independent irritant effect was postulated, as anti-H_1 histamine only slightly reduced the size of the wheal. Ironically, in this work neither glycerine at 2% or less, nor 0.8% phenol produced these reactions, but above that threshold the size of the wheal was directly related to the glycerine concentration. Presumably it causes a greater response in humans when injected intradermally than it does in guinea pigs. No picture is completely straightforward, even for glycerine. For the interested reader, a third article on this same subject [37] seems to show very little difference between purely aqueous and aqueous glycerine intradermal injection of allergens.

C. Safety of Glycerine Versus Other Humectants

Another way to ascertain the safety of glycerine is to consider its toxicology with respect to other humectants. A detailed study in this regard has been done

by Motoyoshi et al. [38] who make several interesting comparisons between
the frequency and degree of erythema observed in closed human patch testing
and the humectant power of the material. Figure 10.3 is adopted from these
authors, and shows erythemal response after 48 h closed patch testing.

Figure 10.4 taken from deNavarre [39], and also quoted in reference 38,
shows the relative humectancy of many of the same materials. Three conclusions can be drawn from these two figures:

1. There is no direct relationship between the degree of irritation produced
    and the humectancy of the material

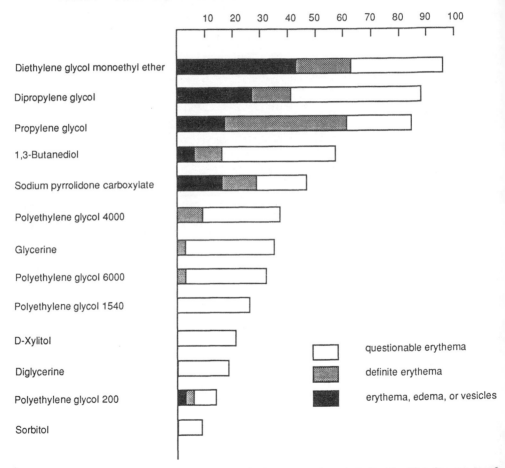

Figure 10.3 Erythemal response from humectants in closed patch tests. Positive frequency of
closed patched tests with 50 (w/w) solutions on 34 healthy males. Site tested: upper back. Duration of applications: 48 h. Reading: 30 minutes after removal (from Ref. 38).

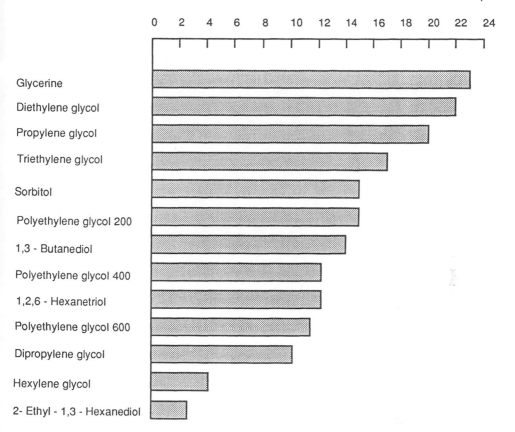

Figure 10.4   Humectant water uptake at 50% RH, 20–27°C   (from Refs. 38 and 39).

2. Propylene glycol, another frequently used material, is much more irritat-
    ing than glycerine
3. Glycerine is clearly the material of choice if one wants to combine high
    water absorption with low irritation potential in an optimum cosmetic in-
    gredient

In this same detailed and comprehensive work, Motoyoshi and co-workers
studied both the degree of clinical irritation over the patch test cycle, as well
as making histological examinations of the skin. Further, these results were
compared with results of animal testing. It was shown that the human skin
reaction to propylene glycol peaked on the Days 15–17, leading to the conclusion that the reaction pattern is a primary irritation. Further, propylene glycol

did not irritate animals without sweat ducts under closed patches. Thus the fact that neither rabbits and guinea pigs, which normally have higher reactions to oil-based cosmetics, nor human skin in open patch testing were irritated, led the researchers to state: "It is concluded that the skin reaction produced when high concentrations of PG (propylene glycol) are used in closed patch tests on human skin includes primary irritation but is thought to be mainly a sweat retention reaction . . . there was considerable leukocyte infiltrate in the vicinity of the sweat glands in the dermis and it is thought that the sweat retention reaction resulted from this."

In short, the irritation caused by propylene glycol seems to occur in the presence of sweat glands.

IV. GLYCERINE AND SKIN CARE

This section deals with the in vivo and in vitro evaluation of the effects of glycerine cosmetics on the skin.

A. Methodology

The subject of cosmetics testing is a wide field in itself, and only brief mention can be made here. For further reading see, for example, the book on cosmetic psychophysical testing by Moskowitz [40]. In outline, however, as a background to this section, one can consider the derivation of data relating to the evaluation of cosmetics on living skin to fall within three general areas.

First is the subjective opinions of the users themselves. To have any validity, it is essential that coded products and nonprejudicial questionnaires are used, so the subjects opinions are free of unintended bias. Further, as with most testing, the inclusion of a control (i.e., a well-known product against which to measure the response) is necessary as an internal calibration in evaluating the results. In this sort of testing it is often difficult to differentiate between responses to actual product performance, and the overall consumer acceptance, as the latter involves appearance, cosmetic feel, perfume, and other such factors.

Second, and most important, is clinical evaluation. Here, independent expert judges evaluate the in vivo effect of materials applied to panelists. One can further divide clinical evaluations into two types. First, normal use tests, such as the affects of creams on dry skin, where the product is applied and evaluated under conditions of intended use. Second, exaggerated use—testing, involving repeated skin insults, to promote and thus gauge the skin irritation/ sensitization of a product. An important factor to promote valid results in clinical testing is the use of "blind" panelists and judges. It is especially important that the judge does not know which product was applied to which test site, as with the best intentions, subjective prejudice could be a factor. Careful control of panelist selection, product application, test sites, and environmental conditions must be maintained. Thus, skin test sites must be selected and rotated so that the test and control products are applied in a balanced manner, to panelists selected for dry skin in the first place. Generally, equal weights of products should be applied. Further, ambient conditions that promote the formation of dry skin are a distinct advantage. This normally means running panels in the winter. Properly conducted clinical tests are the basis for all institutionally accepted performance claims. Of course clinical evaluations can be made using instrumental methods of analysis in addition to judges.

Third is the area of laboratory-controlled instrumental evaluation of cosmetics using in vivo test sites. This allows the accumulation of objective, quantitative data. The drawback often is in relating the data to clinical evaluations. Of course, much work is also done on excised skin using instrumental methods, and some of these studies will also be covered in this section. In general, instrumental tests can be categorized as follows.

1. Visualization of the skin, such as scanning electron microscopy, image analysis, laser-doppier visualization of underlying blood flow, or dansyl chloride staining of dry skin.
2. Mechanical measurement of unperturbed skin, especially profilometry of wrinkles, and ultrasound measurement of skin thicknesses.
3. Mechanical measurements of various skin relaxation phenomena; here, techniques such as the gas-bearing dynamometer quantify the viscoelastic properties of skin.
4. Evaluation of water vapor transmission through the skin. The two major techniques here involve use of either a forced airflow, or static air conditions.
5. Measurement of electrical properties, especially impedence. This technique is an attempt to directly measure stratum corneum moisture content, as the conductivity should go up with moisture. Some complicating factors, however, are surface contamination/electrode contact, electrical passage through different routes, and frequency dependence.
6. Simple mechanical measurements such as in vivo stress/strain techniques and ballistic rebound.
7. Physical or chemical analysis of such factors as sebum concentration using sebum paper, and lipid analysis.

This list does not exhaust the possibilities. In fact, the different instrumental techniques that could be used are virtually unlimited.

With the above thumbnail sketch as a background, some of the various evaluations of glycerine in skin care will be examined.

B. In Vivo Studies Using Direct Observation

Using clinical studies, Bissett and McBride have clearly demonstrated the skin-conditioning effects of glycerine on human and pig skin [41]. In humans, various levels of aqueous glycerine were applied twice daily to dry skin on legs. Evaluation of dry skin was done on a 0 (best) to 5 (worst) scale. Every possible treatment pair, including the water control and four glycerine levels were duplicated 20 times within the study. Table 10.3 is taken from their work.

Starting grade averages were 2.02 to 2.71. Statistically significant differences at the 95% confidence level exist between 5, 10, and 20% glycerine levels. The work on pig and human skin correlated well. In general, the maximum benefit is obtained at a use level of around 29 μmol of glycerine/cm^2 skin per day. Further, in their work with in vivo pig skin, glycerine was compared with petrolatum in skin conditioning. Interestingly, at 40% to 100% glycerine levels no significant difference was found between the skin improvement by the glycerine or 100% petrolatum. This result is fascinating when one considers the almost diametric differences between these two molecules, especially the hygroscopicity of one and the occlusivity of the other.

Bisset and McBride state that the observed benefits are due to more than a temporary masking of the scales, and somehow involves the prevention of skin dehydration.

A 1988 study is another excellent documentation on the skin enhancing benefits of glycerine [42]. A battery of direct and instrumental observations were made, which, in keeping with the arrangement of this chapter, will be discussed both in this and the next section. Visual (photographic) assessment of skin condition was made with 15 female volunteers. A blind crossover study was done using a glycerine-containing lotion and the equivalent nonglycerine placebo on the volunteers. Formulas were not given. Photographs were taken

Table 10.3 Improvement in Human Dry Skin Conditions by Various Concentrations of Glycerine

Treatment	Grade reduction[a] (skin improvement)
5% Glycerine	0.86
10% Glycerine	1.44
20% Glycerine	1.74
40% Glycerine	1.82

[a]Combined results after two weeks treatment using water control.
Source: From Ref. 4.

12 h after treatment. After one week's use, the glycerine-containing lotion improved skin condition in 80% of the subjects, and after two weeks in almost 100% of the subjects tested. The nonglycerine lotion did not show significant improvement in skin condition. This is dramatic evidence of its beneficial effect. Parallel instrumental studies were done on in vivo skin which included transepidermal water loss, skin roughness, coefficient of friction, and electrical impedence, which all demonstrated objectively, the improvements in skin condition due to glycerine. Except for the results on transepidermal water loss (TEWL), they will be discussed in the next section.

Batt et al. [42] measured TEWL on forearms treated with 0.05 ml/20 cm^2 of the appropriate solution. Measurements were done at 60% RH using a Servomed EP1 Evaporimeter (Servomed AB, Stockholm-Vällingby, Sweden). In this technique, ambient humidity without airflow is measured. In 5–15% aqueous solution, glycerine reduced TEWL for 4 h and in a lotion for up to 7 h over simple aqueous control. This certainly makes a correlation between the beneficial effects of glycerine and petrolatum easier to understand, but this result does not correlate well (as the authors themselves state) with the often reported observation that glycerine in fact speeds up TEWL (i.e., 50, 60, and personal observations using forced airflow methods at low RH). As the authors suggest, the amount of water that is measured coming off the skin could be a function of the external RH. At higher external RH's the glycerine could be expected to remove moisture from the atmosphere (especially in the static conditions of the Servomed EP1 Evaporimeter), giving an apparently lower TEWL. With forced airflow methods, at low RH, the extra moisture in the upper skin layers would show up as an increased TEWL. In both instances, glycerine hydrates the skin. The simple occlusive effects of petrolatum are constant regardless of conditions. In their general discussion, the authors state "The data reported here are consistent with the hypothesis that the beneficial effects of glycerol on skin condition are due to its physical effects on the status of water in the outer layers of the stratum corneum. This may be the results of glycerol interactions with stratum corneum lipid structures, or proteins, altering their water-binding and/or hydrophilic properties."

Middleton and Allen [43] have extended the consideration of skin water levels by considering the difference in water layers as one progresses outward through the stratum corneum. Skin becomes especially dehydrated at low temperatures and relative humidity, which, of course affects the outer layers more. They surmise that this will result in a lower extensibility of the surface corneum compared with deeper layers, resulting in a differential flaking and cracking of the surface. In a study in which the main thrust was to show the effectiveness of lactic acid and sodium lactate in skin treatments [44], the efficacy of 5% glycerine (in water) was indeed shown to increase the water-holding capacity and extensibility of solvent damaged guinea pig footpad corneum at 81% relative humidity. The presence of an external glycerine layer in in vivo situations will presumably lessen the diffusion-controlled water gradient across the stratum corneum, thus ameliorating dry skin problems. Another way to look at it is that, depending upon the external RH, a wetter outer layer of skin, also containing glycerine, may or may not lose more water than a dryer, nonglycerine, outer layer. Also, because of glycerine, water replacement is more a displacement than a slow hydrophobic pore controlled diffusion phenomenon. The use of glycerine and other humectants to increase the diffusion of water through hydrophobic membranes has in fact been documented utilizing polyethylene hollow fiber membranes [45]. Using ozone to attach humectants to the 42 µm pores, gave long-term increases in water vapor permeability. Similarly, humectants incorporated into dried polyvinyl alcohol forms, gave a dramatic increase in the water absorption rate when soaked in liquid water [46]. What all this means is that it is not only the humectancy per se that is important, but the humectant effect on the microporous properties of the stiff and relatively hydrophobic keratin chains—which allows a greater rate of water transportation. Kinetic, as well as equilibrium free energy effects, are crucial factors in the consideration of skin hydration. This makes glycerine quite different in its action to the simple water trap that is petrolatum. In spite of all this, however, it is by no means certain that all of the beneficial skin conditioning effects promoted by glycerine and petrolatum are a simple function of skin water level only.

Rather than consider only the relative effects of oily and humectant substances, the cosmetic chemist should also consider their synergistic interactions with each other, as demonstrated, for example, in work done by Ozawa et al. [47]. Thus, with in vivo experiments done in both guinea pig and human skin, a maximum moisturizing effect was obtained with a combination of water, humectants, and oily substances. Similar synergistic effects have also been observed by this author in clinical studies conducted on dry leg skin. Ozawa et al. made the further interesting observation that higher molecular weight humectants, such as hyaluronic acid, and lower molecular weight humectants, such as glycerine, also showed synergistic effects. They claim that an appropriate formulation protects the skin via physiochemical as well as biochemical effects (e.g., promotion of amino acid metabolism), but I think this unlikely. This beneficial mixture of humectant and oil has been underscored in a patent which deals with the alleviation of psoriasis as well as dry skin [48]. Thus a cosmetic composition was prepared containing white petroleum jelly 40, lanolin 40, cocoa butter 10, and glycerine 60 parts. This was applied four times daily to the arms and legs of a female suffering from psoriasis. Within four days the skin cracks had almost healed and most of the scales had disappeared. Within seven days no signs of psoriasis were visible.

The underlying interactions of oils and humectants have been studied using in vitro physical chemical methods [49]. As convention dictates, occlusivity and water-holding capacity of creams were taken as prognosticators of the skin hydrating effects, and this indeed was realized when collagen membrane was used as a skin model. As to be expected, the addition of humectants, such as glycerine to a cream decreased the occlusivity but increased the water-holding capacity. Interestingly, they reported that occlusivity and water-holding capacity depended more on the humectant than the oil content; a finding that gives valuable clues in the formulation of an effective skin treatment. In an extension of the just-reported study, the same team examined the effects of the addition of humectants to water-in-oil creams [50]. In order of increasing effect, the quantity of water migrating through the oil phase was affected by the following humectants: glycerine, urea, and sodium thiocyanate. The effect of humectants in decreasing occlusivity, but increasing the water holding capacity was much more pronounced with water-in-oil than oil-in-water creams.

In other work relating to humectant and oil interactions, Abe has studied the in vitro and in vivo effects of glycerine, liquid paraffin and isopropyl myristate on transepidermal water loss [51]. It is reported that in vitro glycerine does increase transepidermal water loss (as expected) but in vivo, 100% glycerine decreases the same considerably. Thus after 1 h, transepidermal water loss versus the control on forearm skin of healthy adults were reported to be:

Glycerine   41.6%
Liquid paraffin   83.7%
Isopropyl myristate 103.4%

As discussed above, the measured results will depend upon the details of the technique used, and very often glycerine increases measured TEWL, whereas hydrophobic paraffins will invariably reduce it. The author, however, does raise a line of thinking which should be considered. Namely, that the effect of the isopropyl myristate in increasing the water vapor transmission might be related to its solvent effects for the natural skin lipids. This certainly seems plausible. A corollary to this is that we should always consider the system we are dealing with as a whole, a point underlined by Rieger when discussing some of the not so beneficial effects of cosmetics on the skin [52]. Thus, one point made by Rieger is that propylene glycol may be less irritating in a closed patch where it remains dilute than in an open patch where it will concentrate on the skin as the water evaporates. The complete system must be considered. This brings us back to the transepidermal water loss effect of 100% glycerine on the skin as reported above. It is possible that the in vivo transepidermal water loss was measured prior to the establishment of the steady state, so water was still being absorbed by the glycerine itself, or that the relative humidity of the ambient air was very high.

A very interesting patent by Wang extends the use of glycerine/oil combinations to wound healing [53]. An insoluble dextran based film, impregnated with glycerine and oil, was shown to increase the rate of healing of rat wounds that had been caused by either excision or burns. The wound covering material allowed sufficient removal of exuded body fluids, sealed the wound site against bacterial infection, and at the same time kept the wound moist, thus preventing scale formation. Certainly, to the extent that skin cracking can be considered as being very minor skin wounds, this patent can be related to cosmetic function. This only emphasizes the care with which we must interpret in vitro work, as it can never completely simulate the complex metabolism of living skin.

C. In Vivo Observations Using Instrumental/Objective Methods

Skin evaluations based on some type of instrumentation have the ostensible advantage of objectivity, as well as the ability to detect changes not visible to the naked eye. The disadvantage, of course, is that the results may be hard to relate to clinically describable changes in the skin. One study, however, which does show an excellent correlation between clinical and instrumental methods has already been discussed, in part, above [42]. In addition to the TEWL methods already described, Batt et al. studied skin surface topography using a Talysurf 10 Profilometer (Rank Taylor Hobson Ltd., Leicester, England), coefficient of friction by rotational torque of a weighted cylinder, and electrical impedence using proprietary instrumentation. After single applications of glycerine-containing preparations, reductions in the "peak/valley" heights of skin topography showed objective increases in skin smoothness lasting for up to 24 h. Measuring skin rheology is one way to study the effects of water concentration on the stratum corneum. The dynamic coefficient of friction was markedly higher for several hours following glycerine treatment. This result was correlated with an increased friction felt by rubbing moist versus dry skin with a finger which is distinct from the flaky roughness of dry skin. Measuring the impedence or capacitance of the skin is a way to more directly relate to the actual amount of water in the stratum corneum. Again, electrical impedence measurements showed elevated values for at least two hours following the glycerine treatment. Thus all these instrumental methods correlated very well with the clinical observations made by the same authors.

Leveque has written a useful summary of several of the methods used in skin care research [54]. Very briefly stated, the article describes measurement of skin hydration through transepidermal water loss, as well as electrical impedence (as stated, a measure of in situ hydration); skin elasticity through torsional and gas-bearing electrodynamometer techniques; skin smoothness through profilometry and image analysis, as well as instrumental analysis of skin color and skin surface lipid content.

To demonstrate the utility of a skin elasticity device called a "Twistometer," skin treated with either 10% glycerine or urea in a basic oil/water emulsion was measured. Figure 10.5 records the results obtained. Over several

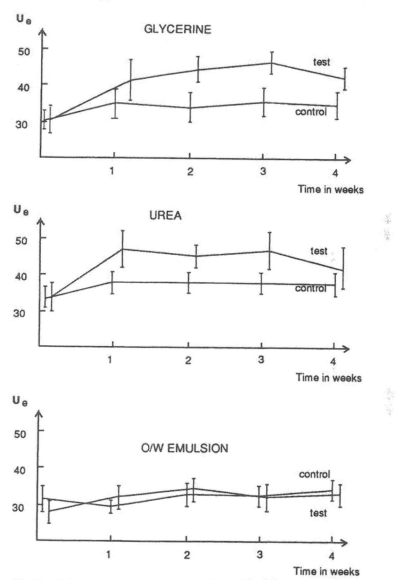

Figure 10.5 Skin extensibility (in vivo) with different treatments. Control and test measurements were taken on separate forearms; glycerine and urea applied at 10% (from Ref. 54).

weeks, glycerine (and urea) significantly increased skin extensibility over the control, whereas the basic emulsion did not. In this technique, a disk is glued onto the skin with adhesive tape and the rotational angle is measured as a function of applied torque. It was determined that the stratum corneum accounted for a large part of the resistance to the recorded skin extensibility. Thus, using this technique, skin flexibility can be shown to be improved by glycerine, a result that would not have been detectable visually under the conditions of the experiment.

In addition to the "Twistometer," viscoelastic properties of the stratum corneum have also been successfully measured with the electrodynamometer. A detailed description of this apparatus has been given by Hargens [55]. Essentially the gas-bearing electrodynamometer (GBE) is a rapid oscillator, of very low intrinsic mass, designed to measure the dynamic stress-strain (hysteresis loops) of easily deformed specimens. Using this apparatus, the elastic and viscoelastic moduli of in vivo stratum corneum can be measured. Christensen et al. have published a detailed study where the GBE was used to measure the effects of moisture on the viscoelastic properties of intact human skin [56]. Compared to normal skin, the hysteresis loops for dry ichthyotic skin were compressed and perpendicular, while those for moistened skin were both fatter and flatter. Figure 10.6, adapted from their work, shows these effects. These same workers showed that skin which had been treated with water vapor alone, returned to normal after several minutes, skin treated with water showed differences for up to 2 hours, and the effect on the hysteresis loop was still visible (31% below baseline) after 3 hours when the skin was treated with an oil-in-water emollient.

Cooper et al. have used the GBE in a detailed in vivo study of dry, normal, and glycerine-treated skin [57]. Figure 10.7 shows typical hysteresis loops as measured by the GBE in their work. In an examination of 28 subjects, these

Dry ichthyotic skin   Normal skin   Highly moist skin

Figure 10.6   GBE hysteresis loops from stratum corneum. Typical shapes from different skin conditions (from Ref. 56).

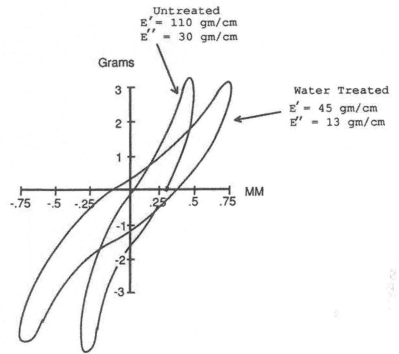

Figure 10.7 Elastic and loss moduli of skin measured by GBE. In vivo human skin (back of hand) at 20% RH. E' = elastic modulus; E" = loss modulus (from Ref. 57).

workers showed a correlation between the elastic modulus and the visually observed skin grading. Visual grading and the elastic modulus were recorded for skin that had been treated with 40% glycerine or just water. The results are shown in Figure 10.8. Five treatments are recorded over a five-day period, showing an impressive correlation between subjective and objective skin softening resulting from the use of 40% glycerine. It is interesting to note that these same workers showed, that as successive layers of the stratum corneum were stripped away the recorded elastic modulus goes down considerably, demonstrating that the measurements indeed originate in the outer layers of the skin.

Serban et al. have measured skin capacitance with respect to initial condition, as well as subsequent treatment with petrolatum and a water-in-oil lotion consisting of water, diisopropyl sebacate, glyceryl monostearate, cetyl alcohol, glycerine and mineral oil [58]. A measurement that they call the Water Diffusion Index (WD$_i$) relates to the increase in capacity (and thus water content) before and after petrolatum occlusion.

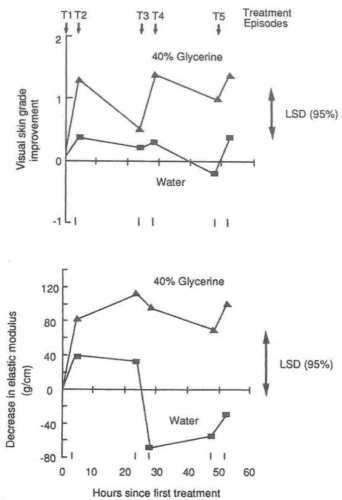

Figure 10.8 Comparison of glycerine effect on elastic modulus, and visual appearance of the skin. Visual skin grade: larger numbers mean less dry (more normal) skin. Elastic modulus: larger numbers mean less tight and more flexible skin (from Ref. 57).

$$WD_i = [C10 - C0]/C0$$

where CO is the capacitance before occlusion and C10 is the capacitance measured at 10 minutes after petrolatum occlusion.

Some of the WD_i scores they obtained are shown in the following Figures 10.9 and 10.10. It is to be noted that dry skin shows a greater increase in capacitance after treatment with petrolatum than normal skin, which reflects

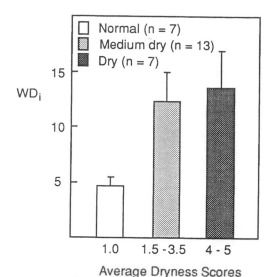

Figure 10.9 Relationship of water diffusion index to skin dryness. WD_i determined electrometrically in skin with dryness subjectively evaluated as: 1 = absent, 1.5 to 3.5 = slight to moderate, 4.5 = intense. (The bars represent standard error.) (From Ref. 58.)

its greater need for additional water. In contrast, in another experiment, dry skin which had been treated with the lotion did not show as much capacitance change on subsequent petrolatum occlusion as dry untreated skin, showing that the glycerine containing lotion had indeed predisposed the skin to contain more water. Furthermore, a comparison between the two figures, shows that treatment of dry skin with lotion makes its apparent water content (as measured by the extent to which petrolatum occlusion can increase the capacitance) similar to untreated normal skin.

As already mentioned, the effects of treatments has been gauged by studying the detailed surface topography of skin. Another example is where Cook and Craft have used skin profilometry to measure the effect of a high glycerine cream on skin condition [59]. At the microscopic level, skin consists of a series of peaks and valleys. In general, the rougher and dryer the skin, the more pronounced these peaks and valleys are. In this work roughness was quantified using a Surfometer SF 101 profilometer. Careful integration of the results yielded a roughness parameter for the number of peaks and the mean peak size. The legs of 10 subjects were studied for 3 weeks using twice-daily application. The high glycerine-containing cream in question had the following in-

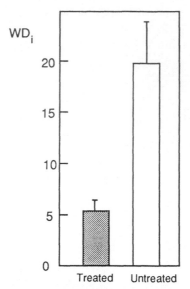

Figure 10.10 Effect of repeated skin lotion treatment on skin water diffusion index. Sites were washed with soap and water prior to measurement. (The bars represent standard error.) (From Ref. 58.)

ingredient: Water, emulsifying wax, glycerine, isopropyl myristate, squalene, beeswax, panthenol, imidazolidinyl urea, allantoin, methyl paraben, propyl paraben, olive oil, retinyl palmitate, fragrance, sodium borate.

Roughness parameters are shown in Table 10.4. The significant increase in the number of peaks, inspite of similar mean peak size, is seen as an overall smoothing of the skin.

D. In Vitro Testing of Skin Condition

Rieger and Deem give a good overview to the use of excised stratum corneum to study the effect of cosmetic ingredients on skin [60]. Although the work discussed above clearly demonstrates the utility of glycerine in increasing the flexibility and condition of the skin when judged in in vivo trials, Rieger and Deem showed that when the elastic modulus was measured in vitro, with complete immersion, the stiffness of the stratum corneum increased with increasing levels of glycerine in a glycerine-water combination. Figure 10.11 demonstrates this effect.

Table 10.4   Effect of a Glycerine-Containing Cream on Skin Roughness

| | Mean ± sd | | |
Roughness parameter	Control	Treated	P value
Number of peaks before treatment	48.52±3.80	48.88±5.33	>0.500
Number of peaks[a] after treatment	51.67±6.29	60.02±5.98	0.025
Mean peak size before treatment	20.22±2.94	20.28±2.99	>0.500
Mean peak size after treatment	20.78±3.49	20.69±3.26	>0.500

[a]This measurement shows a significant improvement when control skin is compared with the treated skin.
Source: From Ref. 59.

It was argued that the interaction between glycerine and water limits the water protein interaction thus making a stiffer protein matrix. An interesting aside was that temperatures below 5°C make the stratum corneum stiffer, but

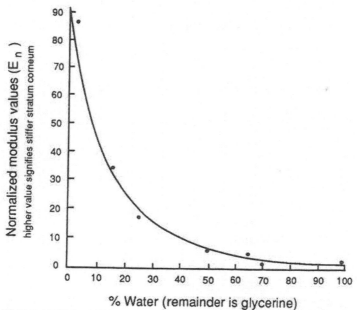

Figure 10.11   Stiffness of stratum corneum v. level of glycerine (from Ref. 60).

above that temperature the elastic modulus is relatively temperature independent. Similar results had been found by Middleton and Allen [43], and later by
Van Duzee [61]. Van Duzee stressed that the modulus is a function of the
water content, not water activity, suggesting no fundamental interactions in
the fibrous protein structure. It should be noted, however, that in this work
urea and lithium bromide were used as the humectants. Although not directly
related to glycerine, the conclusion of Van Duzee with respect to urea and
lithium bromide is interestingly compared with that of Takahshi et al. on a
study of hydroxy acids and in vitro skin [62]. The latter used an oscillating
rheometer rather than the Instron Tensile tester of Van Duzee. They found that
although the uptake of water by sodium lactate is greater at all the relative
humidities studied than it was for lactic acid, the actual skin plasticising effects were much greater with lactic acid. Further, with different hydroxy acids,
the alpha acids were much more effective in plasticizing skin than the beta
acids. They suggest that this is due to a direct interaction between the polar
keratin groups of the stratum corneum, and the alpha hydroxy acid moiety.
This reduces interchain interaction and results in a more pliable substrate.

I have seen no direct evidence to suggest such a direct interaction between
glycerine and keratin chains. However, some interesting studies have been
done on the interaction of aqueous glycerine with collagen which at least suggest some direct interactions with keratin as opposed to just a "water carrier"
role. Interstitial collagens are fibrous proteins consisting of three polypeptides
each of which is a left-handed helix intertwined into a right-handed triple helix. It has been shown that glycerine inhibits the in vitro self-association of
monomeric collagen into these fibrils [63]. Figure 10.12 graphically illustrates
this effect.

On the other hand, glycerine enhances the self-assemblies of globular proteins such as L-asparaginase, actin, and tubulin, and it actually stabilizes the
finished helical collagen against denaturation [64]. This effect is graphically
illustrated in Figure 10.13.

It should be added that both the kinetic and the equilibrium stability of the
collagen is enhanced. The cited works give a complex discussion of these phenomena, but it was concluded that the driving force for the stabilization of the
triple-helical structure of the collagen stems mainly from a favorable interaction with glycerine. It is not hard to imagine that this favorable interaction
could also occur with the monomeric collagen thus hindering the association
pathways. Commenting further upon this interaction, the author states: "Water molecules appear to play a major role in stabilizing the triple-helical structure by serving as bridges of interchain hydrogen bonds and forming chained
structures on the surface of the helix (Privalov 1982). The primary sites of
such water bridges are the carboxyl groups of the second amino acid of each
Gly-X-Y triplet, the amino acid groups and the hydroxy groups of hydroxypro-

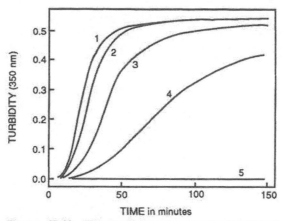

Figure 10.12 Effect of glycerine on the kinetics of collagen self-association: 1 is 0 M, 2 is 0.02 M, 3 is 0.1 M, 4 is 0.2 M., and 5 is 0.4 M glycerine, respectively. 30°C increased turbidity shows increased association (from Ref. 63).

line side chains which occur almost exclusive at the Y position of the triplet (Ramachandran and Ramakrishnan, 1976). It seems possible that glycerol with three hydroxy groups on each molecule can become bound to the collagen molecule through formation of multiple hydrogen bonds replacing the tightly bound water molecules." Exactly what relevance this has to glycerine-keratin interactions is not known. It does suggest, however, a mechanism that involves

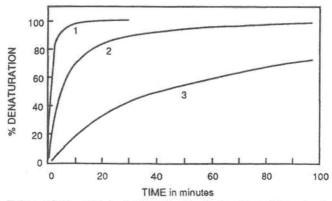

Figure 10.13 Effect of glycerine on the kinetic stability of collagen: 1 is 0 M, 2 is 1.5 M, and 3 is 3.0 M glycerine. Unfolding of triple helix measured at 40.5°C (from Ref. 64).

a direct structural interaction between glycerine and protein, and thus broadens the effect over that of only increasing water content. It further suggests that basing possible skin care benefits on humectancy alone (where many materials, such as inorganic salts and Na PCA, are stronger than glycerine) is very likely to be misleading in gauging actual benefits to skin.

Returning to the Rieger and Deem study [60], under immersion in vitro conditions, 50% glycerine dramatically increases the stiffness when the temperature is dropped to only around 17°C. This is certainly not surprising if one thinks in terms of a glycerine-water interaction. Whereas these results are interesting in increasing our overall understanding, they are not a good in vivo model.

These same workers obtained different results when glycerine and other materials were tested in a 31% RH environment (as opposed to complete immersion). Under these conditions, 4% glycerine has no effect and 50% only a marginal reduction in stiffness. Under these conditions, strong acid and base, as well as sodium pyrrlidone carboxylate, were highly effective agents to increase flexibility, sodium lactate also worked well, whereas oils slightly increased stiffness. The same workers studied stratum corneum and humectant moisture absorption as a function of relative humidity, and certainly glycerine is much more effective at higher humidities. Thus in thinking about in vivo studies, one must consider not only higher external RH conditions, but more importantly the interaction of treatments with the insensible water loss, and the effective high humidity from below the stratum corneum. These factors could dramatically change the in vitro results just described, when moving to an in vivo situation.

These workers also studied in vitro water vapor transmission through stratum corneum as a function of treatment. The results are shown in Table 10.5.

Table 10.5 Effect of Treatment on In Vitro Water Vapor Transmission Through Stratum Corneum

| Material | Rate ($mg\ cm^{-2}\ h^{-2}$) | | Ratio (Treated/Untreated) |
	Untreated	Treated	
25% Glycerine	0.264	0.518	1.96
4% Sodium lactate	0.294	0.372	1.27
4% NaPC[a]	0.142	0.227	1.60
4% Propylene glycol	0.223	0.267	1.20
Mineral oil	0.274	0.205	0.75
Safflower oil	0.309	0.281	0.91

[a]Sodium pyrrolidone carboxylate.
Source: From Ref. 60.

As can be seen, glycerine is very effective in increasing water through-put. Similar in vivo observations have been made. Acknowledging the beneficial in vivo effects of glycerine, these authors suggest that the observed increase in water vapor transmission must be related to the water gradient in living skin, and thus an effective water increase in the higher stratum corneum levels.

Recent workers have made similar observations by studying the flow of tritiated water through hamster ear skin in a diffusion cell [65]. The diffusion cell was held at 32°C, and contained a dessicant on the collection side. Glycerine increased water loss, whereas mineral oil and sesame oil decreased it. It should be mentioned that this author has seen in vivo water diffusion data where, although the increased rate with glycerine was easily observed, the decreased rate with mineral oil was not (but certainly in that work the occlusive effect of petrolatum was seen). Adrangui has measured the water binding capacity of isolated human stratum corneum at 37% to 80% relative humidities, using different humectants in solution and in a cream formula [66]. The use of all the humectants increased the water binding capacity in the order given: sodium pyrollidone carboxylate > sodium lactate > glycerine > propylene glycol ≥ ethylene oxide.

Using the same dynamic rheometer as in reference 62, Takahashi et al. measured the effect on modulus of a number of cosmetic agents added to excised stratum corneum [67]. These workers measured the dynamic elastic, and viscoelastic (loss) moduli of excised stratum corneum at extension 5 μm and cycle speed 30 Hz. This method has the advantage of measuring pre- and posttreatment effects in the same piece of skin. As to be expected, water strongly plasticized the skin, which gradually returned to normal after about 60 minutes at 25°C and 50% RH. Equally to be expected, oils had no effect, and surfactants actually caused a permanent increase in skin stiffness. Unlike water alone, which allowed the skin to return to its original stiffness after 60 minutes, 10% aqueous glycerine stayed about 25% below the baseline as shown in Figure 10.14. At 100%, glycerine had no apparent effect at 25°C and 50% RH, but, of course, this is an in vitro situation and has no underlying water as in living skin. These authors show an interesting graph relating plasticizing effect to water holding capacity of different humectants. (Fig. 10.15). In this in vitro test, it was at least concluded that the skin-softening effects of the low-molecular weight humectants, which themselves have no effect on stratum corneum, may be determined by their water-holding capacities. It must be emphasized, however, that glycerine, of average benefit here, is especially useful to the formulator of refined emulsion cosmetics. Further, as shown above, it has proved to have dramatic cosmetic benefits in the in vivo system, where the interactions are much more complex than in this work.

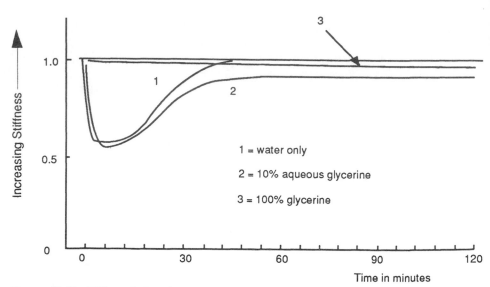

Figure 10.14 Effect of glycerine and water on stratum corneum stiffness (note: numbers on vertical axis are ratio of elastic modulus at time = t : time = 0) (from Ref. 67).

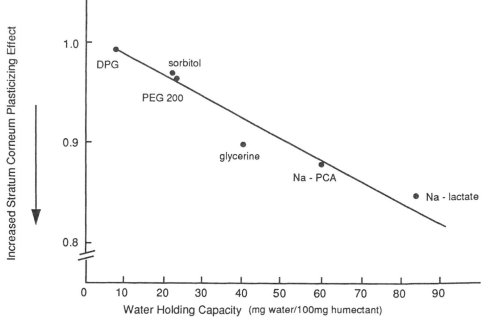

Figure 10.15 Effect of different humectants on stratum corneum plasticity (note: numbers on the verticle axis are ratio of elastic modulus at time ~ 100 min. : time = 0) (from Ref. 67).

V. GLYCERINE AS A PROTECTIVE AGENT AGAINST SKIN TOXINS

To protect against skin toxins, glycerine can be used in two ways. First as a nonsolvent barrier to the penetration of solvents and toxins, and second as a clean-up medium for more polar toxins.

The effects of acetone and kerosene on human epidermis were studied by Lupulescu and Birmingham using ultrastructural and scanning electron microscopy techniques [68]. In particular, the barrier effects of a gel consisting of water, 25% glycerine, 10% cellulose gum, and preservative were examined. Quoting from their paper ". . . light microscopy showed a reduction and disorganization of horny layers with intercellular edema following acetone exposure. Disruption of keratin layers and stratum spinosum were more evident after kerosene administration. None of these signs occurred following a concomitant exposure to the protective agent." Electron microscopy showed details of stratum corneum damage including enlarged intercellular spaces and disrupted desmosomes. Much less cell damage was seen with use of the glycerine gel and it was concluded that this protection results from blocking the absorption and migration of acetone and kerosene through the epidermis.

In related work, a mixture of glycerine, water, and talc was evaluated as a barrier against toluene [69]. In this work, excretion of toluene through the lungs after skin contact was measured. Figure 10.16, clearly shows how toluene uptake through the skin (as measured by breath analysis) is inhibited by the glycerine cream. A deep application where the cream is well massaged into the skin is much more effective than a superficial one. Another closely related mixture which has been shown to protect the skin against organic solvents is an aqueous mixture containing 25–45% glycerine along with 10–20% of nonionic emulsifiers, and minor amounts of cellulose gum [70].

Glycerine is also useful for formulating barrier creams against non-solvents. Thus Glantz et al. showed its effectiveness as part of an ointment against acrylic resins [71]. The experimental ointment in question had the following composition: monolaurin 7%, monomyristin 22%, water 64.8%, glycerine 5%, lecithin 1%, and sorbic acid 0.2%. This ointment had a semicrystalline nature and stayed on the skin for 30 h or more. Very good protection against the irritating effect of acrylic monomers on the skin was obtained. Other workers have shown its usefulness in a soap, borax, mineral oil emulsion as a protective agent against various "deleterious materials" [72]. At 4% it is found as an adjunct in a diaper rash treatment product based on silicone and mineral oils, as well as cetyl alcohol and propylene glycol [73].

In considering whether glycerine will make an effective barrier cream, any change in the irritant or barrier cream must cause its effectiveness to be reexamined. For instance, work has been done which suggests that a glycerine-based product is not effective against m-xylene [74]. Similarly, in an exhaustive study of 55 potential anti-irritants, glycerine, preapplied in a soap

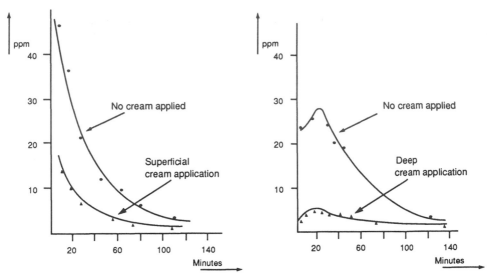

Figure 10.16 Inhibition of toluene uptake through skin by a glycerine cream. Two charts show typical curves. Variation in top control curves is a subject to subject variability. Curves show toluene excretion (ppm) in alveolar air after skin absorption. Deep cream application, i.e. rubbing cream in, typically gives better protection. (From Ref. 69).

emulsion, was not effective in reducing the cutaneous irritation of croton oil [75]. Materials which showed effectiveness in this test were silicone oil, polyacrylamide, collagen, and gelatin. Two reasons why glycerine may not have been effective in this series of tests is that only one irritant, croton oil, was used. Also rabbit was used as the model, which could well give different results from human models. Obviously, in evaluating barrier creams, the individual effectiveness of any combination of formula and solvent needs to be examined, and specific details of the test employed must also be considered.

The work so far in this section uses glycerine in a preapplied formula. Two additional reports described its usefulness when used concurrently with the irritant, as well as when used for postexposure decontamination. In the first case, 0.2% sodium dodecyl sulfate was tested for irritation on 54 women with and without the addition of 5% glycerine [76]. In the cases where glycerine was used, the cases of irritation dropped from 19% to 7% of the women tested. In the second case, glycerine has been shown to be useful in decontaminating skin exposed to phenolic substances [77]. Rats were given a percutaneous exposure to undiluted phenol and swabbed with water, or methylated spirits, or glycerine, or polyethyleneglycol. The latter two materials were much more effective at reducing mortality and convulsions than either water or methylated spirits. This result is in keeping with the use of glycerine as an essential ingredient of phenol eardrops, as it is well known that it reduces the irritation of phenol to the skin.

VI. THE ROLE OF GLYCERINE IN THE PENETRATION OF MATERIALS THROUGH THE SKIN

This section is included to stimulate some thinking about the effect that glycerine and other cosmetic ingredients might play in the transport of solutes through the skin. Different authors deal with limited aspects of the question. Individual results will depend upon solvent/solute interaction, as well as interactions of both with the membrane, and the composition of the receptor fluid. Given that skin is a complex biological substrate, and experimentation being done in this area covers a wide range of in vivo, and in vitro work, as well as the use of different animal models, this review will only serve to present some ideas and trends. A major factor to consider is in the comparison of diffusion cell data with in vivo work. In the former, the water concentration can stay constant as a drug dissolving material, but in the latter the applied solvent evaporates and the drug medium can change rapidly.

A review of two important principles involved is to be found in a study of the effect of glycerine on the percutaneous absorption of methyl nicotinate [78]. Increasing concentrations of glycerine delay the onset of erythema produced by methyl nicotinate. This phenomenon was related to partition coefficients and diffusivity. Taking the partition coefficient idea in its simplest terms, if a drug is applied in an aqueous vehicle, anything which affects the relative solubility of the drug in water versus solubility in the skin and/or skin lipids, will affect skin penetration. In this case the partitioning of glycerine between water/glycerine and isopropyl myristate (IPM) was studied. As glycerine was increased the amount of methyl nicotinate in the IPM decreased. This was related to decreasing methyl nicotinate availability for the skin, and thus a slower onset of erythema. Partitioning is related to available drug concentration.

A second factor to consider is the actual rate of transport, or diffusivity, of a drug both within the solvent and across the membrane. In this work the diffusivity of methyl nicotinate in the solvent (studied across a porcelain barrier) was 270 times faster in pure water than pure glycerine. Figure 10.17 shows these relationships and their effect on erythema, which itself, of course, is directly related to the rate of transport across the skin. It should be noted that the erythema experiments were done by applying test substance-soaked filter paper to the skin.

Other workers have considered partitioning, and the effect of the solvent/solute on the membrane itself. This was done with phenol as the solute, and

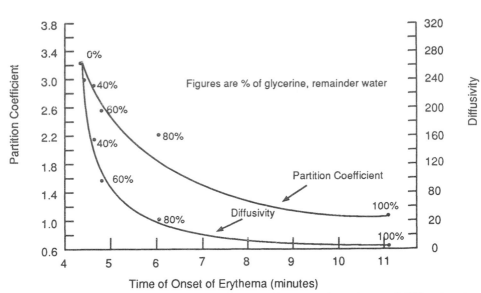

Figure 10.17   Relation of partition coefficient, diffusion, and erythema of different water-glycerine solutions (from Ref. 78).

using an in vitro rat skin diffusivity chamber to measure permeation [79]. Idson [80] has expressed the penetration flux (J_m) of a solute through a membrane by:

$$J_m = Q/A \cdot t = K_m \cdot D \cdot C_v/h$$

where Q is the amount of solute that diffuses across area A in time t; K_m is the membrane : vehicle partition coefficient of the solute; D is the diffusion coefficient of the solute in the membrane of thickness h, and C_v is the concentration of solute in the vehicle.

Scheuplein [81] gives the permeability coefficient (k_p) as

$$k_p = J_m/C_v$$

Figure 10.18 shows the inverse relationship between penetration flux of phenol through excised rat skin and the level of glycerine in the aqueous vehicle. This phenomenon was related to the decreased level of phenol in liquid paraffin when partitioned with glycerine as opposed to water. In addition to this expected partition coefficient result, these authors discuss the effect of the test medium on the membrane itself. Thus, hydration has been shown to increase diffusion through stratum corneum by a factor of 10, and to a certain extent the inhibiting effect of glycerine on penetration could be caused by a partial reduction in the hydration of the stratum corneum caused by the glycerine itself. Another point of interest, mentioned by the author, is that at sufficiently

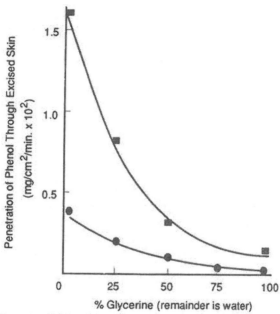

Figure 10.18   Effect of glycerine on phenol transport through skin. Top curve is 5%, bottom curve 2% phenol. (from Ref. 79).

high concentration, the phenol itself could damage the skin and thus auto-promote its penetration flux.

As pointed out by Bronaugh and Stewart, when considering skin penetration, thought must also be given to the solubility of the penetrant in the receptor fluid [82]. This may not be important in in vivo experiments when biological fluids can remove the penetrant, but certainly it can have a great effect in diffusion cell experiments. Among a great deal of other data given, it was shown that when 50:50 glycerine:water was used as a receptor fluid for 1-(3-ethyl-5,6,7,8-tetrahydro-5,5,8,8-tetramethyl-2-naphthalenyl)ethanone, permeation through a rat skin diffusion cell was much less than when a 6% PEG 20 oleyl ether receptor solution was used. This relates to the relative solubility of 1–(3–ethyl–5,6,7,8–tetrahydro–5,5,8,8–tetramethyl–2–naphthalenyl)ethanone in the two receptor media.

Yet another factor in the consideration of drug penetration has been emphasized in a series of papers by Mollgaard and Hoelgaard [83, 84, 85]. As these authors point out, in addition to considering the solubility interactions of drugs and solvents there are occasions where the solvent affects the barrier function of the stratum corneum by its own transport, (or, as suggested above, indirectly by changing the state of hydration of the skin). In this context, in contrast to polyethylene glycol and glycerine, ethylene glycol and propylene glycol show pronounced enhancing properties. Furthermore, in keeping with the hypothesis that the transport of the solvent is important, propylene glycol has good skin penetrating properties but glycerine does not. Figure 10.19 [55] shows the cutaneous permeation rate of estradiol and metronidazole using excised human skin in a diffusion cell.

Not only is the rate of penetration considerably slower with glycerine than with propylene glycol, the 1:1 mixtures are about as slow as with 100% glycerine itself. The large effect that glycerine has in reducing drug transport, seems to be related to the effect that glycerine has in reducing the permeation of propylene glycol itself. Thus, Figure 10.20 [84] shows the permeation rate of estradiol and metronidazole as well as propylene glycol as glycerine is added.

The authors surmise that propylene glycol has the ability to reduce the diffusional resistance of the skin barrier, possibly because the drug is kept dissolved in the glycol as the glycol diffuses through the skin. Certainly, adding the glycol to the skin 48 hours prior to the drug penetration study did not enhance diffusion. The use of propylene glycol to enhance drug penetration is well known in the patent literature. One can speculate why glycerine would so thoroughly inhibit the diffusion of the propylene glycol. Certainly solubility considerations would make the skin-propylene glycol interactions thermodynamically less favorable, but it is possibly a lot more complicated than that.

In an interesting mechanistic tangent, one patent claims a mixture of propylene glycol and a "cell envelope-disordering compound" such as methyl myristate, as a penetration enhancer for a variety of pharmacological agents [86]. They point out that the stratum corneum is an extremely thin surface

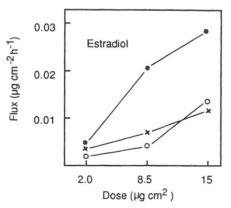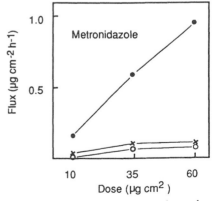

Figure 10.19 Drug penetration through excised human skin in glycerine and propylene glycol. • = Vehicle is propylene glycol; x = Vehicle is 1 : 1 propylene glycol to glycerine; ° = Vehicle is glycerine. (From Ref. 83.)

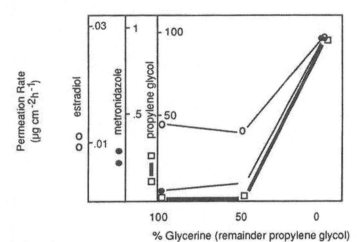

Figure 10.20 Inhibition of propylene glycol and drug transport by glycerine (from Ref. 84).

layer of about 20 microns, where the cell envelopes tend to be mainly polar lipids such as ceramides, sterols, and fatty acids. Despite close packing of the cells, this intercellular, lipid-based layer represents about 15% of the volume. The cytoplasm of the statum corneum cells remains polar and aqueous.

In diffusion measurements, every solvent mixture-membrane system must be studied in its own right. Thus, although an equal amount of propylene glycol added to glycerine did not enhance the penetration of estradiol and metronidazole, an approximately equal quantity of menthol added to glycerine caused a much higher permeation of propanol-HCl through mouse skin than with glycerine alone [87]. Further, and even more directly to the point of uniqueness of every individual system, other workers have shown that propylene glycol : glycerine (1:1-5) makes a good penetrating solvent for guanabenz through postmortem skin in the Franz diffusion cell [88].

In a sense, all these studies indicate that, by itself, glycerine is a poor drug permeation agent. A similar conclusion can be drawn from its effects on tobacco condensate carcinogenicity [89]. Tobacco condensates in acetone containing 0%, 17.5%, and 35% of glycerine were examined for topical carcinogenicity and the ability to produce epithelial hyperplasia in mice. Increasing levels considerably reduced the incidence of tumors and hyperplasia. Among the possible mechanisms advanced, were the increased solubility of the carcinogenic components in the glycerine, and an inhibitory effect exerted at the cellular level by somehow impeding either cell absorption or intracellular transport.

As an aside, Hoelgaard and Mollgaard studied the additional effects of azone (1-dodecylazacycloheptane-2-one) as a drug penetrant enhancer in these systems [85]. Azone seems to act on the skin membrane itself with dramatic results. Thus with 1% azone, 25 times as much metronidazole and propylene glycol were delivered through the skin after a 20 hour period than without.

The skin itself is a very important factor. In the above work, human skin was used and the amount of diffused glycerine was measured using gas chromatography. It was insignificant compared with propylene glycol. The permeability of glycerine (as well as urea, thiourea, and glucose) has also been measured using mouse skin with radioactive techniques in a diffusion cell method [90]. The permeability of glycerine through full thickness mouse skin was 9.5×10^{-5} cm/h. For the dermis alone, however, it jumped all the way up to 0.29 cm/h. This, of course, ties in with the well-known fact that the stratum corneum is, in fact, the diffusion barrier in the skin. In this work, permeability did not correlate with ether-water partitioning.

Not particularly pertinent, but just to show how interesting and complex this world of ours is, some workers have shown that for three-carbon atom glycols (specifically, propylene glycol and glycerine) there is an active transport from the inside to the outside of frog skin pointing to the physiological role of frog skin as an excretory organ. This asymmetric flux does not occur with ethylene glycol [91].

Finally some further insight into the relationship between permeability and partition coefficient was found by studying the permeabilities of glycerine and ethylene glycol through phospholipid- and cholesterol-based bilayers. In cholesterol bilayers the ratio of permeability is similar to the ratio of partition coefficients, as would be expected. However, as indicated in Table 10.6, in phospholipids, glycerine penetrates more than the ratio of partition coefficients would suggest [92].

The authors suggest this discrepancy may be related to the orienting effect of the bilayer itself versus the normal bulk phases. Thus the membranes con-

Table 10.6 Glycerine and Ethylene Glycol Permeability Coefficients (P) in Phospholipid and Cholesterol Bilayer Membranes

Membrane	P Glycerine (cm s^{-1} $\times$ 10^{-6})	P Ethylene Glycol (cm s^{-1} $\times$ 10^{-6})	P Ethylene Glycol/ P Glycerine
Lecithins in n-decane	5.70±1.19	18.30±5.32	3.16
Oxidized cholesterol in n-octane-dodecane	1.57±0.31	14.42±3.57	8.47

Note: Approximate partition coefficient of ethylene glycol : glycerine in hydrocarbon is 7 : 1.
Source: From Ref. 92.

taining the resultant oriented polar groups of the lecithins are more attractive for the hydrophilic molecules than are the corresponding fatty bulk phases.

VII. CRYOPRESERVATION OF SKIN AND TISSUE WITH GLYCERINE

Since 1949, glycerine has been widely known as a protective substance for the freezing and thawing of mammalian cells. Together with dimethyl sulfoxide, it is a key ingredient in frozen cell banking and cryobiology research [93]. A tabular summary of much of the work done in this field, along with summaries of optimum procedures and proposed further research are found in Aggarwal et al. [94]. If viable tissue is frozen quickly the formation of intracellular ice will kill it. If frozen too slowly, on the other hand, the formation of extracellular ice is considered more significant, as the free electrolyte concentration is concurrently elevated and osmotic distortion and death of the cell results. Properly used, cryoprotective agents act both as an antifreeze and to reduce osmotic stresses. Much of the technology in this area, therefore centers around the proper concentrations and use of the glycerine (or dimethyl sulfoxide, DMSO), as well as optimizing the cooling and heating sequences. It might be noted that inspite of the apparent irony of having a noted skin penetrant (DMSO), and a noted nonpenetrant (glycerine) as the two useful agents, cryopreservation does not rely on penetration through the stratum corneum, but through nonbarrier tissues such as the dermis. Nevertheless, as has been pointed out, given that glycerine needs to enter the cell, sufficient pretreatment time must be allowed, as it does penetrate slowly [95].

Optimum conditions noted by Aggarwal et al. include pretreating the tissue with 20-30% glycerine at 4°C for 2 h and the combined use of controlled cooling at 1–5°C/min in the presence of 15% glycerine (or 10% DMSO), with a subsequent rapid warming of 50–70°C/min.

As found in Aggarwal et al., Rheinwald [96] has shown that human keratinocyte cultures and disaggregated single cells can be frozen to minus 196°C in 10% glycerine and reactivated. Whether this has any relationship to the use of glycerine as a skin care agent in cold weather is hard to say, but it does represent an interesting result to the cosmetic chemist. It is noteworthy that skin tissue can be preserved for two to four weeks at noncryogenic temperatures, that is 0 to 4°C, by immersion in 15% glycerine in Ringer's solution [97]. The latter is an electrolyte isotonic with cell tissue.

Different workers have attempted to obtain a deeper understanding of these phenomena by investigation of structural and metabolic changes. In an ultrastructural investigation of frozen human skin, damage was exhibited by increased intercellular spaces, clumping of the nuclear chromatin, and cytoplasmic damage like swollen mitochondria, vacuole formation, and other effects [98]. Rate of freezing, pretreatment, concentration of the cryoprotectant, etc., all have an effect on the damage produced. Glycerine is shown to have a better cryogenic protective effect than DMSO, at concentrations of 15% or less. Another way to study damage is to examine skin metabolic activity after the freeze-thaw cycle. This was done for rat skin following preservation and storage in a glycerine buffer at -196°C [99]. Although glycerine dissolved in buffer medium does protect the skin, it was shown that after storage, the incorporation of 2-[^{14}C]glycine into the proteins, of 6-[^{3}H]thymidine into DNA, and alpha 1-[^{14}C]aminoisobutyric acid transport through the cell membrane are all slowed compared with freshly incubated skin. Two important points noted by these workers was that to preserve the skin tissue, the cyroprotectant must be used in an electrolyte medium. Also, to preserve metabolic activity (which is not an issue in considering the strictly cosmetic use of glycerine on stratum corneum), the skin pre-exposure time needs to be restricted.

It should not be assumed that the glycerine cyropreservation of different cells is equal. This was illustrated in a study of its effects in the preservation of epithelial versus fibroblast cells. Without a cryopreservative, the former are more susceptible to freeze damage, but with 20% or so glycerine, the latter are more susceptible to damage [100]. As an interesting aside, in some experiments these authors pretreated the cells with hyaluronidase to enhance cell penetration.

Whether there is any mechanistic relationship or not with the preservation of epithelial cells versus fibroblasts, other workers have shown that, at the molecular level, glycerine (and other glycols) stabilizes soluble collagen against thermal denaturation [101]. Figure 10.21 shows how the temperature at which soluble collagen is thermally denatured (T_m) steadily increases with increasing concentrations of glycerine as well as to a lesser extent with propylene and ethylene glycols.

Examination of the data shows that the interactions involved are subtle. Thus propane 1,2-diol lowers the denaturation temperature, whereas glycerine and propane 1,3-diol increases it. More recent work, done by measuring the optical rotation of triple helix collagen solutions, has confirmed these findings [102]. Here, 50% glycerine increased the temperature stability by about 10°C, whereas neither sorbitol or propane 1,2-diol had any appreciable effect. The interaction of glycerine and collagen is more fully discussed in Section IV. D.

Presumably glycerine will not affect that part of the denaturant process that is due to increased thermal disordering, but will affect that due to the decreased intramolecular bonding of water, and thus maintain, at increasing temperatures, a relatively increased attraction between the collagen and the solvent water molecules (and/or as discussed in Section IV.D between collagen and glycerine itself as a partial substitute for structural water). In the most precise mechanistic sense, this stabilization of collagen by glycerine with in-

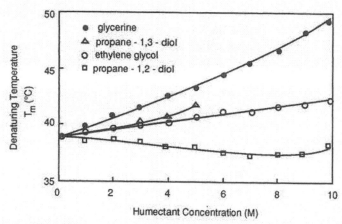

Figure 10.21 Humectant effect on acid-soluble collagen denaturization. Experiments done in 0.15 M potassium acetate buffer, pH 4.7. (From Ref. 101.)

creasing temperature, may have no direct relationship to either its cryogenic properties or its cosmetic benefits. In a general sense, however, one can certainly correlate the positive benefits of glycerine with its effects on the water structure of complex water/protein interactions in general.

A vivid example of the beneficial replacement of water by glycerine is seen in its use to dehydrate and preserve pig skin at normal temperatures [103]. Quoting from Basile, "In fact, glycerine dehydrates tissue by replacing most of the intracellular water, without, however, altering the ionic concentration of the cells, and for this reason its acts as an efficient protector of cell integrity. In addition, at concentrations of more than 48%, glycerine acts as a powerful antiseptic." Thus using 96% glycerine, Basile was able to dehydrate and sterilize skin that had not lost its cellular characteristics, and that on rehydration, regained its original flexibility. With the lower levels of glycerine used at cryogenic temperatures its preservative action is not an issue. Certainly, however, the room temperature preservation of skin acts as a powerful reminder of the various benefits that it can impart and the interrelationships of the medical work, taken as a whole, with the use of glycerine as a cosmetic.

VIII. SKIN LIPIDS: GLYCERINE AS A NATURAL METABOLITE

As fats are broken down, or formed in the body, glycerine can be either taken up or released. Thus, at very small levels, it is already a constituent of the epidermal environment. This section will merely document this fact because of its relevance to the subject of glycerine and the skin, and is not intended to be a treatise on this large and important subject.

Two very distinct types of lipids must be considered in dealing with the subject of lipids and the skin. These are sebum, as secreted by the sebaceous glands, and the epidermal lipids themselves. A good review on this subject has been written by Downing, Wertz, and Stewart [104]. The major constituents of human sebum consist of triglycerides (60%), wax esters (25%), squalene (12%), and cholesterol esters (2%) [105]. Epidermal lipids are quite different, consisting principally of ceramides (50%], cholesterol (20%), and free fatty acids (25%) [106]. With the composition of human sebum in mind, as well as its well-known increase in secretion at puberty, it is not surprising that skin surface glycerine levels have been studied in relationship to acne vulgaris [107]. In this study, both glycerine and free fatty acid levels were measured, as both should be derived equally from fat breakdown. Three types of patients were evaluated, those with untreated acne vulgaris, those treated with oral tetracycline, and control patients without the disease. Figure 10.22 shows the measured and calculated glycerine levels taken from these three groups.

The fact that the free fatty acid levels were, in fact, as predicted, was taken to mean that although extra glycerine is formed in acne patients as a result of breakdown of the excess fatty sebum secretion, the glycerine itself may be a substrate for *Propionibacterium acnes*.

The above report is fairly conclusive that glycerine is a natural, albeit extremely minor, component of skin due to the breakdown of sebum fats. In fact, corroboration has been obtained in the study of water extracts of normal skin where low levels of glycerine, presumably originating from fat breakdown, were observed [108].

In contrast, in studies of rat skin exaggeratedly washed with soap [109], radioactive glycerine (as well as other substrates) was shown to be incorporated into incubated skin sections. In this study, when skin was irritated with exaggerated washing, DNA synthesis was stimulated and this was followed by increased incorporation of both labeled glycerine and acetate into glycerol lipids (phospholipids and triglycerides). In fact, soap irritation caused a large increase in epidermal triglyceride production, as opposed to the prior irritation condition where most of the radioactivity is taken up in the phospholipid species. The germinative cells (stratum basale) of the epidermis were the locus of the observed effect. In fact, no changes in lipid production caused by soap irritation were observed in the sebaceous glands of the dermis, or the dermis itself. As might be expected, under normal circumstances in the epidermis, phospholipids and sterols are the major radioactive lipids synthesized, whereas in the sebaceous glands labelled triglycerides, sterol esters, and wax esters predominate [110].

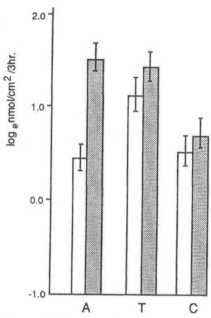

Figure 10.22 Naturally occurring glycerine levels in acne patients. A = acne
vulgaris patients; T = treated acne vulgaris patients; C = control subjects.
(From Ref. 107.)

In another study, this time of rat skin in essential fatty acid deficiency
[111], marked increases in the uptake of glycerine and acetate into all lipid
classes was observed particularly being predominant in phosphatidylcholine.
This report includes an interesting discussion on the role of linoleic acid on
skin barrier function, which is beyond the scope of this work. An interesting
stated hypothesis, however, is that phosphatidylcholine containing linoleic acid
is the key lipid for barrier function, and alternate species are synthesized in
the absence of metabolically obtainable linoleic aid.

The observant reader will have noticed that in this section, dealing with
glycerine as a metabolite in the dermis/epidermis, phospholipids were cited as
a predominant species in the basal layer, whereas ceramides (or sphingolipids)
have been cited as the major epidermal lipid species. The reason for this is
that the lipid type in the human epidermis changes as one goes from the basal
layer to the statum corneum. Lampe et al. [112] have a good graphical representation of these changes (Fig. 10.23).

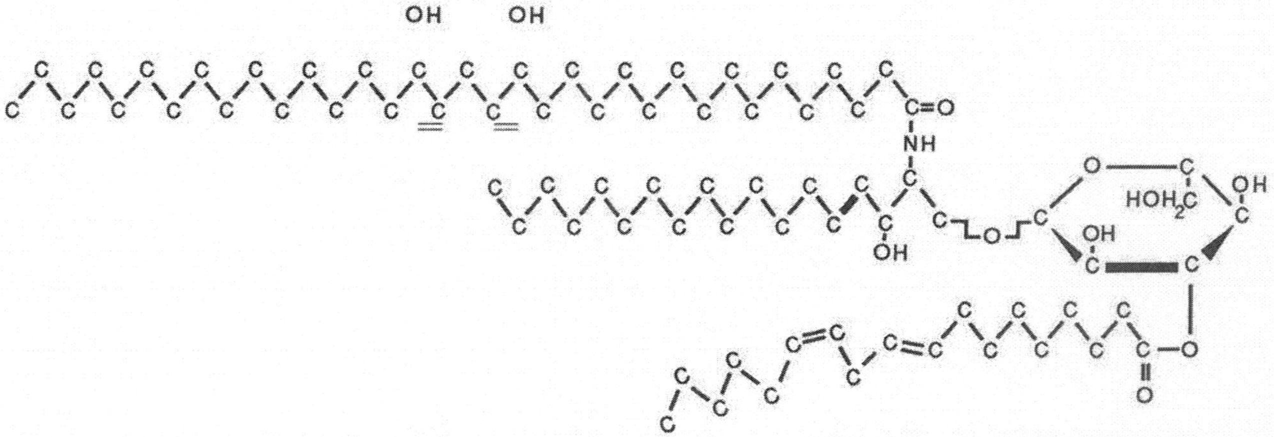

Figure 10.24 Problemin structure: possible natural epidermal water barrier. 1-(3′-*O*-linoleyl)-β-glucosyl-*N*-dihydroxypenta-triacontadienoyl sphingosine. (From Ref. 114.)

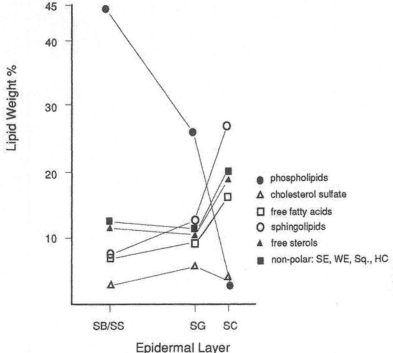

Figure 10.23 Variation of lipid species in human epidermis. SB = stratum basale; SS = stratum spinosum; SG = stratum grandulosum; SC = stratum corneum. (From Ref. 112.)

For the interested reader, two detailed reviews on the subject of epidermal lipids are written by Elias [113] and Yardley [114], where of special interest is the identification by Yardley of ceramide "problemin" as having key water barrier functions (see Fig. 10.24). This is to be compared with the conclusions of Takahashi et al. [111] just mentioned.

REFERENCES

1. Blank, I. H. Factors which influence the water content of stratum corneum, J. Invest. Dermatol. 18: 433–440, 1952.
2. Henkel KGaA, Oleochemicals Division, Triacetin, Glycerine, Diacetin, TEGA, 1988.
3. Dow Chemical U.S.A., Synthetic Glycerine, Form No. 115-601-R87, 1987.
4. Crow, L. and Kerrigan, R., Backgrounder—high quality synthetic glycerine from the Dow Chemical Company finds wide use, Dow Chemical U.S.A., 1987.

5. U.S. Soap & Detergent Association, 1986 totals, as quoted in ref. 4.
6. Moxey, P., The role of glycerine in cosmetic and toiletry products, Mfg. Chem. Aerosol News, November: 27–30, 1967.
7. Glycerine Producers Association, Uses of glycerine, 8–9, received from The Soap and Detergent Association, New York.
8. Newman, A. A., with additional chapters by L. V. Cocks, Glycerol, C.R.C. Press, Cleveland, 1968, pp. 176–179.
9. Vaughan, C. D., Using solubility parameters in cosmetics formulation, J. Soc. Cosmet. Chem., 36: 319–335, 1985.
10. German Patent 3,445,749, Stable gel-like shaving cream, 1985.
11. British Patent 1,416,296, Blass, J. M., Styptic preparations for use on human skin, 1975.
12. U.S. Patent 3,034,966, Williams, E. W., Nail hardening composition and method of making same, 1962.
13. Japanese Patent 8627917, Kao Corp, Aqueous eye lotion containing indomethacin and glycerols, 1986.
14. Japanese Patent 85115509, Sunstar Inc., Transparent cleansing gel compositions for the skin, 1985.
15. Japanese Patent 85116622, Pentel Co., Ltd., Pen-type liquid cosmetics, 1985.
16. Courtney, D. L., Polyols in creams and lotions, Cosmet. Toiletries, 95: 27–34, 1980.
17. Griffin, W. C., Behrens, R. W., and Cross, S. T., Hygroscopic agents and their use in cosmetics, J. Soc. Cosmet. Chem., 3: 1–15, 1952.
18. Strianse, S. J., Comparative value of polyols in cosmetic emulsions, Am. Perf. Cosmet., 77: 10, 31–39, 1962.
19. Segur, J. B., and Miner, C. S., Jr., Hygroscopic and viscosity effects of glycerine in cosmetics, Proceedings of the Scientific Section of the Toilet Goods Assn., No. 19, May 1953.
20. Henny, G. C., Evanson, R. V., and Sperandio, G. J., An evaluation of humectants in cosmetic emulsions, J. Soc. Cosmet. Chem., 9: 329–335, 1958.
21. Lo, Y-C., Froning, G. W., and Arnold, R. G., The water activity lowering properties of selected humectants in eggs, Poultry Sci. 62: 971–976, 1983.
22. Schwartzkopff, U., Ohl, K., Heilmann, T., and Brecht, G., Nondrying cleaning cloth, German Patent 3,447,499, 1986.
23. CRC Handbook of Chemistry and Physics, 61st ed., CRC Press Inc, Boca Raton, FL, 1980–1981, D-239, p. 240.
24. Zanker, A., Nomograph to maintain constant relative humidity, Chem. Ind. 5: 568–569, 1975.
25. Adapted from Refs. 16 and 23.
26. Sloan, A. E., and Labuza, T. P., Humectant water sorption isotherms, Food Prod. Dev. 9(10): 68, 1975.
27. Sakamoto, K. and Suzuki, E., A new method to evaluate the hygroscopicity of humectants by broad-line pulsed NMR, Fureguransu Janaru, 10(5): 59–63, 1982.
28. Kang, Y. K., Gibson, K. D., Nemethy, G., and Scheraga, H. A., Free energies of hydration of solute molecules. 4. Revised treatment of the hydration shell model, J. Phys. Chem. 92(16): 4739–4742, 1988.

29. Fisher, A., Reactions to popular cosmetic humectants. Part III. Glycerine, propylene glycol, and butylene glycol, Cutis, 26: 243–244, 269, 1980.
30. Hannuksela, M., Allergic and toxic reactions caused by cream bases in dermatological patients, Int. J. Cosmet. Sci. 1: 257, 1979.
31. Guillot, J. P., Martini, M. C., Giauffret, J. Y., Gonnet, J. F, and Guyot, J. Y., Safety evaluation of some humectants and moisturizers used in cosmetic formulations, Int. Cosmet. Sci. 4: 67–79, 1982.
32. U.S. Patent 4,605,555, Sato, M., Katsuragi, Y., Sakano, Y., Sugihara, K., and Aimoto, K., Composition and method for treating keratosic disorder of skin and mucosa, 1986.
33. Harrison, H. N., Shuck, J. M., and Caldwell, E., Studies of the pain produced by mafenide acetate preparations in burns, Arch. Surg., 110: 1446–1449, 1975.
34. Waltz, D., Irritant action due to physico-chemical parameters of test solutions, Fd. Chem. Toxicol., 23(2): 299–302, 1985.
35. Radford, E. and Berkowitz, E. C., The use of intradermal tests and relevance of negative control in patients with negative or equivacal modified rast test scores to inhalant allergens, Laryngoscope, 97: 675–677, 1987.
36. Menardo, J. L., Bousquet, J., Bataille, A., Restagny, G., and Michel, F-B., Effects of diluents on skin tests, Ann. Allergy, 51: 535–538, 1983.
37. Imber, W. E., Allergic skin testing: a clinical investigation, J. Allergy Clin. Immunol. 60(1): 47–55, 1977.
38. Motoyoshi, K., Nozawa, S., Yoshimura, M., and Matsuda, K., The safety of propylene glycol and other humectants, Cosmet. Toiletries, 99: 83–91, 1984.
39. deNavarre, M. G., The Chemistry and Manufacture of Cosmetics, Vol. II, U.C.C. Information, 1962, p. 169.
40. Moskowitz, H. R., Cosmetic Product Testing, Science and Technology Series, Vol. 3, (E. Jungermann, ed.), Marcel Dekker, Inc., New York and Basel, 1984.
41. Bissett, D.L., and McBride, J. F., Skin conditioning with glycerol, J. Soc. Cosmet. Chem., 35: 345–350, 1984.
42. Batt, M. D., Davis, W. B., Fairhurst, E., Gerrard, W. A., and Ridge, B. D., Changes in the physical properties of the stratum corneum following treatment with glycerol, J. Soc. Cosmet. Chem. 39: 367–381, 1988.
43. Middleton, J. D., and Allen, B. M., The influence of temperature and humidity on stratum corneum and its relation to skin chapping, J. Soc. Cosmet. Chem., 24: 239–234, 1973.
44. Middleton, J. D., Development of a skin cream designed to reduce dry and flaky skin, J. Soc. Cosmet. Chem., 25: 519–534, 1974.
45. Ito, H., Hasegawa, A., and Yoshihara, T., Hydrophilization of porous hydrophobic membranes, Japanese Patent 8719208, 1987.
46. Japanese Patent 8779240, Osawa, T., Ideshita, R., Matsuno, D., and Hyodo, S., Cosmetic poly(vinyl alcohol) foams, 1987.
47. Ozawa, T., Nisiyama, S., Horii, I., Kawasaki, K., Kumano, Y., and Nakayama, Y., Humectants, and their effects on the moisturization of skin, Hifu, 27(2) 276–288, 1985.
48. Johnson, Z., Treating psoriasis with cosmetic compositions, U.S. patent 4,454,118, 1984.

49. Nishijama, S., Komatsu, H., and Tanaka, M., A study on skin hydration with cream. Influence of its components on skin hydration, J. Soc. Cosmet. Chem. Japan, 16: 136–143, 1983.

50. Nishijama, S., Komatsu, H., and Tanaka, M., A study on skin hydration with w/o type cream (II), J. Soc. Cosmet. Chem. Japan, 17: 116–120, 1983.

51. Abe, T., Biopharmaceutical studies on the effect of some topical vehicles on human skin, Chem. Pharm. Bull., 27(2): 386–391, 1979.

52. Rieger, M. M., Skin irritation: the continuing enigma, Cosmet. Toiletries, 99: 63–65, 1984.

53. Wang, P. Y-C., Wound coverings, U.K. Patent Application 2,099,704A, 1982.

54. Leveque, J-L., Physical methods to measure the efficiency of cosmetics in humans, Cosmet. Toiletries, 99: 43–52, 1984.

55. Hargens, C. W., The gas-bearing electrodynamometer (GBE) applied to measuring mechanical changes in skin and other tissues, in Bioengineering and the Skin (R. Marks and P. A. Payne, ed.) Proceedings of the European Society for Dermatological Research Symposium, Cardiff, 1979, MTP Press Ltd., Boston, pp. 113–122.

56. Christensen, M. S., Hargens, C. W., Nacht, S., and Gans, E. H., Viscoelastic properties of intact human skin: Instrumentation, hydration effects, and the contribution of the stratum corneum, J. Invest. Dermatol. 69: 282–286, 1977.

57. Cooper, E. R., Missel, P. J., Hannon, D. P., and Albright, G. B., Mechanical properties of dry, normal and glycerol-treated skin as measured by the gas-bearing electrodynamometer, J. Soc. Cosmet. Chem., 36: 335–348, 1985.

58. Serban, G. P., Henry, S. M., Cotty, V. F., Cohen, G. L., and Riveley, J. A., Electrometric technique for the in vivo assessment of skin dryness and the effect of chronic treatment with a lotion on the water barrier function of dry skin, J. Soc. Cosmet. Chem., 34: 383–394, 1983.

59. Cook, T. H., and Craft, T. J., Topographics of dry skin, non-dry skin, and cosmetically treated skin as quantified by skin profilometry, J. Soc. Cosmet. Chem., 36: 143–152, 1985.

60. Rieger, M.M., Deem, D. E., Skin Moisturizers. II. The effect of cosmetic ingredients on human stratum corneum, J. Soc. Cosmet. Chem., 25: 253–262, 1974.

61. Van Duzee, B. F., The influence of water content, chemical treatment and temperature on the rheological properties of stratum corneum, J. Invest. Dermatol., 71: 140–144, 1978.

62. Takahashi, M., Machida, Y., and Tsuda, Y., The influence of hydroxy acids on the rheological properties of stratum corneum, J. Soc. Cosmet. Chem., 36: 177–187, 1985.

63. Na, G. C., Butz, L. J., Bailey, D. G., and Carroll, R. J., In-vitro collagen fibril assembly in glycerol solution: Evidence for a helical cooperative mechanism involving microfibrils, Biochemistry, 25: 958–966, 1986.

64. Na, G. C., Interaction of calf skin collagen with glycerol: Linked function analysis, Biochemistry, 25: 967–973, 1986.

65. Lieb, L. M., Nash, R. A., Matias, J. R., and Orentreich, N., A new in-vitro method for transepidermal water loss: a possible method for moisturizer evaluation, J. Soc. Cosmet. Chem., 39: 107–119, 1988.

66. Adrangui, M., Evaluation of the water retention capacity of the human stratum corneum. Effect of humectants incorporated in Sedefos 75 based creams, Bull. Tech./Gattefosse Rep., 79: 57–62, 1986.

67. Takahashi, M., Yamada, M., and Machida, Y., A new method to evaluate the softening effect of cosmetic ingredients on the skin, J. Soc. Cosmet. Chem., 35: 171–181, 1984.

68. Lupulescu, A. P., and Birmingham, D. J., Effect of protective agent against lipid-solvent-induced damages, Arch. Environ. Health, January/February: 33–36, 1976.

69. Guillemin, M., Murset, J. C., Lob, M., and Riquez, J., Simple methods to determine the efficacy of a cream used for skin protection against solvents, Br. Jo. Ind. Med., 31: 310–316, 1974.

70. Romanian patent 68155, Dumitrescu, N., Dumitriu, R., and Muresan, D., Composition for protecting the skin against organic solvents and petroleum derivatives, 1979.

71. Glantz, P. O., Larsson, K., and Nyquist, G., A new skin protecting agent against acylic resins, Odont. Revy, 265–272, 1976. Vol 27.

72. Indian Patent 73CA2453, Chloride India, Barrier creams having protective action against deleterious materials, 1973.

73. U.S. Patent 3,567,820, Sperti, G. S., Compositions and treatment for the alleviation of diaper rash, 1971.

74. Lauwerys, R., Lachapelle, J. M., Buchet, J. P., Tennstedt, D., Kivits, A., Bertrand, F., and Triest, A., Cutaneous absorption of industrial solvents; evaluation of the effectiveness of barrier creams, Can. Med. Trav., 19(1): 3–6, 1982.

75. Guillot, J. P., Martini, M. C., Giauffret, J. Y., Gonnet, J. F., and Goyot, J. Y., Anti-irritant potential of cosmetic raw materials and formulations, Int. J. Cosmet. Sci. 5: 255–265, 1983.

76. Japanese patent 828650, Shiseido Co. Ltd., Cleaning compositions with low skin irritation, 1983.

77. Brown, V. K. H., Box, V. L., and Simpson, B. J., Decontamination procedures for skin exposed to phenolic substances, Arch. Environ. Health, 30: 1–6, 1975.

78. Hadgraft, J., Hadgraft, J. W., and Sarkany, I., The effect of glycerol on the percutaneous absorption of methyl nicotinate, Br. J. Dermatol., 87, 30–36, 1972.

79. Roberts, M. S., and Anderson, R. A., The percutaneous absorption of phenolic compounds: the effects of vehicles on the penetration of phenol, J. Pharm. Pharmac., 27: 599–605, 1975.

80. Idson, B., Biophysical factors in skin penetration, J. Soc. Cosmet. Chem., 22: 615–634, 1971.

81. Scheuplein, R. J., Advances in biology of the skin, Vol. XII, Pharmacology and the Skin, (W. Montagna, E. J., Van Scott, and R. B. Stroughton, eds.) Appleton Century Crofts, New York, 1972.

82. Bronaugh, R. L., and Stewart, R. F., Methods for in-vitro percutaneous absorption studies III: Hydrophobic compounds, Pharm. Sci.,73(9): 1984. P. 1255–1258

83. Mollgaard, B., and Hoelgaard, A., Vehicle effect on topical drug delivery. I.Influence of glycols and drug concentration on skin transport, Acta Pharm. Suec., 20: 433–442, 1983.

84. Mollgaard, B. and Hoelgaard, A., Vehicle effect on topical drug delivery. II. Concurrent skin transport of drugs and vehicle components, Acta Pharm. Suec., 20: 443–450, 1983.

85. Hoelgaard, A. and Mollgaard, B., Dermal drug delivery-improvement by choice of vehicle or drug derivative, J. of Controlled Release, 2: 11–120, 1985.

86. Wickett, R.R., Cooper, E.R., and Loomans, M.E., Penetrating topical pharmaceutical compositions, European Patent 0043738 A2, 1982.

87. European Patent 14146 A2, Tsuk, A.G., Enhancement of transdermal drug delivery, 1985.

88. Japanese Patent 8617513, American Home Products, Transdermal pharmaceuticals containing propylene glycol and glycerine as solvents for the enhancement of drug transport, 1986.

89. Wilson, J., Clapp, M.J.L., and Conning, D.M., Effect of glycerol on local and systemic carcinogenicity of topically applied tobacco condensate, Br. J. Cancer, 38: 250–257, 1978.

90. Ackerman, C. and Flynn, G.L., Ether-water partitioning and permeability through nude mouse skin in vitro. I. Urea, thiourea, glycerol, and glucose, Pharm., 36: 61–66, 1987.

91. Storelli, C., Svelto, and Lippe, C., Transport of polyhydric alcohols across frog skin, in Proceeding of the First Europen Biophycics Congress, Vol. III, 14–17 Sept., 1971, Baden Austria.

92. Lippe, C., Galluci, E., and Storelli, C., Permeabilities of ethylene glycol through lipid bilayers, membranes, and some epithelia, Archives de Physioligie et de Biochemie, 79: 315–318, 1971.

93. Sherman, J.K., Pretreatment with protective substances as a factor in freeze-thaw survival, Cryobiology, 1(4): 298–300, 1965.

94. Aggarwal, S. J., Baxter, C. R., and Diller, K. R., Cryopreservation of skin: An assessment of current clinical applicability, J. Burn Care Rehabil., 6(6): 469–476, 1985.

95. Berggren, R. B., Ferraro, J., and Price, B., A comparison of cryophylactic agents for pretreatment of preserved frozen rat skin, Cryobiology, 3(3): 272–274, 1966.

96. Rheinwald, J. G., Serial cultivation of normal human epidermal keratinocytes, Methods Cell Biol, 21A: 299–254, 1980.

97. Horova, J., and Samohyl, J., Conservation of skin transplants by chilling, by means of glycerol, Scripta Medica, 54(1): 53–60, 1981.

98. Biagini, G., Poppi, V., Cocchia, D., Ruboli, G., Damiani, R., and Laschi, R., Skin storage in liquid nitrogen. An ultrastructural investigation, J. Cutaneous Pathol., 6: 5–17, 1979.

99. De Loecker, De Wever, F., Jullet, R., and Stas, M. L., Metabolic changes in rat skin during preservation and storage in glycerol buffer at −196 C, Cryobiology, 13: 24–30, 1976.

100. Athreya, B. H., Grimes, E. L., Lehr, H. B., Greene, A. E., and Coriell, L. L., Differential susceptibility of epithelial cells and fibroblasts of human skin to freeze injury, Cryobiology, 5(4): 262–269, 1969.

101. Hart, G. J., Russell, A. E., and Cooper, D. R., The effects of certain glycols, substituted glycols, and related organic solvents on the thermal stability of soluble collagen, Biochem. J., 125: 599–604, 1971.

102. Linder, H., Collagen in cosmetics, Parfuem. Kosmet., 65(6): 340, 342–3, 346, 1984.
103. Basile, A. R. D., A comparative study of glycerinized and lyphilized porcine skin in dressings for third-degree burns, Plastic Reconstruct. Surg., June: 969–972, 1982.
104. Downing, D. T., Wertz, P. W., and Stewart, M. E., The role of sebum and epidermal lipids in the cosmetic properties of skin, Int. J. Cosmet. Sci. 8: 115–123, 1986.
105. Downing, D. T., and Stewart M. E., Methods in Skin Research, (D. Skerrow and C. J. Skerrow, eds.), John Wiley & Sons Ltd., Chichester, 1985, pp. 349–379.
106. Yardley, H. J., and Summerly, R., Lipid composition and metabolism in normal and diseased epidermis, J. Pharm. Ther., 13: 357, 1983.
107. Rebello, T., and Hawk, J. L. M., Skin surface glycerol levels in acne vulgaris, J. Invest. Dermatol., 70: 352–354, 1978.
108. Padberg, G., Uber die Kohlenhydrate im wassrigen Eluat der menschlichen Hautoberflache, Arch. klinische u. experimentelle Dermatologie, 229: 33–39, 1967.
109. Prottey, C., and Hartop, P. J., Changes in glycerolipid metabolism in rat epidermis following exaggerated washing and soap solutions, J. Invest. Dermatol., 61: 168-179, 1973.
110. Prottey, C., Hartop, P. J., and Ferguson, T. F. M. Lipid synthesis in rat skin, Br. J. Dermatol. 87: 586–607, 1972.
111. Takahashi, M., Sato, T., and Akino, T., Metabolic changes in scaly lesions of rat skin produced by essential fatty acid deficiency, Tohoku J. Exp. Med., 138: 261–274, 1982.
112. Lampe, M. A., Williams, M. L., and Elias, P. M., Human epidermal lipids; characterization and modulations during differentiation, J. Lipid Res. 24: 131–140, 1983.
113. Elias, P. M., Epidermal lipids, membranes, and keratinization, J. Dermatol. 20: 1–19, 1981.
114. Yardley, H. J., Epidermal lipids, in Biochemistry and Physiology of the Skin, (A. Goldsmith, ed.), Oxford University Press, Oxford, 1983, pp. 363–381.
115. Yardley, H. J., Epidermal lipids, Int. J. Cosmet. Sci. 9: 13–19, 1987.
116. Bowser, P. A., Nugteren, D. H., White, R. J., Houtsmuller, U. M. T., and Prottley, C., Identification, isolation, and characterization of epidermal lipids containing linoleic acid, Biochim. Biophys. Acta, 834: 419–428, 1985.

11

Methods for Evaluating the Efficacy of Cosmetics Containing Glycerine

Yohini Appapillai

Neutrogena Corporation, Los Angeles, California

I. INTRODUCTION

Glycerine is incorporated into cosmetics primarily with the aim of "moisturizing" and preserving the suppleness of skin. It does, however, offer other advantages to the formulator. In some cases, adding glycerine to a cosmetic product is like buying an insurance policy for the formula. This chapter focuses on ways of assessing the potential skin care benefits to the consumer. Emphasis is given to the interpretation and relevance of the data obtained by use of popular noninvasive biophysical techniques. To this end, case studies gleaned from the literature are reviewed.

Before discussing "how" to evaluate effectiveness, we need define "what" parameters to evaluate. Only then can we have a basis for measuring the degree of success achieved.

A. The Meaning of "Moisturization"

"Moisturization" remains a nebulous concept. Complications arise because the expression implies that water content of the skin is increased. In most cases this is an inadequate, and at times an inaccurate description of what the product delivers, or ultimately what the consumer desires; namely, healthy, soft, supple skin.

"Moisturization" encompasses a wide range of biological/biophysical changes in the uppermost layer of the skin, the stratum corneum (SC). Kligman defines a moisturizer in operational terms as "a topically applied substance or product that overcomes the signs and symptoms of dry skin" (1). He does not simply equate moisturization with increasing the water content of the skin because it would be a "gross oversimplification." You will hear this sentiment echoing throughout this chapter. Indeed, the main message here is that it is necessary to use more than one parameter to assess the overall benefit accruing from the use of a product.

B. What Is "Dry Skin"?

Most of us are familiar with the symptoms of everyday types of dry skin (xerosis) for which we seek solace from cosmetic preparations. The skin surface looks and feels rough, and dull ashen patches develop due to the scattering of light by scales (1, 2). Skin feels taut and inflexible, often itches, and may even develop deep cracks due to poor extensibility.

Little is known about the nature and the pathogenesis of cosmetic dry skin. The working hypothesis is that it is a subtle disorder of how the skin desquamates (1). The significant changes occur near the surface. The SC continually sheds clusters of corneocytes (squames) from the surface (desquamation), while fresh squames replace them from the underlying layers. Normally the clusters are very small and the process is more or less invisible. In dry skin, larger aggregates of squames are shed in the form of flakes or scales. Anatomically the epidermis appears unaltered, even though the root of the problem must reside therein (1).

The basic tenet among skin care professionals, cosmetic formulators, marketeers, and consumers alike is that dry skin results from a low water content. Clinical observations link low relative humidity (RH) and absolute humidity (dew point) of the environment and low temperatures with a higher incidence of dry skin (3, 4). The prevalence of dry skin in exposed body sites rather than in protected areas supports this tenet (2).

In vitro experiments pioneered by Blank in the 1950s (5, 6) showed that isolated brittle callous could be made soft and flexible by immersion in water, whereas months of immersion in oils, petrolatum, or anhydrous lanolin failed completely. Ever since, rightly or wrongly, cosmetic industry professionals have focused on the lack of water as the cause of dry skin. In actuality dry skin belongs to a family of disorders and manifests itself as rough, brittle, scaly skin, which may or may not lack water (1, 7). One thing is clear, oily skin, rich in sebum, does not represent the opposite of dry skin. There is no correlation between the two (7).

The SC normally contains 10–30% water (8–11). This water content is regulated by the barrier properties and the water-holding capacity of the stratum corneum. Blank showed that when the water content dropped below 10%, isolated callous lost its suppleness (5).

D. The Function of a "Moisturizer"

In practical terms, a moisturizer will maintain, or, restore, the elasticity and the flexibility of the stratum corneum (SC), a function referred to as plasticizing the SC, while decreasing surface roughness. Water alone may well accomplish this, but the benefit is fleeting due to rapid evaporative loss, and percutaneous absorption.

Formulators attempt to slow down the evaporative loss of water by using occlusive materials such as petrolatum which have high diffusional resistance to water, or humectants such as glycerine which bind water. Most cosmetic moisturizers contain emollients, such as mineral oil, and fatty alcohols, which have little or no effect on water transport across the skin surface, unless applied heavily. These emollients provide temporary relief to dry skin by filling in the gaps between loosened flakes and gluing them back to the surface, making the surface look and feel less rough (4).

II. THE ROLE OF GLYCERINE IN SKIN CARE

Glycerine improves the skin condition in vivo. This irrefutable fact is well established on the basis of generations of consumer acceptance and more recent clinical documentation. The preceding chapter by Mast provides an overview of in vivo and in vitro test results reported in the literature.

Nevertheless, skepticism still prevails regarding glycerine's moisturizing ability. The argument is that it acts mainly by humectancy and hence tends to scavenge water from around itself, which may not be desirable when it is on the skin. First of all, it should be recognized that the use of a humectant in an effective moisturizer does not establish humectancy as its principal or only mode of action. In fact, glycerine has been shown to plasticize the skin without increasing its water content (12). Its use in the leather industry to increase the suppleness of leather (13) and its use in the cryopreservation of skin and tissue and red blood cells (see Chapter 10) may be related to its efficacy in maintaining the suppleness of intact skin.

When a glycerine film is present on the skin, the region of water evaporation shifts to this film. When ambient humidity is low, the film will increase the throughput of water across the SC by drawing up water from the lower layers and giving it up to the atmosphere. In high-humidity environments, the film would absorb moisture from the air, as well as retain moisture from sweat glands and underlying epidermis. Whether this moisture is released to the skin or not is a moot point. I believe that a good part of the moisture would be available to the SC, considering: (1) the intrinsic hygroscopicity of the SC

(5, 6), and (2) that aqueous solutions of glycerine continuously lose water to the atmosphere at ambient humidities lower than 95% (14). Thus, a wet glycerine layer on the skin, would plasticize the SC by virtue of the water it holds, and its own intrinsic ability.

Glycerine is not the only material shown to plasticize skin without hydrating it. A modified triglyceride from soybean oil (15) and alpha-hydroxy acids such as lactic acid (16, 17) and 2-hydroxyoctanoic acid (18, 19) have been shown to plasticize skin outside the humectancy pathway. Using elasticity measurements of normal and delipidized skin samples, Leveque demonstrated the plasticizing effect of SC lipids (2). These are highly significant findings which attest to the fact that water is *not* the only plasticizer of skin, as the popular dogma since Blank's classic experiments would have you believe.

Glycerine is a natural metabolite present in the skin, liberated during the enzymatic hydrolysis of glycerides and phospholipids. Perhaps it influences the biochemical processes in ways we have yet to determine (21). Could it be feeding the skin exogenously? Or, does it play a purely structural role in the water ecology of the SC, interacting with SC lipids and/or proteins as some authors have suggested?

A. Glycerine and the Preservation of the Lipid Barrier to Transepidermal Water Loss

The stratum corneum (SC) serves as a barrier to the evaporation of water from the viable epidermis within, while protecting it from external chemical and physical insult. In its absence, water will evaporate from the skin surface as if from a free water surface (22). The SC is made up of about twenty layers of flat, keratin-filled, cornified cells interdigitated in a lipid matrix, much like a wall built of bricks set in mortar (23). Intercellular lipids (ICL) arranged in multiple bilayers constitute the water barrier (24–27) (see Fig. 1).

Recent studies by Imokawa et al. (28, 29), and by Yamamura and Tezuka (30) suggest that these intercellular lipids play a critical role in the water-holding properties of the SC, along with the natural moisturizing factors (NMF). Glycerine may be playing a vital role in preserving and enhancing the function of these key lipids.

In an important new publication, Froebe et al. present convincing evidence to support the hypothesis that glycerine maintains the intercellular lipid cement in a fluid, liquid crystalline state (31). Earlier, Friberg and Osborne proposed that optimal barrier to water loss is achieved when the intercellular lipids are in a liquid crystalline phase (32). In this recent study, glycerine was incorporated into Friberg's physical model of SC lipids at concentrations of 0–15%, under low and high relative humidity (6% RH and 92% RH, respectively).

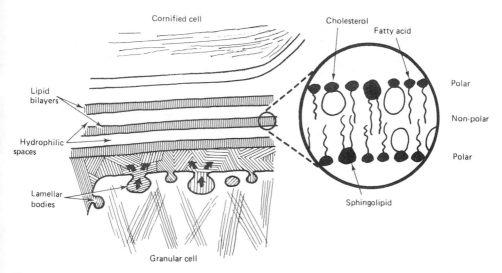

Figure 11.1 Diagram showing formation of intercellular substance of the stratum corneum. Lamellar bodies containing neutral lipids are secreted from cells of the granular layer. The contents of lamellar bodies reorganize into sheets of lipid bilayer. (From Ref. 84.)

The dramatic effect of glycerine on SC lipid structure is evident in Figures 2 and 3. Figure 2 shows at the outset, both, a control sample without any glycerine, and a sample containing 10% glycerine exhibiting a pattern characteristic of lamellar liquid crystals. Figure 3 documents the changes occurring over time in the same samples maintained at 6% RH. After 6 hours the control has begun to form fine crystals, while the 10% glycerine sample is largely liquid crystalline. By 96 hours the control has undergone a phase transition to solid crystal, while the glycerine containing sample is a mixture of solid and liquid crystal, despite almost complete dehydration. At high RH (92%) a similar contrast was observed between control and test samples, although the phase transition proceeded more slowly than at low RH. The glycerine containing sample remained predominantly liquid crystal even at 96 hours.

Against this backdrop, we will review some of the more popular instrumental methods capable of reliably monitoring skin function. When using instruments it is imperative that the investigator be aware of exactly what is being measured. Otherwise, spurious sensitivity will be recorded.

III. EVALUATION TECHNIQUES

Skin bioengineering is a rapidly growing field. This article is not intended to be a treatise on the multitude of noninvasive techniques that are being used

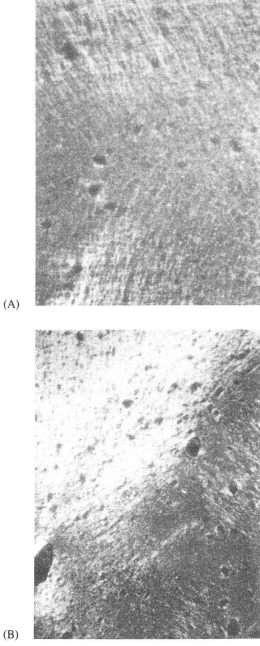

(A)

(B)

Figure 11.2 Model lipid with 0% glycerine (A) and 10% glycerine (B) viewed under polarized light at 400× magnification at the initial time (from Ref. 31).

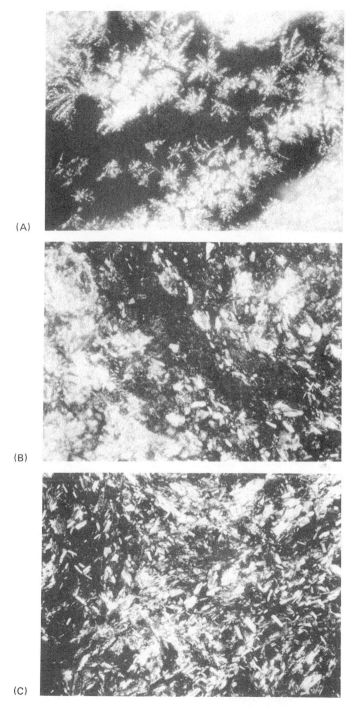

Figure 11.3 Model lipid with 0% glycerine (A–C) and 10% glycerine (D–F), after 6, 24, and 96 h of exposure to 6% RH, viewed under polarized light at ×400 magnification (after 6 h) or 100 × magnification (after 24 and 96 h) (from Ref. 31).

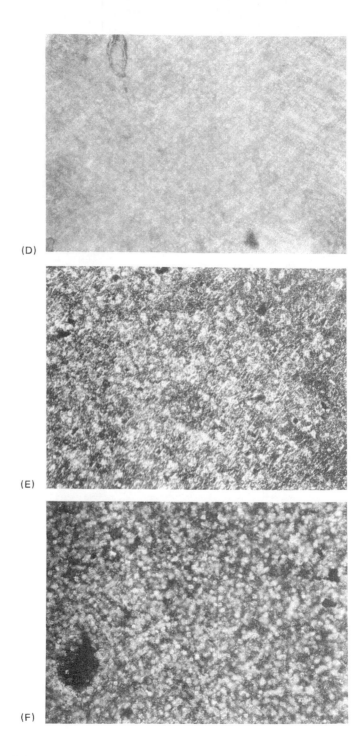

(D)

(E)

(F)

Figure 11.3 Continued.

to evaluate skin care product performance. The interested reader is referred to several excellent reviews (33–39) spanning this dynamic field.

The usefulness of noninvasive instrumental techniques for the evaluation of skin function has been increasingly documented (40–43). Biophysical methods offer several advantages over subjective assessments by trained practitioners and human volunteers: (1) they generate objective data amenable to statistical analysis; (2) they can be performed quickly, on larger panels, thereby compensating for the marked individual differences exhibited by panelists, (3) "nonvisible" or "clinically nonquantifiable" skin pathology can be evaluated, (4) they enable comparisons between treatments made at different times, and in different locations (multicenter studies for example).

To be meaningful, testing regimens should as far as possible mimic intended product usage. Ideally, efficacy testing and claims substantiation should involve a three-pronged approach, in which subjective, visual, and sensory judgements by trained practitioners, and consumer panels are correlated with the instrumental readings (34). This is the only way to ensure that the benefits measured using laboratory equipment are relevant in the "real world."

For examples of this integrated approach to efficacy testing of glycerine-containing products, see Dunlap's evaluation of hand and body lotions (44), Dahlgren et al.'s paper on the effects of bar soap constituents (12), and Batt et al.'s report on skin conditioning with glycerine (45).

A. Similarity of Glycerine and Water

For the purposes of this discussion, it is important to stress the intrinsic water-like behavior exhibited by glycerine. Glycerine is a small molecule with three hydroxyl groups on a hydrocarbon back bone [$CH_2(OH)$ $CH(OH)$ $CH_2(OH)$]. It has a relatively high dielectric constant of 42.5, compared with most organic fluids, which have dielectric constants below 7 (46). Water has a dielectric constant of 78.5 (46). Glycerine could be expected to facilitate the passage of electricity by its plasticizing action on the SC. These similarities coupled with its hygroscopicity make it difficult to quantify the water content in the SC following application of moisturizers containing glycerine. Most techniques currently used to measure moisturization in vivo cannot distinguish between the individual contributions of glycerine and water.

B. Design of Performance Evaluation Tests

Plasticization of SC is universally regarded as the major function of its water content. It is this plasticity and not the water content per se which is important for the SC in its resistance to flaking and cracking (16). In evaluating efficacy of a moisturizer, what matters is the level of improvement attained and the duration of the benefit. It does not matter whether it is water, glycerine, or a blend of the two, that produces the observed improvement.

Even if it were possible to quantify the increase in water content of the SC following application of a cosmetic containing glycerine, it would be an incomplete measure of "moisturization." To get a true picture, we must use a battery of techniques capable of objectively evaluating the health of the SC based on indicators such as barrier competency, elasticity, electrical permitivity, etc. Relevance of the findings to the real-life situation should also be confirmed by appropriately designed clinical studies, and subjective panel assessments. This issue is of fundamental importance to the cosmetic industry, and should be understood and appreciated by everyone associated with it.

A basic problem associated with in vivo assessment of biophysical properties is that variation in the quantity measured is often caused by factors other than the factor of interest. It is imperative that environment-related variables, instrument-related variables, and subject-related variables be strictly controlled so that worthwhile correlations can be made.

C.   General Precautions for In Vivo Performance Evaluation

A climate-controlled facility provides the best environment for conducting comparative product testing. If such a facility is not available, a separate room should be set aside where the testing may be conducted in peace and quiet. Only those involved in the testing should be in the room. By recording ambient humidity and temperature over a period of time, it may be possible to select a window of time when fluctuations are minimal. Any room that can be maintained at a steady temperature range of 20–25 C, and a stable RH range of 20–50% would suffice. If such control is not possible, the number of participating subjects should be increased to accommodate the increased statistical variation.

The subject should be resting quietly. The test operator should avoid any significant conversation aside from remarks intended to put the subject at ease. Subjects should be fully briefed *before* measurements are carried out. An accommodation period of 30 min or longer is necessary for a subject's body and skin temperature to come to equilibrium with ambient conditions.

Even under the best circumstances, there is day-to-day variation within a given subject, therefore, using the subject as his/her own internal control is recommended. This is done by including a control site of uninvolved skin adjacent to the test site. The volar area of the forearm is most frequently used for comparative product testing, because it shows high homogeneity with respect to properties such as sweat rate, corneocyte size, and cohesion (47). Even here, the sites show some variation (48). It is recommended that sites within approximately 4 cm of the elbow or the wrist are avoided.

Contralateral similarity is often assumed; however, this may not be the case. The dominant arm may give significantly different readings (49), especially if it had been used in a physical endeavor just prior to the study.

D. Transepidermal Water Loss Measurement

Transepidermal water loss (TEWL) is defined as insensible water loss through the skin separate and distinct from active perspiration (50). In this technique water flux through the SC is measured, providing a direct measure of the barrier integrity and an indirect assessment of water content. Measurement of TEWL is widely used to assess changes in hydration of the SC (51, 52) and the effect of topically applied moisturizers (53–55), and irritants (41). Its popularity notwithstanding, assessing moisturizer efficacy using TEWL measurements alone has its limitations. This very important point will be discussed at length in this section.

The different methods used in measuring TEWL have been extensively reviewed in the literature (50, 51, 56, 57). For in vivo measurements the best commercially available measuring device is the Evaporimeter EPI (Servomed, Stockholm, Sweden, Fig. 4), which has replaced the earlier cumbersome methods. For a comprehensive review of precautions to take when measuring TEWL with the Evaporimeter, the reader is referred to a recent report issued by the Standardization Group of the European Society of Contact Dermatitis (58). The key points made there are summarized below.

1. Principle

The basis for the Evaporimeter is that in the absence of air currents, a boundary layer of air develops near the skin surface. Steady-state water exchange from the skin surface creates a water vapor pressure gradient which extends from the skin surface to this boundary. The value of the gradient can be computed from partial pressure and temperature calculated at two fixed points on the linear gradient lying perpendicular to the skin surface. The rate of evaporation is proportional to the vapor pressure gradient, and TEWL in square grams per hour can be derived from the following equation (59):

$$\text{TEWL} = D' \frac{dP}{dx}$$

where D' is a constant $= 0.670 \times 10^{-3}$g (mhP^{-1}) and $dP/dx =$ vapor pressure gradient (Pa/m).

The depth of the boundary depends on the site and air circulation. In general, a mean depth of 10 mm may be assumed (59, 60). Accordingly, the Evaporimeter probe has two sets of sensors stacked vertically 3 and 9 mm above the surface measured. The sensors are a pair of transducers, one for measuring relative humidity and the other for measuring air temperature. They are housed in a small Teflon cylinder open at both ends (Fig. 5). Air currents disrupt the linear gradient and give rise to wide and rapid fluctuations in measurement. This very sensitive technique is capable of measuring fluxes as

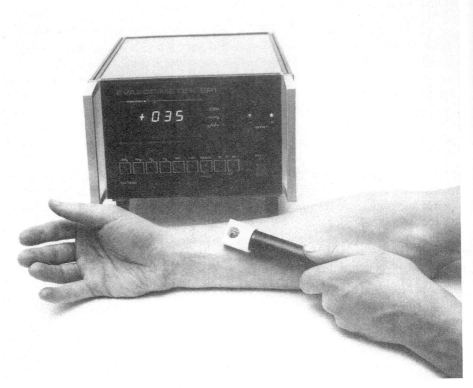

Figure 11.4   Servomed Evaporimeter EPI for measurement of transepidermal water loss.

small as $6 \times 10^{-8} g/cm^2 s$. Therefore, the importance of controlling exogenous and endogenous variables cannot be overemphasized.

2. Practical Guidelines

The probe should be applied with light, even, pressure, parallel to the skin surface. It is best to maintain the surface measured in the horizontal plane.

Figure 11.5   Cross-sectional view through Teflon shield of the Evaporimeter probe, showing sensor arrangement (from Ref. 56).

The probe handle houses amplifiers which are temperature sensitive, therefore it should not be held directly by hand, especially when making repeated measurements. The probe is best handled with an insulating glove or the calibration rubber stopper supplied with the equipment.

Contact time of the probe on the skin should be as short as possible to avoid occlusive build up of moisture. It helps to start the measurements without the use of the damping filters. Usually, stabilization of the TEWL value is reached within 30–45 seconds of the time the probe is placed on skin (61, 62). Longer stabilization time may be required for test sites exhibiting excessive water loss. Even after stabilization, the readings often continue to fluctuate very rapidly, due to extraneous disturbances, making it difficult to get an exact reading. Recording the measurements by means of a strip chart recorder makes it easier to follow the course of the measurement, and measure TEWL accurately (63). Alternatively, the output from the Evaporimeter may be interfaced with a microcomputer programmed to acquire any number of data points and statistically analyze the data (54).

The built-in electronic filters giving time constants of 10 and 20 seconds (s) may also be used for damping the fluctuations. The following procedure is recommended by the manufacturer (64, 65). When the reading appears to have stabilized, the 10 s button should be pressed, there should follow a 5 s pause before the 20 s button is pressed. The reading should become stable during the next 30 s period and this is considered the measured value (61).

Again, air currents are the greatest impediment to accurate determination of TEWL using this technique. Air-conditioning systems operating in the test area, drafts caused by opening and closing of doors, and movement of people are some typical disturbances. In environmental chambers where temperature and humidity are controlled by air-conditioning systems that circulate air rapidly, these fluctuations are impossible to avoid unless some form of draught shield is used. Two devices have been adapted: (1) a chimney can be affixed to the probe head, extending the column of static air above the sensors by 1–2 cm (66), and (2) a measuring box can be used, having an open top and holes for the placement of forearms of both the subject and the investigator (58).

Ambient air temperature should be maintained at a constant 20–22°C for two very important reasons: (1) to prevent thermal sweating, which is elicited in most people at ambient temperatures over 28–32°C, and (2) to maintain a steady skin surface temperature. Skin surface temperature has a profound influence on TEWL (67, 68). For example, in one study, an approximate twofold increase in TEWL was noted with an increase in skin temperature from 23 to 33°C. Therefore, skin surface temperature should be recorded as a routine part of TEWL measurements. Mathias et al. have derived an equation for correcting TEWL to a standard skin surface temperature of 30°C (68). However, at ambient air temperatures around 20–22°C, skin surface temperature normally stays within 28–32°C, and TEWL values do not require correction (58).

Changes in ambient relative humidity (RH) alter TEWL in a complex manner. On one hand, as ambient RH increases, the vapor pressure gradient between the skin and the surroundings decreases, resulting in lower TEWL. At higher RH, on the other hand, the water content of the skin is higher, and the skin becomes more permeable (diffusion and partition coefficients increase). Ideally, RH should be regulated to about 40% (58). The investigator should record ambient RH as a standard routine.

3. Interpretation of Results

The application of an occlusive material to the skin surface will physically block the surface (at least partially) and lower TEWL for a time. This reduction in TEWL is traditionally taken as a direct measure of the moisturizing efficacy of the material. Occlusive moisturization is one of the best ways to treat dry skin (1), as it increases the water content from within and keeps it there. However, it should be realized that after several hours, increased hydration may actually cause TEWL to rise above the baseline value, because the diffusion coefficient of the skin increases with hydration (69).

Thus, there are opposing mechanisms at work, and some occlusive materials may increase TEWL depending on what point in time the measurement is made. Furthermore, lower TEWL in theory could result from dehydration of the skin, as might result from topical application of a desiccant. Although unlikely, it illustrates the not-so-simple relationship between water content and flux in the SC.

Therefore, making a moisturization claim based solely on TEWL measurements is neither good laboratory nor good clinical practice. This is particularly true in the case of commercial moisturizers which typically contain a mix of occlusive materials, humectants, and emulsifiers or surfactants; all of which have the potential to affect TEWL in different ways.

In dealing with glycerine, or products containing glycerine or similar humectants, we face an added complication. Unlike petrolatum, glycerine is not a simple water trap. It plays a dominant role in regulating SC water flux due to its sensitivity to ambient RH changes. Even while measuring TEWL we induce these changes, and should therefore, pay attention to the way TEWL measurement was made. Let me illustrate this point with two reports in the literature.

Reitschel and De Villez tested the moisturizing efficacy of a hand cream with a high glycerine content on 10 volunteers with normal skin (70). A 0.02 ml aliquot of the product was applied to a 6.25 cm^2 area (3.2 µl/cm^2) on the forearm. An adjacent untreated site served as control. The Meeco electrolytic moisture analyzer was used to measure TEWL (50, 56). This device employs

Table 11.1 Increase in Transepidermal Water Loss Following the Application of a Hand Cream with High Glycerine Content[a]

Time after application (min)	Control (mg/cm^2/h)	Increase over control (mg/cm^2/h)	% Increase
65	0.122	0.229	188
130	0.099	0.198	200
195	0.095	0.215	226
Average	0.105	0.214	204

[a]Measured using the ventilated chamber technique.
Source: From Ref. 70.

the ventilated chamber technique of measuring TEWL, in which a continuous stream of dry nitrogen gas is passed through a sampling chamber placed over the test site. The moisture content of the gas is measured before and after passing through the chamber using electrolytic water sensors. Note the direct contrast to the Servomed, which measures the water gradient within the static boundary layer immediately adjacent to the skin surface. Readings were taken 1, 2, and 3 hours after product application. The results are displayed in Table 1. At all three readings, TEWL of the test site was greatly increased compared with the untreated control site (204% on average, $p \leq 0.001$).

The authors conclude that since adequate time was allowed for excess water present in the cream to evaporate before the first measurement, the increased amounts of water detected were due to the interaction between the moisturizer and the stratum corneum. The net effect being more water made available to the stratum corneum.

Their results show that a large amount of moisture was being held on or near the surface of the skin. Exactly how much of it was available to the SC cannot be determined. It can be argued that since glycerine, and mixtures of glycerine and water plasticize the skin, and since SC has a great affinity and capacity for absorbing water (5, 20), the existence of a wet film of glycerine within the upper layers of the SC is indicative of the product's moisturization potential (see Section II). Therefore, it is fair to conclude that the results support the moisturizing ability of the product over a period of 3 hours. Ideally, this should have been corroborated by clinical evidence or by at least one other technique such as skin elasticity, or conductance.

As an aside, it should be mentioned that this hand cream was tested in a separate 14-day home-use study conducted by independent clinical researchers, and 50–78% improvement was registered in the categories of redness, roughness, and dryness. Improvement of skin surface roughness as evidenced by image analysis was 23–46% (71).

Wilson et al. explored the relationship between TEWL, skin surface water loss (SSWL), and water content of the SC, as it pertains to the assessment of moisturization and soap effects (55). SSWL represents the sum of excess water loss plus steady-state TEWL. For occlusive moisturizers, SSWL measured immediately after product removal correlates positively with efficacy (72, 73).

TEWL and SSWL were measured using a Servomed Evaporimeter. SC water content was measured with the dielectric water content probe (74). This device transmits microwave radiation of a selected frequency targeted at the water molecules. The amount of energy absorbed by the water molecules is proportional to their concentration. The frequency of oscillation of a molecule is related to its dielectric constant. Because glycerine has a sufficiently high dielectric constant, the probe directly senses the glycerine, and a distinction between water and glycerine content cannot be made.

Moisturizers were applied to sites on both forearms; a control site was left untreated on each arm. Assuming contralateral anatomical similarity, water loss and dielectric probe response (DPR) were measured on separate forearms. Products were applied to premarked test sites; an hour later TEWL and DPR measurements were made over the applied films. Following each measurement the site was wiped eight times with a tissue wipe and remeasured; final readings were recorded one hour later.

The SSWL data from 10 subjects treated with glycerine and petrolatum are presented graphically in Figure 6 (mean±SD). One hour following application, SSWL measured over the glycerine film shows a marked depression, while that over petrolatum shows a smaller depression. This might suggest that glycerine was a better occlusive moisturizer. However, immediately after wiping, the glycerine site appears unchanged from the control site, while the petrolatum site shows a jump in SSWL indicating release of moisture trapped underneath the film.

The authors offer a possible explanation for these observations: "The Evaporimeter probe imposes a slight increase in humidity over the test film at the moment of measurement which places the glycerol film in transition, absorbing moisture from the air and acting against TEWL."

We found that Evaporimeter measurements made over glycerine films applied on glass slides record an immediate, sharp, drop which gradually, over 1 hour, returns to baseline. However, similar depressions were not observed in preliminary experiments with several moisturizers containing 5–40% glycerine. Moisturizers were applied to glass slides, allowed to dry for 1 hour, and the Evaporimeter readings were taken (75). Therefore, to ensure that the change in TEWL is not an artifact of the measuring technique, it is recommended that the behavior of product films on glass slides be monitored.

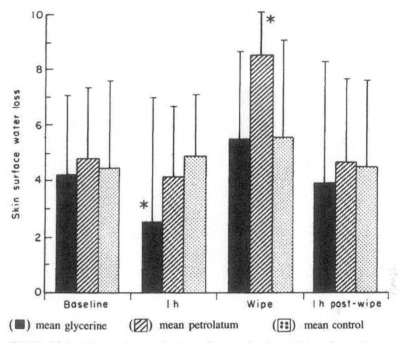

(■) mean glycerine   (▨) mean petrolatum   (▦) mean control

Figure 11.6 Skin surface water loss after application of glycerine and petrolatum. Glycerine (1 h) and petrolatum (post wipe) are significantly different from baseline ($p \leq 0.05$) (from Ref. 55).

There is a second explanation for the lower SSWL at the glycerine site: any water evaporating from the skin surface will be trapped as liquid water within the film, lowering the measured TEWL. It should become clear from the data presented below that wiping removes only part of the glycerine. The residual glycerine may have effectively offset the SSWL increase.

Thus, in the case of moisturizers based on humectants such as glycerine, measurements taken after wiping the site are not useful. However, TEWL measured over the intact film may correlate with moisturizing ability.

The DPR data from the above experiment are plotted in Figure 7. Glycerine significantly increases DPR (compared with control) both in the prewipe measurement made 1 hour after application as well as in the subsequent wiped measurement. The high DPR recorded is due to a composite of water and glycerine. The glycerine value remains elevated 1 hour post wipe, probably due to residual glycerine adhering to skin. In the case of petrolatum, little

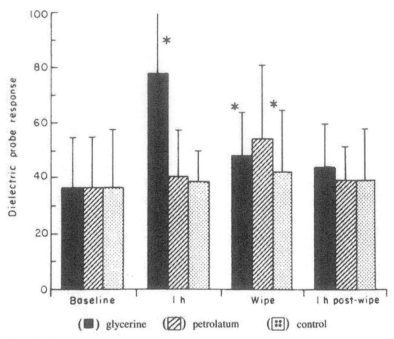

Figure 11.7 Dielectric probe response after the application of glycerine and petrolatum. Glycerine (1 h) and petrolatum (post wipe) are significantly different from baseline (p≤0.05) (from Ref. 55).

difference is seen at 1 hour past application, but a significant increase is noted immediately after wiping away the film.

These results indicate that penetrating, adherent, materials like glycerine resist "wipe-off." We have confirmed this hypothesis (75) by repeating the experiment, and examining the forearm by Attenuated Total Reflectance (ATR) (76) Fourier transform infrared (FTIR) spectroscopy, a technique which permits direct scanning of the skin surface and a few microns below. Figure 8 shows the reflectance spectrum of bare skin at the test site prior to application of glycerine. Figure 9 shows the same site 1 hour after application of glycerine. Figure 10 shows the spectrum of the site after wiping 16 times with a tissue wipe applied to the skin with considerable pressure. There appears to be a significant amount of glycerine still left on the skin. It should be added that the ATR spectrum of the site scanned nearly 8 hours later was very similar to Figure 10, but washing with water removed the glycerine. Thus glycerine, and moisturizers formulated with glycerine have the potential to resist rub-off and provide all-day moisturization.

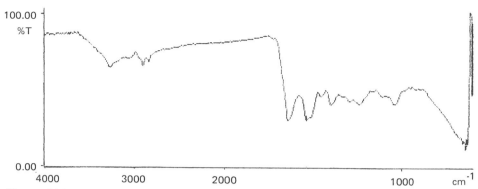

Figure 11.8   Attenuated total reflectance FTIR spectroscopy of "clean" skin of forearm, in vivo.

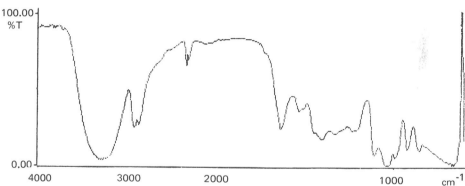

Figure 11.9   Attenuated total reflectance FTIR spectroscopy of forearm (same test site as in Fig. 8), 1 h after the application of a film of glycerine, in vivo.

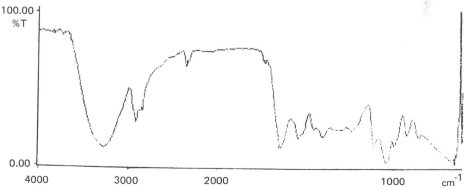

Figure 11.10   Attenuated total reflectance FTIR spectroscopy of the same test site (compare with Figs. 8 and 9), following 16 wipes with a folded tissue wipe, in vivo.

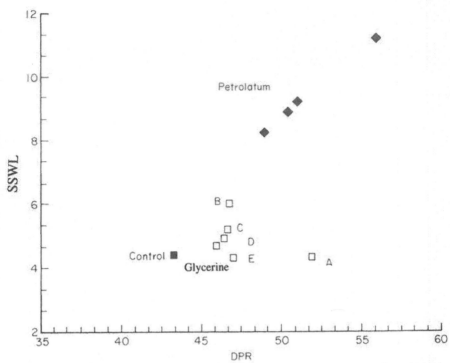

Figure 11.11 Skin surface water loss and dielectric probe response post wipe. Comparison of commercial products with glycerine and petrolatum (from Ref. 55).

Wilson et al. extended their study of commercial moisturizers. Included in the investigations were a model glycerine moisturizer (Product A, the hand cream used in the preceding study), a model nonglycerine moisturizer (Product B), and three others containing mixtures of fatty materials and humectants (Products C, D, and E).

The wiped SSWL and DPR results of five separate studies are correlated together in Figure 11. Note that the four petrolatum means lie along a straight line extending from the untreated control. Only Product B lies close to this conceivable occlusive hydration course. Glycerine, Product A and the control site generate a separate vector which is almost horizontal, showing moisturization without a complimentary SSWL increase. Petrolatum and Product A exhibit higher moisturizing ability than pure glycerine. Like glycerine, Product A remains on the skin all day, and is not removed by wiping, as evidenced by ATR-FTIR spectroscopy (75).

E. Electrical Measurements

The skin exhibits very complex electrical properties, which could in principle be related to its physical and chemical state. The phenomenological similarity between Fick's law and Ohm's law has led to the acceptance of electrical measurements for assessing effects connected with skin permeability (77). More specifically, many a moisturizing claim has been made by relating measured values of skin conductance (or impedance) to skin permeability increase induced by hydration of the SC (78–83). Attesting to the relevance of this technique to cosmetic scientists and dermatologists, is the finding of Leveque et al. that skin conductance is inversely related to the degree of typical winter xerosis of the face (84). The influence of ambient humidity on SC hydration was demonstrated using skin conductance measurements (85).

The main advantages to using electrical measurements, aside from the noninvasive nature of the tests, are: (1) high sensitivity—electrical changes with function are much greater as a percentage of the baseline than is the case with other biophysical techniques (86); (2) ease of measurement—they are quick and very simple to perform, permitting investigations on a large sample size.

Leveque and De Rigal have identified three possible means of conduction within intact skin, which are influenced by its water content (87). (1) Keratin chains in the SC have a dipolar moment. The plasticizing action of water facilitates the movement of the dipoles, resulting in a marked increase in dielectric permitivity. (2) Mobility of ions in the intercellular space is dependent on the viscosity of the SC, which is lowered by hydration. (3) Water molecules themselves are able to form a continuous network of hydrogen bonds permitting proton transfer between adjacent molecules.

Water content of the SC has been shown to exist in three different environments, tightly bound up to 10%, loosely bound from 10 to about 30 or 50%, and anything over being present as free water (88). The stronger the water is bound to the keratin, the less it increases the conductivity of the system (89). Therefore, plasticizers such as glycerine which alter the nature of water binding to the keratin proteins will have a direct influence on the readings.

1. Principle

In order to relate electrical measurements made on the skin to its physical condition, two kinds of models have been employed. The more common model, based on construction of an "equivalent circuit," assumes that certain skin components act like electrical devices. The widely accepted electrical analogue of the skin is illustrated in Figure 12 (48). This equivalent circuit assumes one electrode is placed on the skin surface, while the other is within the epidermis. However, Salter (77) and Edwards (91) caution that the true electrical behavior

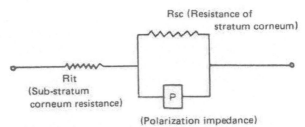

Figure 11.12 Equivalent electrical circuit of the skin. This represents the situation for one surface electrode and one subdermal electrode. Here R_{it} is the resistance of the substratum corneum tissue, and R_{sc} is the stratum corneum resistance (from Ref. 48).

of skin is too complicated for any such representation to be accurate beyond a single set of physical constraints on the measurement conditions.

According to this model, the impedance of intact skin, that is, the total electrical opposition to the flow of alternating current (AC) is comprised of two components, resistance (R) and capacitance (C) according to the following relationship: where f stands for the frequency of the applied AC (92). Direct current measurements are avoided because they cause polarization of the electrodes.

$$Z^2 = R^2 + [\tfrac{1}{2}\pi f C]^2$$

The operating frequency (f) is an important consideration, as the impedance of the SC is high at low frequencies (0–100 Hz) requiring the use of aqueous contact electrodes which them selves directly affect the hydration state of skin. Higher frequency impedance measurements allow the use of dry electrodes. For the assessment of skin hydration, frequencies in the 10Khz to low Mhz zone are recommended, as the influence of ions is very much smaller than that of water in this range (see Fig. 13) (93).

The Tagami Single Frequency Impedance Device. Among commercially available single frequency devices for measuring skin moisturization, the instrument used most often, about which sufficient technical and experimental information is available, is the Skicon Skin Surface Hydrometer (IBS instruments, Japan) developed by Tagami (78–80).

The Skicon-200, shown in Figure 14, measures electrical conductance of the skin surface. The probe is composed of two concentrically arranged brass electrodes separated by a cylindrical insulator. The diameter of the central electrode, and the inner diameter of the outer cylindrical electrode are 1 and 4 mm, respectively. AC of 3.5 Mhz frequency flows between the two electrodes via skin tissue and the conductance is read automatically on the digital recorder as reciprocal impedance in micro-ohms.

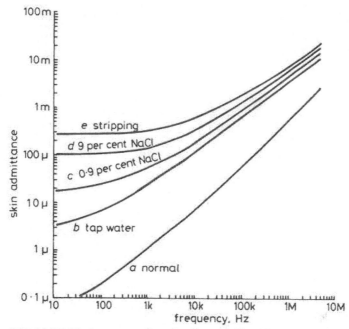

Figure 11.13 Frequency characteristics of skin admittance or conductance (from Ref. 93).

The spring-loaded probe tip has a simple sliding device which enables the probe to be applied to the skin with light, consistent pressure (30 g) throughout the study. If the probe is pressed too hard or pressed with differing amounts of pressure, water content at different levels of the epidermis will be sampled, leading to inconsistent results. Using cellophane tape stripping of serial layers of the SC, Tagami showed that the principal hydration detected by Skicon is from the most superficial layers of the SC. The probe should be held perpendicular to the surface measured for 3 s. This timing interval is adequate to allow stabilization of the electronic circuits while avoiding occlusive moisture buildup.

The Salter Multifrequency Impedance Device.   The second approach relating observed electrical behavior of the SC to its physical and chemical state was suggested by Rosenberg (94), and developed by Pethig (95) and Salter (96). It is based on the theories of solid-state physics. The SC is seen as a disordered, principally ionic, semiconductor or solid electrolyte. The activation energy for semiconduction is reduced by increased permittivity induced by hydration of the keratins in skin.

For a detailed analysis of this complex theory refer to two articles by Salter (86, 89). Simply put, the keratins in the SC are considered protonic

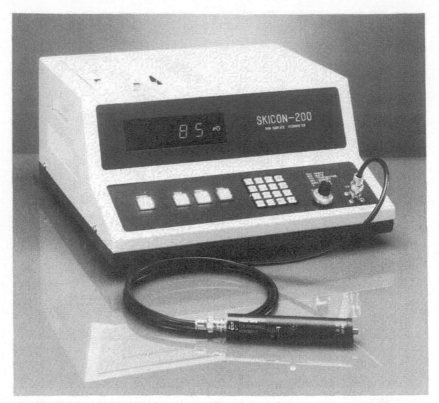

Figure 11.14   Skicon-200 for examination of electrical conductance of skin.

semiconductors, exhibiting small energy gaps between their conduction and valence bands. Electrons can be excited to energy levels between the two bands, where they remain for a short time, giving rise to charge transfer and capacitance, as in the case of transistors.

The theory, and supporting experimental evidence generated by Salter show that the alternating current complex impedance of the SC (Z) closely obeys the Cole equation (97):

$$Z = R_s + jx_s = R_\infty + \frac{R_0 - R_\infty}{(1 + j_\omega \tau_p)^{1 - m}}$$

where R_s is the (real) series resistance varying from 0 to ∞ and X_s is the equivalent (imaginary) series reactance; R_0 and R_∞ are the limiting values of R_s as the frequency tends to 0 or ∞, respectively; ω is defined as $2\pi f$, where f is the measurement frequency; τ is the mean relaxation time; m is related to width distribution curve of relaxation times, and takes values between 0 and 1. It is this parameter m that is sensitive to SC moisture content. Salter asserts

Figure 11.15 The Solartron 1170 frequency response analyzer is shown on the left and the Apple II± with dual disc drive and thermal printer on the right.

that for adequate characterization of the skin, electrical behavior must be measured over as wide a range of f as possible.

The Salter Complex Impedance Device (SCID) shown in Figure 15 is based on this principle. It consists of a Solartron 1170 Frequency Response Analyzer (FRA) (Solartron Electronics Group Ltd., England), which can sweep through a large frequency range (e.g., 0 to 4,000Hz) quickly, interfaced to a microcomputer. Measurements are made using a spring-loaded probe tipped with a pair of dry Ag/AgCl electrodes, each 10 mm in diameter and with 10 mm spacing between them. The FRA is programmed to measure impedance at 23 separate frequencies. The measurement cycle takes about 30 seconds, at which time the probe is removed from the site. Data manipulation and recycling of the device for a subsequent measurement require an additional 30 seconds. The output is a unitless relative hydration index, which is claimed to be impervious to changes in skin temperature, and sweating (77). The index ranges in value from 20 (very dry) to 80 (very hydrated) (85).

2. Practical Guidelines

For either approach it is imperative that all experimental conditions be carefully controlled. The precautions mentioned in Section III.C. should be heeded if meaningful correlations are to be drawn from the data. This is true especially for single frequency devices, since such measurements are not unique descriptors of skin properties (77, 90, 91). It is impossible to separate the effects of hydration and skin temperature using a single frequency, as the suppleness of skin is related to both its water content and its temperature. Measurement of skin surface temperature at each test site should alert the investigator of any undue change. An untreated control site should always be included, allowing the subject to be his/her own internal control. Provided such strict measures are taken, single frequency devices such as the Skicon are useful for relative measurements of product performance (91).

3. Interpretation of Results

As pointed out above, glycerine by its plasticizing action and ability to hydrogen bond would affect skin conductance independently of the water content, though not as strongly as water. Therefore, it would be an incorrect assumption to attribute the observed changes in conductance following topical application of a product containing glycerine entirely to changes in water content. This is true of other plasticizers as well, and ingredients active on proteins such as urea, which can alter the ratio between water molecules tightly bound to the protein, loosely bound, and "free," without changing the water content. This very sensitivity makes electrical measurements valuable for evaluating the relative merits of glycerine-containing products. Devices which accurately measure water content only would not reflect the total effect on skin.

One complicating factor is that inflammation of the skin is accompanied by an increase in conductivity (87). Nole et al. demonstrated that, as the barrier function of the skin was damaged by repeated soap insult while subjects were monitored under constant environmental conditions, both TEWL, and conductance values increased (85). This underscores my point that basing efficacy evaluations on any one technique is inadequate, and may lead to erroneous conclusions. Use of two complementary techniques such as TEWL and electrical measurements or skin elasticity measurements will provide a mechanism fingerprint for any test substance.

IV. CONCLUSION

Within the past decade great strides have been made in objective evaluation of the performance of skin care products. We now have several excellent tools, a few of which were discussed here, to help assess the condition of the skin. These devices can provide increased sensitivity which was lacking in classical clinical studies (33). However, successful use of the techniques requires strict control of extraneous sources of variability, and operator skill.

It should be appreciated that each instrument usually measures a single property, whereas human perception is the result of multiple sensory inputs processed by the brain. Therefore, a much better understanding of skin's response can be gained by using a battery of tests than relying on any one technique.

Reduction of consumer terminology to physically measurable attributes is a major challenge not to be taken lightly. For example, "feels smooth" is not adequately defined by surface topography alone; surface friction and skin hardness should be included for unambiguous correlation (47). Similarly, "moisturization" should not be assessed by mere measurement of water content, particularly when skin plasticizers other than water are involved.

Certain moisturizers deliver more than a transient benefit through fundamental modifications in the desquamation process (4, 98). Such products are labeled "therapeutic" or "functional" moisturizers. Kligman's classic regression method (98) is the most widely used test to identify and discriminate between such products. An abbreviated version of this clinical method, incorporating instrumental techniques has been developed by Prall et al. (4). Their version compresses the treatment and regression phase into a single week, instead of the six required by the original method, and is better suited to the time frame within which much of product development and claim substantiation is carried out.

As instrumentation becomes more powerful and sensitive, the question of relevance to the real world looms larger. The concern is that the instruments may be measuring subtle differences not perceivable by the consumer (34). Such differences may, in fact, be harbingers of skin effects eventually discernible by the user. Thus, they can be valuable guides for the formulator and the dermatologist. Correlating instrumental measurements with panelist self-assessment and/or expert evaluation, in an expanded study simulating in-use conditions is ultimately the only sure way to establish a claim.

ACKNOWLEDGMENTS

I am grateful to Dr. Donald Orth of Neutrogena Corporation, for his enthusiastic support and valuable suggestions. I would like to thank Dr. Donald Wilson of Cygnus Research Corporation for several helpful discussions regarding TEWL measurements.

REFERENCES

1. Kligman, A. M., Lavker, R. M., Grove, G. L., and Stoudemayer, Some aspects of dry skin and its treatment, in Safety and Efficacy of Topical Drugs and Cosmetics (A. M. Kligman and J. J. Leyden, eds.), Grune and Stratton, New York 1982, pp. 221–238.
2. Chernosky, M. E., Clinical aspects of dry skin, J. Soc. Cosmet. Chem., 27, 365–376, 1976.

3. Gaul, L. E., and Underwood, G. B., Relation of dewpoint and barometric pressure to chapping of skin. J. Invest. Dermatol., 18, 9–12, 1951.

4. Prall, J. K., Theiler, R. F., Bowser, P. A., and Walsh, M., The effect of cosmetic products in alleviating a range of skin dryness conditions as determined by clinical and instrumental techniques, Int. J. Cosmet. Sci., 8, 159–174, 1986.

5. Blank, I. H., Factors which influence the water content of the stratum corneum, J. Invest. Dermatol., 18, 433–440, 1952.

6. Blank, I. H., Further observations on factors which influence the water content of the stratum corneum, J. Invest. Dermatol., 21, 259–271, 1953.

7. Pierard, G. E., What does "dry skin" mean?, Int. J. Dermatol., 26(3), 167–168, 1987.

8. Bulgin, J. J., and Vinson, L. T., The use of differential thermal analysis to study the bound water in stratum corneum membranes, Biochim. Biophys. Acta, 136, 551–560, 1967.

9. Foreman, M. I., A proton magnetic resonance study of water in human stratum corneum, Biochim. Biophys. Acta, 437, 599–603, 1976.

10. Hansen, J. R., and Yellin, W., NMR and infrared spectroscopic studies of stratum corneum hydration, American Chemical Society Meeting, Abstracts, March–April 1971.

11. Walkley, K., Bound water in stratum corneum measured by differential scanning calorimetry, J. Invest. Dermatol., 59, 225–227, 1972.

12. Dahlgren, R. M. et al., Effects of Bar Soap Constituents on Product Mildness, Proceedings Second World Conference on Detergents, pp. 127–134.

13. Mannheim, P., Structure modifiers in cosmetics, Soap Perfumery and Cosmet, July 713, 720, 1959,

14. Courtney, D. L., Polyols in creams and lotions, Cosmet. Toiletr., 95, 27–34, 1980.

15. Osborne, D. W., The skin softening properties of maleated soybean oil, Cosmet. Toiletr., 103, 57–70, 1988.

16. Middleton, J. D., Development of a skin cream designed to reduce dry and flaky skin, J. Soc. Cosmet. Chem., 25, 519–534, 1974.

17. Takahashi, M., Machida, Y., and Tsuda, Y., The influence of hydroxy acids on the rheological properties of stratum corneum, J. Soc. Cosmet. Chem., 36, 177–187, 1985.

18. Hall, K. J., and Hill, J. C., The skin plasticization effect of 2-hydroxyoctanoic acid, (1) The use of potentiators, J. Soc. Cosmet. Chem., 37, 397–407, 1986.

19. Hill, J. C., White, R. H., Barratt, M. D., and Mignini, E., The skin plasticization effect of a medium chain 2-hydroxy acid and the use of potentiators, J. Appl. Cosmetol., 6, 53–68, 1988.

20. Leveque, J-L, and Rasseneur, L., Mechanical properties of stratum corneum: influence of water and lipids, 155–161, in The Physical Nature of the Skin (R. M. Marks,. S. P. Barton, and C. Edwards, eds.), MTP Press, Lancaster, England, 1988, pp. 155–161.

21. Rieger, M., Skin, water and moisturization, Cosmet. Toiletr. 104, 41–42, 44–50, 1989.

22. Idson, B., Dry skin moisturizing and emolliency, Drug Cosmet. Ind., 41–43, 1980.

23. Williams, M. L., and Elias, P. M., CRC Crit. Rev. in Therapeutic Drug Carrier Systems, 3, 95–122, (1987).
24. Elias, P. M., Epidermal lipids, barrier function, and desquamation, J. Invest. Dermatol. 80 (suppl), 44s–49s, 1983.
25. Elias, P. M., and Friend, D. S., The permeability barrier in mammalian epidermis, J. Cell. Biol., 65, 180–191, 1975.
26. Elias, P. M., Epidermal lipids, membranes, and keratinization, Int. J. Dermatol., 20, 1–19, 1981.
27. Golden, G. M., Guzek, D. B., Kennedy, A. H., McKie, J. E., and Potts, R. O., Biochemistry, 26, 2382–2388, 1987.
28. Imokawa, G., and Hattori, M., A possible function of structural lipids in the water-holding properties of the stratum corneum, J. Invest. Dermatol., 84, 282–284, 1985.
29. Imokawa, G., Akasaki, S., Hattori, M., and Yoshizuka, N., Selective recovery of deranged water-holding properties by stratum corneum lipids, J. Invest. Dermatol., 87, 758–761, 1986.
30. Yamamura, T., and Tezuka, T., The water-holding capacity of the stratum corneum measured by ^{1}H-NMR, J. Invest. Dermatol., 93, 160–164, 1989.
31. Froebe, C. L., Simion, F. A., Ohlmeyer, H., Rhein, L. D., Jairajh Mattai, Cagan, R. H., and Friberg, S. E., Prevention of stratum corneum lipid phase transitions in vitro by glycerol—an alternative mechanism for skin moisturization, 41, 51–65, 1990.
32. Friberg, S. E., and Osborne, D. W., Small angle x-ray diffraction patterns of stratum corneum and a model structure for its lipids, J. Disp. Sci. Technol., 6, 485–495, 1985.
33. Grove, G. L., Techniques for substantiating skin care product claims, in Safety and Efficacy of Topical Drugs and Cosmetics (A. M. Kligman and J. J. Leyden, eds.), Grune and Stratton, New York, 1982, pp. 137–176.
34. Grove, G. L., Design of studies to measure skin care product performance, Bioeng. Skin, 3, 359–373, 1987.
35. Leveque, J. L., Physical methods for skin investigation, Int. J. Dermatol., 22, 368–375, 1983.
36. Grove, G. L., Noninvasive methods for assessing moisturizers, in Clinical Safety and Efficacy Testing of Cosmetics (W. C. Waggoner, ed.), Marcel Dekker, New York, 1990, pp. 121–148.
37. Grove, G. L., and Grove, M. J., Objective methods for assessing skin surface topography noninvasively, in Cutaneous Investigation in Health and Disease (J. L. Leveque, ed.), Marcel Dekker, New York, 1988, pp. 1–32.
38. Potts, R. O., Stratum corneum hydration: Experimental techniques and interpretations of results. J. Soc. Cosmet. Chem., 37, 9–33, 1988.
39. Leveque, J. L., Grove, G., de Rigal, J., Corcuff, P., Kligman, A. M, and Saint Leger, D., Biophysical characterization of dry facial skin, J. Soc. Cosmet. Chem., 82, 171–177, 1987.
40. Maibach, H. I., Bronaugh, R., Guy, R., Turr, E., Wilson, D., Jacques, S., and Chang, D., Noninvasive Techniques for Determining Skin Function, Cutaneous Toxicity (V. A. Drill, and P. Lazar, eds.), Raven Press, New York, 1984, pp. 63–97.

41. Berardesca, E., and Maibach, H. I., Bioengineering and the patch test, Contact Dermat., 18, 3–9, 1988.
42. Alexander H., and Miller, D. L., Determining skin thickness with pulsed ultrasound, J. Invest. Dermatol., 72, 17–19, 1979.
43. Nilsson, E. G., Tenland, T., and Oberg, P. A., Evaluation of a laser doppler flowmeter for measurement of tissue blood flow, EEE Trans. Biomed. Eng., 27, 597–604, 1980.
44. Dunlap, F. E., Clinical evaluation of a highly effective hand and body lotion, Curr. Ther. Res., 35, 72–77, 1984.
45. Batt, M. D., Davis, B., Fairhurst, E., Gerrard, W. A., and Ridge, B. D., Changes in the physical properties of the stratum corneum following treatment with glycerol. J. Soc. Cosmet. Chem., 39, 367–381, 1988.
46. CRC Handbook of Chemistry and Physics, 57th Ed., CRC Press, Columbus, 1976–1977, pp. E-55, 56.
47. Prall, J. K., Instrumental evaluation of the effects of cosmetic products on skin surfaces with particular reference to smoothness, J. Soc. Cosmet. Chem., 24, 693–707, 1973.
48. Archer, W. I., Kohli, R., Roberts, J. M. C., and Spencer, T.. S., Skin impedance measurement. In Methods for Cutaneous Investigation (R. L. Rietschel and T. S. Spencer, eds.), Marcel Dekker, New York, 1990, pp. 121–142.
49. Malczewski, R. M., and Phillips, R. N., The use of statistical process control to analyze moisturization study data, presented at the 16th I.F.S.C.C. Congress, New York, October 8–11, 1990.
50. Spencer, T. S., Transepidermal water loss: methods and applications, in Methods for Cutaneous Investigation (R. L. Rietschel, T. S. Spencer, eds.), Marcel Dekker, New York and Basel, 1990, pp. 191–217.
51. Idson, B., In vivo measurement of transepidermal water loss, J. Soc. Cosmet. Chem., 29, 573–580, 1978.
52. Berube, G. R. Messinger, M., and Berdick, M., Measurement in vivo of transepidermal water loss. J. Soc. Cosmet. Chem., 22, 361–368, 1971.
53. Rietschel, R. L., A skin moisturization assay, J. Soc. Cosmet. Chem., 30, 369–373, 1979.
54. Wu, M-S, Yee, D. J., and Sullivan, M. E., Effect of a skin moisturizer on the water distribution in human stratum corneum, J. Invest. Dermatol., 81, 446–448, 1983.
55. Wilson, D., Berardesca, E., and Maibach, H. I., In vivo transepidermal water loss and skin surface hydration in assessment of moisturization and soap effects, Int. J. Cosmet. Sci., 10, 201–211, 1988.
56. Wilson, D. R., and Maibach, H. I., A review of transepidermal water loss, in Neonatal Skin, Structure and Function (Maibach, H. I., and Boisits, E. K., eds.), Marcel Dekker, New York-Basel, 1982, pp. 83–100.
57. Grice, K. A., Transepidermal water loss, in The Physiology and Pathophysiology of the Skin, Vol. 6 (A. Jarrett, ed.), Academic Press, London, 1980, pp. 2115–2127.
58. Pinnagoda, J., Tupker, R. A., Agner, T., and Serup, J., Guidelines for transepidermal water loss (TEWL) measurement, Contact Dermat. 22, 164–178, 1990.

59. Nilsson, G. E., Measurement of water exchange through skin, Med. Biol. Eng. Comput., 15, 209–218, 1977.
60. Wheldon, A. E., and Monteith, J. L., Performance of a skin Evaporimeter, Med. Biol. Eng. Comput., 18, 201–205, 1980.
61. Pinnagoda, J., Tupker, R. A., Coenraads, P. J., and Nater, J. P., Comparability and reproducibility of the results of water loss measurements: a study of 4 evaporimeters, Contact Dermat., 20, 241–246, 1989.
62. Blichmann, C. W., and Serup, J., Reproducibility and variability of transepidermal water loss measurements, Acta Dermato-venereologica, 67, 206–210, 1987.
63. Pinnagoda, J., Tupker, R. A., Smit, J. A., Coenraads, P. J., and Nater, J. P., The intra- and inter-individual variability and reliability of transepidermal water loss measurements, Contact Dermat., 21, 255–259, 1989.
64. ServoMed Evaporimeters Operation Handbook, ServoMed, Vallingby, Stockholm, Sweden, 1981.
65. A Guide to Water Evaporation Rate Measurement, ServoMed, Vallingby, Stockholm, Sweden.
66. Seitz, J. C., and Spencer, T. S., Use of capacitive evaporimetry to measure effect of topical ingredients on transepidermal water loss, J. Invest. Dermatol., 78, 351, 1982.
67. Grice, K. A., Sattar, H., Sharratt, M., and Baker, H., Skin temperature and transepidermal water loss, J. Invest. Dermatol., 57, 109–110, 1971.
68. Mathias, C. G. T., Wilson, D. M., and Maibach, H. I., Transepidermal water loss as a function of skin surface temperature, J. Invest. Dermatol., 77, 219–110, 1981.
69. Cooper, E. R., and Van Duzee, B. F., Diffusion theory analysis of transepidermal water loss through occlusive films, J. Soc. Cosmet. Chem., 27, 555–558, 1976.
70. Rietschel, R. L., and De Villez, R. L., Skin moisturization without occlusion, Cutis, 27, 543–544, 1981.
71. Neutrogena Skin Care Research Institute Report, Clinical testing to verify claims of skin improvement made for five leading hand moisturizers, 1985.
72. Jacques, S., Water content and concentration profile in human stratum corneum, Ph.D Thesis, University of California, Berkeley, 1984.
73. Rietschel, R. L., A skin moisturization assay, J. Soc. Cosmet. Chem., 30, 369–373, 1979.
74. Jacques, S. L., A linear measurement of the water content of the stratum corneum of human skin using a microwave probe, presented at the 32nd Annual Conference on Engineering in Medicine and Biology, Denver, Colorado, 1979.
75. Appapillai, Y., Lewis, L., unpublished data.
76. Baier, R. E., Noninvasive, rapid characterization of human skin chemistry in situ, J. Soc. Cosmet. Chem., 29, 283–306, 1978.
77. Salter, D. C., Instrumental methods of assessing skin moisturization, Cosmet. Toiletr., 102, 103–109, 1987.
78. Tagami, H., Ohi, M., Iwatsuki, K., Kanamaru, Y., and Ichijo, B., Evaluation of the skin surface hydration in vivo by electrical measurement, J. Invest. Dermatol., 75, 500–507, 1980.

79. Tagami, H., Electrical measurement of the water content of the skin surface, Cosmet. Toilet., 97, 39–47, 1982.

80. Tagami, H., Kanamuru, Y., Inoue, K., Suehisa, S., Inoue, F., Iwatsuki, K., Yoshikuni, K., and Yamada, M., Water sorption-desorption test of this skin for in vivo functional assessment of the stratum corneum, J. Invest. Dermatol., 78, 425–428, 1982.

81. Wortzman, M., and Grove, G. L., Assessment of long-lasting moisturizers by skin surface electrical hydrometry, presented at the 7th International Symposium of Bioengineering and the Skin, Milwaukee, June 16–18, 1988.

82. Clar, E. J., Her, C. P., and Sturelle, C. G., Skin impedance and moisturization, J. Soc. Cosmet. Chem., 26, 337–353, 1975.

83. Serban, G. P., Henry, S. M., Cotty, V. F., and Marcus, A. D., In vivo evaluation of skin lotions by electrical capacitance and conductance, J. Soc. Cosmet. Chem., 32, 421–435, 1981.

84. Friedmann, P. S., The skin as a permeability barrier, in Scientific Basis of Dermatology (A. J. Thody and P. S. Friedmann, eds.), Longman Group UK Ltd., 1986, p. 28.

85. Nole, G. E, Boisits, E. K., and Thaman, L., The Salter complex impedance device as an instrument to measure the hydration level of the stratum corneum in vivo, Bioeng. Skin, 4, 285–296, 1988.

86. Salter, D. C., Quantifying skin disease and healing in vivo using electrical impedance measurements, in Non-Invasive Physiological Measurements, Vol. 1 (P. Rolfe, ed.), Academic Press, London-New York-San Francisco, 1979, pp. 21–64.

87. Leveque, J. L., and De Rigal, J., Impedance methods for studying skin moisturization, J. Soc. Cosmet. Chem., 34, 419–428, 1983.

88. Walling. P. L., and Dabney, J. M., Moisture in skin by near-infrared reflectance spectroscopy, J. Soc. Cosmet. Chem., 40, 151–171, 1989.

89. Salter, D. C., Electrical phenomenology of the keratins of human skin, Special Session on Biological Materials, Institution of Electrical and Electronics Engineers Annual International Conference on Electrical Insulation and Dielectric phenomena, Buck Hill Falls, PA, October 16–20, 1983, published in the IEEE Annual Report of the CEIDP, 1983, pp. 397–402.

90. Kohli, R., Archer, W. I., Roberts, J. M. C., Cochran, A. J., and Li Wan Po, A., Impedance measurements for non-invasive monitoring of skin hydration—a reassessment, Int. J. Pharmaceutics, 26, 275–287, 1985.

91. Edwards, C., The electrical properties of skin, in Nature of the Skin (R. M. Marks, S. P. Barton, C. Edwards, eds.), MTP Press Ltd., Lancaster, England, 1988, pp. 209–213.

92. Tregear, R. T., Physical functions of skin. Academic Press, London-New York, 1966, pp. 53–72.

93. Yamamoto, Y., Yamamoto, T., and Ozawa, T., Characteristics of skin admittance for dry electrodes and the measurement of skin moisturization. Med. Biol. Eng. Comput., 24, 71–77, 1986.

94. Rosenberg, B., Electrical conductivity of proteins. II. Semiconduction in crystalline bovine hemoglobin, J. Chem. Phys., 36, 816–823, 1962.

95. Pethig, R., Dielectric and Electronic Properties of Biological Materials, John Wiley and Sons, Chichester, 1979.
96. Salter, D. C., Studies in the Measurement, Form and Interpretation of Some Electrical Properties of Normal and Pathological Human Skin In Vivo (3 volumes), D. Phil. thesis, University of Oxford, 1981.
97. Cole, K. S., Permeability and impermeability of cell membranes for ions. Cold Spring Harbour Symposium, Quant. Biol. 4, 110, 1940.
98. Kligman, A. M., Regression method for assessing the efficacy of moisturizers, Cosmet. Toilet., 93, 27–35, 1978.

12

Chemistry and Biology of Monoglycerides in Cosmetic Formulations

Jon J. Kabara

Lauricidin, Inc., Galena, Illinois

I. INTRODUCTION

Although the use of animal and vegetable fats (triglycerides) was recorded in ancient times [1 and references therein], their breakdown products, mono and diglycerides have a shorter history. A self-emulsifying glyceryl stearate (Tegin) was introduced commercially around 1927. Much of what we know about this group of products comes from their use in the food industry as emulsifiers. Their functional importance derives from their ability to make oil and water phases mix to form an emulsion. It is estimated that about 65% of the emulsifier used in the food industry are mono and diglycerides. It should be mentioned that it is primarily the monoester which is responsible for emulsification activity. While historically, monoglycerides were used as emulsifiers, present-day studies suggest wider application for this unique group of modified fats. Depending on their fatty acid chain length, monoglycerides in cosmetics may function as emollients, deodorants, preservatives, and/or transdernal agents. Despite their multi-functional characteristics, these products are the safest group of chemicals known and indeed are food substances per se. The monoglycerides of edible fatty acids are approved for food use and are regarded as Generally Recognized As Safe (GRAS). This recognition can be found under 121.101 of the Code of Federal Regulations (2 Dec. 1964 *Federal Register*). The latter designation (GRAS) is given only to a handful of chemicals whose safety is hardly in question.

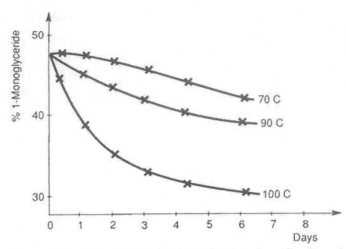

Figure 12.1 Reduction in monoglyceride content as a function of time and temperature (Ref. 3).

II. PREPARATION OF MONOGLYCERIDES

Monodiglycerides were first synthesized in 1853, but were not produced and used commercially until the 1920s.

If a triglyceride is reacted with glycerol at a high temperature, for example, at 200°C. with alkaline catalysis, the resulting product is a mixture of mono-, di-, and triglycerides as well as traces of unreacted glycerol [2]. Provided that the reaction temperature is high enough, the composition of the reaction product will depend only on the relative ratio of fat/glycerol. For instance, a normal commercial monoglyceride may be obtained as a result of the reaction between ca. 1 mol triglyceride and 2 mol glycerol. At lower temperatures, however, a product like this is no longer in an equilibrium state; and, as a consequence, the monoglyceride content tends to decrease as a function of temperature, time, and catalyst traces (Figure 12.1). It is thus important to avoid extended heating of monoglycerides [3].

Normally the composition of commercial so-called 40% monodiglyceride varies considerably:

Monoglyceride	35–40%
Diglyceride	45–50%
Triglyceride	10–15%
Free glycerol	1–7%

In practice the highest monoester content that can be achieved without distillation is about 60%. A detailed and critical review of the process was presented in 1982 by Sonntag [4].

Monoglycerides exist in two isomeric forms, beta or the 2-form and the alpha or 1-form.

$$
\begin{array}{ccc}
\text{H} & & \text{H} \\
| & & | \\
\text{H—C—OR} & & \text{H—C—OH} \\
| & & | \\
\text{H—C—OH} & \rightleftarrows & \text{H—C—OR} \\
| & & | \\
\text{H—C—OH} & & \text{H—C—OH} \\
| & & | \\
\text{H} & & \text{H} \\
\text{alpha } (\alpha)\text{-form} & & \text{beta } (\beta)\text{-form} \\
90\text{–}95\% & & 5\text{–}10\%
\end{array}
$$

The relative amount of the two isomers is a result of an equilibrium which is temperature dependent. At 200°C, the equilibrium content of 1-monoglyceride is 82%, but at 20°C, it is 95%. Since the distilling temperature is relatively high, low (80%) one-monocontent might be obtained even though the total monoglyceride content is in the order of 95% in commercially distilled products (Table 12.1). Only the alpha form responds to periodic oxidation, which is the basic reaction in the normally applied chemical determination of mono-glyceride content.

Distilled monoglycerides are produced batchwise or continuously, by a two-step production process. The first step results in a monodiglyceride mixture, which can either be finished and brought into the trade, for instance, in the form of spray crystallized powder or flakes. In the second step, the monodi-glycerides are further concentrated through molecular distillation.

The essential equipment for the production of distilled monoglycerides is the molecular still. It operates under extremely low pressure (0.001 mmHg). At higher pressures undesireable products are obtained. It is characteristic of the process that a large proportion of the di- and triglyceride as well as excess of glycerol is recycled.

Table 12.1 Composition of Commercially Distilled Monoglyceride

Monoglycerides	94.2%
Diglycerides	3.6%
Triglycerides	0.7%
Fatty acid	0.8%
Glycerol	0.7%

III. ANALYSIS OF MONOGLYCERIDES

The need for analysis of these lipids may arise under several conditions of their production and use. Analysis provides for production control and the detection of batch to batch variation; products purchased from different suppliers can be compared; especially where the monoglyceride is used as a preservative [i.e., distilled monolaurin (Lauricidin)].

Glycerides may be mixtures of individual chemical species which differ by degree of esterification, by chain length of the fatty acid, by degree and type of unsaturation, and by the positional isomer (α or β) of the fatty acid.

Numerous analytical methods are available. The method most commonly used where the "pure" chemical specie (alpha monoglyceride) is available is by periodic acid oxidation. The quantitative oxidation of the 1-monoacylglycerol by excess periodic acid and the back-titration of the excess periodic acid [5].

This procedure shows good producibility and accuracy, but it also has several disadvantages: (a) the method is tedious and time consuming, (b) free glycerol interferes unless it is extracted by water or salt solution, and (c) the

Table 12.2 Fatty Acid Composition of Lauricidins (wt % as Methyl Esters)[a]

Fatty acid	Old (1977)	New (1980)	Powder (1985)
8:0	0.1	—	0.1
10:0	0.5	0.1	0.7
12:0	92.5	94.5	95.8
14:0	6.4	3.6	2.8
16:0	0.3	0.9	0.2
18:0	0.1	0.6	0.2
18:1	0.1	0.2	0.2
18:2	—	0.1	—

[a]Methyl esters were prepared by the following procedure: The lauricidin sample (10 mg) was dissolved in 0.5 N sodium methoxide in methanol (1 ml) in a screw cap vial. The solution was maintained overnight at room temperature then hexane (2 ml) was added. An aliquot of the hexane layer was taken for GC analysis. GC conditions: Column, glass WCOT SILAR-5CP (64 m × 0.28 mm ID); column temp = 110–170°C at 1°C/min; injector and detector temp = 220°C; carrier gas (N_2) flow rate = 0.78 ml/min.

Table 12.3 Compositions of Monoacylglycerals in Lauricidin[a]

Acyl group	Position	Old (1977)			New (1980)			Powder (1985)		
8:0	2	—	—	—	—	—	—	—	—	—
	1(3)	0.1	0.1	100	—	—	—	0.1	0.1	100
10:0	2	—	—	—	—	—	—	—	—	—
	1(3)	0.4	0.4	100	0.1	0.1	100	0.7	0.7	100
12:0	2	0.1	0.2	0.2	5.2	5.4	5.6	0.3	0.3	0.3
	1(3)	92.7	92.8	99.8	88.1	90.6	94.4	95.9	96.1	99.7
14:0	2	—	—	—	0.2	0.2	5.7	—	—	—
	1(3)	6.3	6.3	100	3.1	3.2	94.3	2.6	2.6	100
16:0	2	—	—	—	—	—	—	—	—	—
	1(3)	0.2	0.2	100	0.5	0.5	100	0.1	0.2	100
Others[b]		0.2	—	—	2.8	—	—	0.3	—	—

[a]The first, second, and last columns in each lauricidin indicate peak area % of all components emerged on chromatograms[1], wt % of monoacylglycerols only,[2] and ratio of 1 (3) to 2-monoacylglycerols,[3] respectively. TMS ethers were prepared according to the procedure described by Kuksis (A. Kuksis, in Lipid Chromatographic Analysis, 2nd Edition, vol. 1, Edited by G. V. Marinetti, Marcel Dekker, New York and Basel, 1976, p. 215). GC conditions: Column, glass WCOT SILAR-SCP (34 m $\times$ 0.28 mm); column temp = 160–180°C at 0.5°C/min; injector and detector temp = 220°C; carrier gas (N_2) flow rate = 0.85 ml/min.
[b]Does not include monacylglycerols.

method determines merely the alpha monoacylglycerol content, and equilibrium between the mono- and 2-monoacylglycerols is assumed. Moreover, this equilibrium is dependent on temperature, and thus a number of factors have to be ascertained before an analysis using iodine can be performed.

More sophisticated methods developed by lipid chemists are available. These are usually less time consuming and more accurate, because isomers (α and β monoglycerides) can be quantified. Thin-layer chromatography (TLC), although not usually used to quantify lipid mixtures can and has been used [6–8]. TLC is particularly suited to large numbers of samples where semi-quantitative data are required to distinguish α and β isomer changes in a sample and/or for quality control. The author has used silica gel G plates impregnated with 10% boric acid and $CHCl_3$:MeOH (98:4, v/v) as the developing solvent (courtesy of Prof. Toru Takagi, Japan).

In our experience gas chromatographic separation and analysis of glyceride mixtures provided by Itabashi and Takagi has been very satisfactory with monoglycerides [9]. Analysis on batches of glycerol monolaurate (Lauricidin) are presented in Tables 12.2 and 12.3. The monoester content of this commercial product was always 94 $\pm$ 2%; the lauric acid greater than 92%.

One problem of gas-liquid chromatography (GLC) is the need to make volatile derivatives of the chemical species being analyzed. In the above procedure either acetate and trimethylsilyl (TMS) derivatives were prepared prior to gas chromatography. Because of the nonquantitative silylation of monoglycerides, correction factors need to be employed. While yields of ethers are high, they decrease with increasing chain length of the fatty acid (Table 12.4).

An analytical method which does not depend on derivative formation is high-performance liquid chromatography (HPLC). The procedure is easy and rapid to carry out. The method has been shown to give reproducible results and quantitative data with standard deviations comparable or better than the periodic acid oxidation method. In an early (1978) report by Riisom and Hoffmeyer [11], saturated mono-, di-, and triglycerides were separated on 10μM LiChrosob DIOL by eluting components isocratically with isooctane 95%: isopropanol 5% (v/v) [11]. The eluted components were monitored by ultraviolet (UV) absorption at 213 nm. Limitations of this method are that the 1- and 2-monoester are not separated sufficiently to allow quantification of the individual peaks; free fatty acids and diglycerides elute together. Also, the ultraviolet detection is not quantitative for mixtures containing a variety of unsaturated components.

HPLC utilizing infrafed (IR) detection reportedly overcomes some of the shortcomings of earlier methods [12]. To prevent the tailing of free fatty acids into the diglyceride area, esterification of the free fatty acids was necessary (Editor's note: use of a fluoricil column would achieve separation without esterification). Another method purported to be simple and convenient was presented by Takano and Kondoh [13]. This method depends upon reversed-phase HPLC for the simultaneous determination of the homologous distribution and the ratio of positional isomers of monoglycerides in commercial products (Fig. 12.2).

Table 12.4 Results Obtained in Extraction Experiments with 0.10% Each of Monolaurate, Monomyristate, Monopalmitate and Monostearate Dissolved in Soya Oil[a]

Monoglyceride	Extraction yield (%)		
	Mean	±	SD
Monolaurate	94	±	5.9
Monomyristate	91	±	4.7
Monopalmitate	86	±	3.4
Monostearate	82	±	6.6

[a]The results are based upon five experiments.
Source: From Ref. 10.

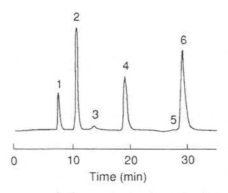

Figure 12.2 Analysis of commercial margarine with the glyceride-selective detector. 1, glycerin; 2, 1-monolaurin (internal standard); 3, 1-monomyristin; 4, 1-monopalmitin; 5, 2-monostearin; 6, 1-monostearin. [13]

IV. TOXICOLOGY AND SAFETY OF GLYCERIDES

The nutritional effects of mono-, di-, and triglycerides are similar since fats are metabolized to mono- and diglycerides during the normal process of fat absorbtion. Therefore the safety assessment of partial glycerides is based on the rationale that amounts of material that are formed in the body naturally, continuously and abundantly are nontoxic.

Numerous feeding experiments have been carried out with mono- and diglycerides dating back to early (1941) studies by Braun and Shrewsbury [14]. The only deleterious effect noted from the feeding of 8–24% monostearin or monolinolein in the diets of rats for 8 weeks was a somewhat slower growth among the monostearin-treated rats.

Hine et al. [15] reported that no measurable irritation was produced by the application of glycerin to the skin and eyes of rabbits. Subcutaneous injections of mono- and diacetin were noted by Li et al. [16] to occasionally cause local irritation in mice and rats. In the rabbit eye, 50% monacetin caused only a slight degree of irritation, while diacetin and triacetin in similar concentrations caused marked congestion and moderate edema. Daily application, for 45 days, of a 30% acetoglyceride emulsion to the skin of albino guinea pigs was reported by Ambrose and Robbins [17] to result in no local irritation of systemic reactions.

In a study in which weanling male and female Holtzman rats were fed diets with or without 50% saturated, partially acetylated monoglycerides, Hertig and Crain [18] noted the appearance of a foreign body-type reaction in the body fat occurring within 8 weeks of initiation of the test.

In the only reported feeding study using chicks, ten 1-day-old chicks were fed diets containing diacetyl tartaryl glycerol monostearate, glycerol lacto-palmitate, or succinylated monoglyceride at dosage levels of 570 mg/kg, lacto-palmitate, or succinylated monoglyceride at dosage levels of 570 mg/kg, 2.85 g/kg, and 8.55 g/kg for a period of 90 days [19]. Diacetyl tartaryl glycerol monostearate caused growth depression at all levels at 7 weeks, slight hyper-emia of the duodenum and ileum at the lower levels, and moderate to severe hyperemia at the highest concentrations. Gut erosion was observed at the high level, with only slight changes at the intermediate level. Glycerol lactopalmi-tate caused growth depression at 90 days. Liver weight was slightly increased and there were instances of very slight hyperemia and gut erosion at the inter-mediate level at 90 days. Succinylated monoglyceride caused very slight growth retardation at 90 days. Cecal size was increased at each level, and slight hy-peremia and one instance of gut erosion occurred at the intermediate level.

When triglycerides of different chain lengths (C_2-C_{11}) were evaluated, there was a great difference in the toxicity of the various triglycerides [20]. The most toxic value (97 ± 4 mg/kg) was noted for triisovalerin (C_5). With longer-chain (C_9-C_{11}) fats, the median lethal dose (LD_{50}) was more than 10 g/kg and could not be determined. The LD_{50} was lowest for triisovalerin and increased when the length of the fatty acid chain was less or greater than C_5.

For a period of two years, Fitzhugh et al. [21] maintained 24 rats on a basal diet with a 25% supplement of Myverol (glyceryl monostearate). Compared with a similar number of control animals, growth and longevity of the test rats was normal and detailed microscopic pathological examinations of all major organs and tissues revealed only a single change: an increase in the number of calcified renal tubular casts attributable to the treatment.

In a study by Mattson et al. [22], in which 12 groups of 10 weanling male Sprague-Dawley rats were fed diets containing various pure mono-, di-, and triglycerides at a level of 25% for a period of 10 weeks, growth of all groups was normal and autopsies revealed no peculiarities. Ames et al. [23] reported that the feeding of monoglycerides (derived from the fatty acids of cottonseed oil) to rats for three generations disclosed no untoward effects attributable to the ingestion of the compounds. On the basis of a feeding study in which rats were fed mono-, di-, and triglycerides (prepared from a mixture of partially hydrogenated soybean and cottonseed oils) at levels of 15 or 25% of the diet for 70 days, Harris and Sherman [24] stated that these compounds exhibit no differences in caloric efficiency, nor do they produce any differences in body weight gain.

Growth, organ weights, and mortality were normal for all groups of 8 young male rats which were maintained for 608 days on diets containing 10% glyceryl lactopalmitate with low monoglyceride content, 10% glyceryl lacto-oleate with high or low monoglyceride content, 10% polyglycerol lactooleate with high monoglyceride content, or 10% acetylated tartaric acid ester of glycerol monostearate [25].

In a study by Orten and Dajani [26], in which male weanling hamsters were maintained for 28 weeks on diets containing 5–15% glyceryl monostearate, the animals of the 15% groups showed a slight weight loss, while the 5% hamsters exhibited a higher weight gain than the controls; no consistent pathological changes were observed.

In 1978, the entire subject of glyceride safety was reviewed, and reassuringly found that these types of chemicals are free from adverse systemic topical effects [27].

V. EMULSIFIERS

While fats and oils have been used as cosmetics since antiquity, there has been a shift in modern times to products which are nongreasy. The production of stable water/oil or oil/water emulsions realized this objective. Among the earliest derivatives used for this cosmetic purpose were "monoglycerides" such as glyceryl monosterate (GMS). It should be noted that commercial monoglycerides are actually mixtures of mono- (35–60%), di- (35–50%) glycerides with the balance comprising of triglycerides, glycerol, and free fatty acids. It is only as recently as 1969 that distilled monoesters have been made commercially available to the food industry. These glycerides are 94–96% monoester content. Their application to cosmetic formulae is only now being appreciated. Preparations of distilled monoesters are almost odorless, colorless, and among their remarkable properties, they produce smaller particle size emulsions not attainable using commercial monoglyceride preparations, and possess antimicrobial activity and transdermal effects.

The ingredient declaration which is necessary or which appears on cosmetic products is written in CTFA-ese (Cosmetic Toiletry and Fragrance Association). Even for an organic chemist, the label has to be decoded. For a chemical to be listed, the purity of the ingredient in question is not an issue in the *Dictionary of Cosmetics Ingredients* compiled by the CTFA. Thus "glyceryl monolaurate" may indicate a partial glyceride of lauric acid but give no information to the formulator as to the purity of the lipid derivative. This serious situation probably requires appropriate action by the CTFA committee involved.

Glyceryl monostearate functions as an emulsifier, stabilizer, thickener, opacifier, or emollient in products such as skin creams and lotions, antiperspirants (creams and roll-ons), shampoos, cream hair rinses and conditioners, as well as suntan creams and lotions.

Monoesters of saturated fatty acids are important for use in cosmetic preparation even though the highly purified glycerides are somewhat expensive. While monoglycerides of unsaturated fatty acids have unique properties, it is only the form of the double bond that is biologically active. In large-scale commercial production of unsaturated monoglycerides, some trans form of the unsaturated fatty acids becomes evident. Saturated fatty acids obtained from animal or vegetable fats (C_{14}, C_{16}, C_{18}) predominate, with glycerol monostearate being the most popular.

However, fatty acids (C_8–C_{12}) obtained from coconut and palm kernel oils have unique properties. Glyceryl monolaurate (GML) is the most unusual and has multifunctional properties. The distilled version of GML has the following properties.

Stability to air oxidation
Chemically neutral in the presence of active ingredients
Low adour and pure white
Outstanding emulsification properties
Excellent emolliency
Minimal changes due to pH
Low or no skin irritation
Excellent spread and "feel" on the skin
Penetrating (transdermal) properties
Antimicrobial against bacteria, fungi/yeast, and lipid-coated viruses (herpes I and II)

Saturated monoglyceride can be used in a variety of preparations such as lotions, creams, and oils. Because of their excellent spreading and penetrating properties, these monoglycerides do not give a greasy feel to the skin. Their skin affinity allows greater penetration of active ingredients into skin. It is suggested that creams made with these saturated monoglycerids would be useful in pharmaceutical preparations.

Monoglycerides in contrast to di- and triglycerides are excellent emulsifiers because they contain a hydrophilic and hydrophobic portion. The hydrophilic portion of the emulsifier extends itself into the aqueous phase of a cosmetic and creates a "membrane" between the water-oil boundaries. Fat particles are encapsulated, and an emulsion is formed. Distilled monoglycerides in contrast to mixed glycerides, form emulsions with smaller particle size and therefore are more stable and easier to preserve [28]. Neutralization of small amounts of free fatty acid present in the monoglyceride preparation can be carried out by adding an equivalent amount of sodium hydroxide. The sodium salt of the free

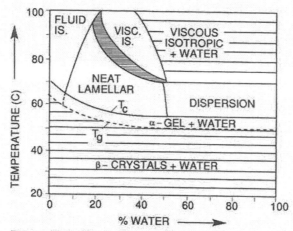

Figure 12.3 Phase diagram of an aqueous system of a saturated distilled monoglyceride (fatty acid composition as fully hydrogenated lard).

fatty acid will help stabilize the α-crystalline gel structure formed when the monoglyceride-water dispersion is cooled to room temperature.

Optimum stability of pure oil/water (o/w) emulsions are obtained by blends of emulsifiers of medium HLP (hydrophilic-lipophilic balance). Monoglycerides which have HLB less than 4 must be combined with emulsifiers which are more hydrophilic (i.e., HLB of 8-12). Such mixtures form more condensed monolayers at the interface.

Precautions should be taken during formulation with monoglycerides to avoid overheating the mixture. Overheating causes phase changes and separation. Figure 12.3 shows the separation of a highly viscous isotropic phase.

An emulsion is thermodynamically unstable and will separate into two liquid phases. Before complete separation, dispersed droplets within an emulsion will coalesce to form larger sized particles. Emulsions that have smaller particle size are more stable than those with larger particles. Distilled monoglycerides are more effective than nondistilled commercial monoglycerides because they facilitate formation of smaller droplets within the emulsion.

Due to isomerization of the monoglyceride, the usual mixture at thermodynamic equilibrium contains a weight ratio of about 88:12 of the 1:2 isomer [29]. While the biological properties of the two isomers are similar, there are differences in their phase properties. These differences have been described as being due to the cross-section of the polar group and hydrocarbon chain [30]. Thus, the transition lamellar → cubic and cubic → reversed hexagonal phases in 1-monolein occurs at lower temperatures than for the 2-isomer. These liquid crystal states, from which monoglycerides can exit, are important to the formulator, since it can control the solubility of other lipophilic substances in the formula.

A more complete discussion of phase changes for monoglycerides is presented below.

VI. PHASE STATE OF MONOGLYCERIDES

How well a monoglyceride performs in a cosmetic emulsion depends on the purity and the type of fatty acid in its composition. The latter determines the physical and chemical properties of the emulsifier.

Monoglycerides are known to be *polymorphic* and are classified according to sub-α, α, β' and β crystalline forms. These are arranged in order of increasing melting point. Thermodynamically the β-crystalline form is the most stable. Monoglycerides of saturated fats are marketed in the β-crystalline form.

Lipid crystals of saturated monoglycerides can be produced by mixing the lipid (10–50%) in water and heating the mixture. The mixture is heated to slightly above the "conversion" temperature. The "conversion" temperature is considered the lowest temperature at which lipid particles in contact with an excess of water absorb water and are converted to spherical particles (liposomes) with marked birefringence. In the case of 1-monolaurin, the conversion temperature is 45°C. The mixture is therefore heated to 50–60°C. This temperature is maintained until equilibrium has been achieved and the mixture is allowed to cool at a rate of 0.5–5.0°C/min. Cooling is continued to room temperature with constant stirring. The *hydrophilic* lipid produced in this manner are thin leaf-shaped crystals with two hydroxyl groups inclined toward the surface of the crystal.

Lipophilic surface crystals can be produced by crystallization of the molten phase.

Early (prior to 1950) phase diagrams dealing with monoglyceride were not very useful, because the purity of the nonionics was always in question.

When distilled monoglycerides (C_{14} or higher) are melted and added to water at the same (melting) temperature, a firm "gel" is formed [31]. The gel, usually containing about 15–25% by weight of water, is not dispersible in water and is not easily soluble in oil. If the excess water is decanted, the gel can be worked mechanically into an optically clear solid of a consistency resembling very heavy stopcock grease.

When a gel is cooled below the melting point of the monoglyceride, crystals begin to form. If crystallization is appreciable, the gel structure deteriorates and a white creamy paste is the result.

Thus the lower temperature limit of stability is related to melting point; 1-monostearin (m.p. 81°C) gels are stable at about 80°C., and 1-monoolein (m.p. 35°C.) gels are stable at about 20°C. Mixed monoglycer-

Table 12.5 Gel Points of Monoglycerides

Monoglycerides	m.p. (°C)	Gel point (°C)
Distilled monolaurin	54–57	No gel to 99°C
Distilled monomyristin	62–65	85
Distilled monopalmitin	67–68	68
Distilled monostearin	70–72	70
1-Monomyristin	70.5	No gel to 99°C
1-Monopalmitin	74–74.5	82
1-Monostearin	81–81.5	81
Distilled monolein	35	<20

Distilled products usually contain about 90–92% 1-monoglyceride, 5–8% of 2-monoglyceride, and 1–3% of glycerine, diglycerides, and free fatty acids.
Source: From Ref. 31.

ides, such as those from natural fats and oils yield mixtures of gel and paste which are not readily dispersed in water.

The upper temperature limit of stability is very close to 100°C. In most cases, if a portion of gel is heated on a spatula, the water begins to boil out before the solid structure deteriorates. Highly unsaturated monoglycerides (e.g., from cottonseed oil or bean oil) form gels which change at about 98°C.

At about 10–15°C below the melting point, the monoglyceride, became smooth creamy dispersion. Finally, the monoglyceride coagulated into a non-dispersible lump which was translucent to transparent. This last conversion was called the "gel point." The exact temperature of gelatin is dependent upon molecular weight of the fatty acid (monolaurin does not gel, but mono-palmitin does) and upon the purity of the monoglyceride (Table 12.5).

Additives can prevent gelation, with triglycerides (15–20% required) about twice as effective as diglycerides (30–40% required). Highly hydrophilic co-emulsifiers prevent gelation, resulting in one of three types of emulsions at least two of which are thixotropic.

Effective coemulsifiers are the strongly hydrophilic types, which are partially water soluble, and may be anionic, nonionic, or cationic. Examples which have been used successfully include ordinary sodium soaps, sodium lauryl sulfate, polyoxy compounds (Tween-80), sucrose monopalmitate, quaternaries, and certain protein hydrolysates. The more hydrophilic materials are more effective, and less is required to accomplish the same result. An optimum quantity of coemulsifier exists, at least when sodium oleate is the coemulsifier.

Water is partially soluble in monoglycerides and in monoglyceride-containing blends. By this technique many water-soluble materials can be incorporated into an oil solution.

Mono- and diglyceride mixtures have been marketed in a "self-emulsified" or soap-containing form for many years. The relatively weak, water-in-oil emulsifying properties of mono- and diglyceride mixtures permit oil-in-water emulsions to be formed by this technique. It is also possible to disperse distilled monoglycerides in water, using soap; the appearance of the emulsion is different.

Preparation of such a coemulsified blend can be effected by mixing water, powdered soap, and molten monoglyceride and applying vigorous agitation. The temperature of agitation must be above the gel point. High agitation, such as that from a Waring Blender or a high-pressure dairy homogenizer provides the greatest degree of homogeneity and stability of the resultant emulsion.

Three types of monoglyceride coemulsified blends exist. The first (type A) exists at or above the normal gel temperature of the monoglyceride. It is amorphous by x-ray diffraction (as is the gel), is quite translucent, in some cases transparent, and has a consistency resembling egg white or a fresh dispersion of starch in hot water. It is thixotropic (Table 12.6). A dilute type A coemulsion, when whirled in a container, has been observed to back up as much as 180°, demonstrating an extremely elastic nature.

Type B emulsions result when type A emulsions are cooled appreciably below the gel temperature of the monoglyceride. Type B is a crystalline dispersion a consistency of translucent pudding, not quite as viscous but still thixotropic (Table 12.6). It resembles a cooled or aged dispersion of starch in water. It has lost most of the elasticity and filamentous appearance. The crystal structure is alpha.

Upon standing, most type B emulsions change to type C. The latter is a white, opaque dispersion in water which resembles a normal dispersion of hard fat.

A characteristic feature of distilled monoglycerides is their ability to form lyotropic mesophases. This is observed when the emulsifiers, which are normally insoluble in water, are dispersed in water and heated. At a certain tem-

Table 12.6 Thixotropy of Type As and B Emulsions[a]

Brookfield, viscometer speed (rpm)	Measured viscosity (cp)	
	Type A (65°C)	Type B (35°C)
20	430	280
10	700	400
4	1600	600
2	2300	1000

[a]Ten g distilled monoglycerides of hydrogenated lard, 200 g water, and 0.5 g sodium oleate.

perature close to the melting point, presumably hydrocarbon chains of the molecules become fluid, and at the same time the water will penetrate between the polar groups, resulting in the formation of liquid crystalline structures, which can be of lamellar, hexagonal, or cubic types (Fig. 12.4).

The polar groups of the emulsifiers are oriented toward the water, which in the lamellar phase is present in films, alternating with the lipid bilayers. In the hexagonal and cubic lyotropic mesophases, the molecules are arranged in cylinders and spheres, respectively. The structure of such mesophases can be studied by x-ray diffraction, and structural parameters such as thickness of the water layer and the lipid bilayer can be determined.

The mesomorphic behavior of distilled monoglycerides in water is demonstrated in a binary phase diagram (Fig. 12.3).

At temperatures slightly above the melting point of the saturated monoglyceride, a lamellar mesophase is formed. At higher temperatures, a cubic viscous isotropic mesophase is produced, and anyone who has worked with distilled monoglycerides knows how important it is to avoid the formation of this phase because of its stickiness and the difficulty of removing it, especially from long product pipes. While it is thus important to have a practical knowledge of the mesomorphic phases of monoglycerides and other emulsifiers, it is even more important for their proper application.

V. PHASE BEHAVIOR OF MONOGLYCERIDES

In order to properly formulate monoglycerides into cosmetic products, it is necessary to understand the various phase behavior of monoglycerides. Briefly, monoglyceride–H_2O systems exhibit a variety of mesomorphic (between liquid and crystal) states analogous to those of soap and H_2O systems.

Water is unable to dissolve more than a small (ca. 100 ppm) fraction of a monoglyceride. However, near and above their melting point, monoglycerides incorporate (in various mesomorphic phases) an extra 20% of water. Water lowers the melting point of crystalline monoglyceride, but not before a neat phase is encountered. Neat is the only mesomorphic phase of monolaurin and it occupies a large part of the phase diagram. "Neat" is recognized to be lamellar in structure. With trans unsaturated monoglycerides, the phase diagram is similar to monostearin, but with the temperature scale shifted 30°C downward. Monoolein (unsaturation) shows a further drop in the neat region and an outcropping of middle phase area. Details on phase diagrams of various monoglycerides from C_{12} to C_{22} can be found in a 1965 report by Lutton [32]. The report by Lawrence and McDonald details monolaurin—water systems [33].

The best and most complete account of phase behavior for saturated distilled monoglycerides was described in 1969 by Krog and Larsson [34] and in

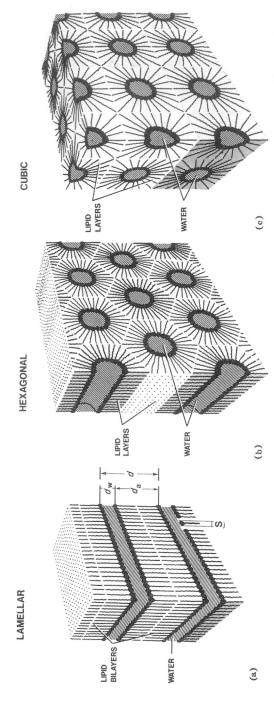

Figure 12.4 Schematic structure models of (a) the lamellar mesophase, (b) the hexagonal II mesophase and (c) the cubic mesophase (courtesy of A.V.I. Publishing Co.).

1975 by Krog [35]. These workers presented details on how monoglycerides behave in aqueous systems under different conditions with regard to concentration temperature, pH, etc., and the relevance this may have in their usage. Much of the discussion below is taken from Krog [35].

When emulsifier crystals are mixed with water and the temperature raised above the Kraft point (T_c temperature where hydrocarbon chains will transform from the solid state to a liquid like state), water penetrates through the layers of the polar groups and a lamellar mesomorphic phase is formed. If the temperature is raised higher, the lamellar structure changes and other mesophases may exist (i.e., cubic or hexagonal). This can be noted in Figure 12.4. On cooling below the Kraft point, the hydrocarbon chains crystallize again, normally in the α-crystalline form with hexagonal chain packing. The same volume of water may still be located between the lipid bilayers and, in this case, a gel structure results (Figure 12.5 c). Different crystal forms are designated α, β', or β due to variation in chain packing. More details can be obtained from Larsson [36] or Chapman [37].

The lamellar phase is most dominant in systems containing saturated monoglycerides while the cubic phase is dominant in systems containing unsaturated monoglycerides.

Phase diagrams of monoglycerides with lower chain lengths have been reported by Larsson [38]. The lamellar neat phase mesophase exists in these systems [38]. At the same time Larsson described a new phase, to which he applied the term "dispersion," in monoglyceride/water systems. The dispersion state exists in the water-rich region diagram and is obtained by adding water to the lamellar neat phase of all saturated monoglycerides, or by heating a mixture of monoglyceride–water (i.e., 10:90 w/w) above the Kraft point. Within temperature limits the dispersion phase is a stable state consisting of spherical aggregate of monoglycerides in equilibrium with water.

Monoglycerides with chain length shorter than C14 do not form a gel structure on cooling from a lamellar phase or a dispersion; instead they crystallize directly in the β-form.

Distilled monoglycerides were made commercially available to the food industry in 1960. Various problems arose when the distilled monoglycerides were handled in the same manner as the mixtures of mono-, di-, and triglycerides. For example, in the baking industry it was common to make emulsions of the lipid mixtures in water by mixing the emulsifier in boiling water. When this same procedure was attempted with monoglycerides of more than 90% monoester content, the formation of cubic viscous isotropic phase caused separation of the aqueous mixture into lumps. Knowledge from phase diagrams caused the industry to begin to use these products in the form of aqueous preparations. The reader with more than casual interest in this subject is urged to read Krog [35].

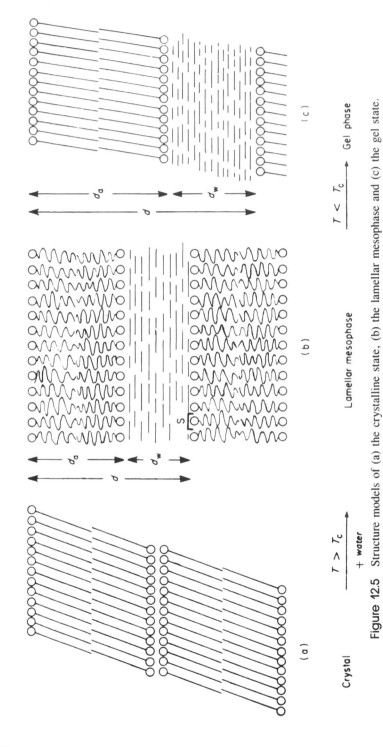

Figure 12.5 Structure models of (a) the crystalline state, (b) the lamellar mesophase and (c) the gel state.

The above discussion can be considered on the basis of a more general phenomenon of fat derivatives, namely polymorphism. Polymorphism is the ability of a single chemical specie to exist in different crystalline forms. This is a phenomenon that takes place during the manufacturing, storage, and use of a cosmetic. The material may change from one phase state to another. The rate at which these changes can take place is a direct function of the temperature at which the material is held following manufacture. A classic example is Noxema skin cream. This product after manufacture must be "aged" under strict conditions of time and temperature before being released for consumption.

Lipophilic derivatives exhibit this behavior. The progression of transformation is always in the same direction, namely, increased melting temperature resulting in a more stable crystalline structure

$$\rightarrow \beta' \qquad\qquad \rightarrow \beta$$

unstable form   stable form

The β crystalline form has the highest melting point, and thermodynamically, is the most stable.

VI. MULTIFUNCTIONAL ROLE OF MONOGLYCERIDES

The most popular commercial fatty acid ester is glycerol monstearate (GMS). In 1971, about 3.5 million pounds were sold, and it is estimated that the 1977 total will be close to 5 million [39]. GMS functions as an emulsifier, stabilizer, thickener, opacifier, or emollient in products such as skin creams and lotions, antiperspirant creams and roll-ons, shampoos, cream hair rinses and conditioners, as well as suntan creams and lotions.

Monoglycerides of saturated fats have the following major physical properties.

Stability against air oxidation
Soluble in alcohol
Non irritating to skin
Excellent spreading on the skin
Good penetrating properties
Outstanding emolliency
Chemical neutrality

In addition to the above properties, pure (>90% monoester content) monolaurin has good activity against gram-positive organisms as well as mold, yeast, fungi, and viruses. Commercial monolaurin (<60% mono content) is not antimicrobial. In the presence of chelating agents distilled monolaurin is active against gram-negative organisms, especially *P aeruqinosa*.

VII. THE ROLE OF MONOGLYCERIDES IN COSMETICS

Fats and oils have played an important role in the composition of cosmetics providing emolliency, moisturizing, grooming, and acting as solvent and/or vehicle to carry "active" agents. Dating back to antiquity lipids extracted from material sources were found to be useful. While the list includes oils from numerous plants and vegetables, some animal sources, little or no use was made of marine oils. Perhaps the odor from oils derived from fish severely limited its use. A short history of fats and oils used in cosmetics was presented by DeNavarre [10]. Many of the oils and fats in use today are the same ones dating from antiquity. However, technical progress has helped develop cosmetic products which have are perceived by the user to be less greasy. Derivatives of fats and oils have been prepared with a wide divergence of properties.

Perhaps the most important attribute required of an oil to be used in a skin lotion is its emollience. The term emollient is applied to those substances which help maintain a smooth, soft, and pliable texture to the skin. In part, this is accomplished by affecting the hydration of skin, namely by preventing or relieving skin dryness. Normal, healthy skin retains its hydration by controlling the transfer of water. With increasing age or in certain disease states where the skin is dry, the phenomenon is associated with increased water loss. Blank has shown that the water content of skin is directly related to slip and flexibility [40].

VIII. PREPARATION OF A NOVEL CREAM BASE

Most cosmetic bases prepared with commercial mixed glycerides (mono-, di-, and tri-) exhibit hydrophobic properties, i.e., water-repellent crystals. However, by understanding mesomorphic phases which distilled monoglycerides undergo, it is possible to form a unique cosmetic preparation. Larsson obtained a patent in 1969 for such a novel ointment base [41]. The best example of said invention is the use of monolaurin. While bases using monolaurin were known in prior art these α-monoglyceride molecules are disordered in an emulsion or gel state. In making this new cream base from a monolaurin and water mixture, the emulsion contains crystals which have hydrophilic surfaces. The importance of such a crystalline phase is that the ointment base can form a film over the skin. This film forming ability is important in pharmaceutical preparations used in open wounds and discharging skin areas. The ointment base can be used for dermatologically active substances. The crystals can directly absorb polar substances or solubilize amphiphilic agents added *before* crystal formation. What this means in practice is that preservatives like parabens would be entrapped in the nonpolar core of the base. Consequently, less preservative would be available.

Because of the unusual properties of the dispersion of monolaurin [41], active substances soluble in the ointment should move transdermally into the skin at a high rate. The base has films of water alternating with films of lipids. Thus, monolaurin can act as a transdermal agent or an enhancer of transport into the skin.

Dispersion of distilled monoglycerides is more difficult because of their tendency toward clumping. One method for making more stable dispersions of monoglycerides is described in a patent of Top-Scor Products, Inc. [42]. The monoglyceride is dispersed in water without the use of a coemulsifier by homogenizing the lipid at the transition temperature. This procedure is similar to technology disclosed by Larsson.

IX. MONOGLYCERIDES AS ANTIMICROBIAL AGENTS

Recent interest in monoglycerides is due to structure–function studies eminating from the laboratories of Kabara [43–46] and Shibasaki [47–49]. Fatty acids as soaps have been known to be germicidal from antiquity. Basic studies showed that monoesters of fatty acids were more antimicrobial than fatty acids themselves [50, 51]. The above reports all indicated glyceryl monolaurate to be the most active commercially available monoglyceride (Table 12.7) while some unsaturated fatty acid monoglycerides were also shown to be active. However, their preparation on a large scale cause trans isomerization of the active cis form. The trans isomers were usually found to be inactive.

The above statement concerning structure–function relationships is true as far as specific organisms and cosmetic products are concerned. However, the effects of monoglycerides can be neutralized by a variety of polymers, including starch and protein, and because binding seems to follow biological activity, several Japanese workers have reported that the volatile fatty acids esters (C_8, C_{10}) were more active than the higher (C_{12}) homologs [52, 53]. These reports, negating the generalizations found, must be viewed as exceptions. Beuchat [54] and Shibasaki [49] have supported the finding that the lauric acid derivative is the most active monoglyceride even in the presence of other complexing stuffs. Beuchat [54] compared the effects of glycerides, sucrose esters, benzoate, sorbic acid, and potassium sorbate against *Vibrio parahaemolyticus*. His results again indicated that the C_{12} monoglyceride was more active than lower (C_8, C_{10}) or higher (C_{14}) chain length derivatives. Also, the low MIC value for monolaurin (≤ 5 µg/ml) indicated it to be more effective than sodium benzoate (300 µ g/ml) or sorbic acid (70 µg/ml).

Because these esters were regarded not to be active against gram-negative organisms [55], the inhibition of *V. parahaemolyticus* growth was somewhat surprising, even though Kato and Shibasaki [48, 56] reported effects on gram-negative bacteria. In these latter cases, the Japanese workers demonstratedthat chelating acids (citric and polyphosphoric acid) were necessary to produce an enhancing effect on the biocidal action of monoglycerides against several gram-negative organisms. Their interpretation was that acids which are

Table 12.7 Minimum Inhibitory Concentrations (g/ml) for Fatty Acids and Their Corresponding Monoglycerides

Organism[a]	Unde-canoic acid	10-Unde-cednoic acid	10-Unde-cenoyl mono-glyceride	10-Unde-canoyl mono-glyceride	11-Dode-cenoic acid	Dode-canoic acid	Dodecanoyl monoglyc-eride	12-Tride-cenoic acid	Tride-canoic acid	Tride-canoyl mono-glyceride
Streptococcus faecalis	NI[b]	NI	500	500	NI	500	NI	1000	NI	NI
Streptococcus pyogenes	125	1000	125	125	250	62	8	125	1000	62
Staphylococcus aureus	1000	1000	500	500	NI	500	250	1000	NI	NI
Corynebacterium sp.	31	31	62	62	125	31	16	31	NI	NI
Mocardia asteroides	62	125	125	62	62	62	16	125	1000	125
Candida albicans	1000	1000	250	100	1000	1000	500	1000	NI	NI
Saccharomyces cerevisiae	500	500	250	100	500	1000	250	500	1000	NI

[a]*Escherichia coli* and *Pseudomonas aeruginosa* were not affected.
[b]NI = MIC > 1000 g/ml.
Source: From Ref. 45.

Table 12.8 Comparison of the Antibacterial Activities of Fatty Acid Esters and Some Commonly Used Preservatives

Food additive	Minimum inhibitory concentration (g/ml)[a]		
	Bacillus subtilis	*Bacillus cereus*	*Staphylococcus aureus*
Monocaprin	123	123	123
Monolaurin	17	17	17
Butyl-*p*-hydroxybenzoate	400	200	200
Sodium lauryl sulfate	100	100	50
Sorbic acid	4000	4000	4000

[a] By the agar dilution method.
Source: From Ref. 48.

chelators release a significant amount of cellular lipopolysaccharide from bacterial walls and cause the organism to behave like a gram-positive bacteria. Apparently the presence of chelators is not always a prerequisite for demonstrating inhibitory action of monoglycerides against all gram-negative bacteria, since *V. parahaemolyticus* is affected in a manner similar to gram-positive organisms [54].

In order to better appreciate the antimicrobial effects of these nontoxic agents, data from Shibasaki's laboratory is presented (Tables 12.8 and Table 12.9). In these tables the antibacterial and antifungal properties of monocaprin (C_{10}) and monolaurin (C_{12}) are compared with some well-known cosmetic

Table 12.9 Comparison of the Antifungal Activities of Fatty Acid Esters and Some Commonly Used Preservatives

Food additive	Minimum inhibitory concentration (g/ml)[a]		
	Aspergillus niger	*Candida utilis*	*Saccharomyces cerevisiae*
Monocaprin	123	123	123
Monolaurin	137	69	137
Butyl-*p*-hydroxybenzoate	200	200	200
Sodium lauryl sulfate	100	400	100
Sorbic acid	1000	1000	1000
Dehydroacetic acid	100	200	200

[a] By the agar dilution method.
Source: From Ref. 48.

Table 12.10 Preservative Effect of Food-Grade Chemicals in a Lauricidin Lotion Oil-Water System[a]

Preservative	At 24 h		At 48 h		At 1 wk	
	Pseudomonas aeruginosa	*Escherichia coli*	*Pseudomonas aeruginosa*	*Escherichia coli*	*Pseudomonas aeruginosa*	*Escherichia coli*
EDTA						
0.1%	+	+	+	+	−	−
0.2%	−	±	−	±	−	−
0.3%	−	±	−	±	−	−
Methylparaben						
0.1%	+	+	+	+	+	+
0.2%	+	+	+	+	+	+
0.3%	+	+	+	+	+	−
EDTA + methylparaben						
0.1% + 0.1%	+	+	±	+	−	−
0.2% + 0.2%	−	+	−	+	−	−
0.3% + 0.3%	−	±	−	±	−	−

[a] +, growth; ±, slight growth; −, no growth.
Source: From Ref. 57.

preservatives, especially butyl parabens. Monolaurin was shown to be the most active agent.

As previously discussed, a serious limitation of the monoglycerides is their lack of activity against gram-negative strains. This limitation can easily be overcome by the combination of a chelating agent, i.e., EDTA, citric or lactic acid. In a model cosmetic preparation where glyceryl monolaurate served as both preservative and emulsifier, no other preservative (methyl parabans) was needed. The cosmetic preparation was challenged with 10^7CFU of *P. aeruginosa* (Table 12.10).

It has also been shown that EDTA lowers the minimum inhibitory concentration of monolaurin against yeasts. Shibasaki and Kato [58] found that EDTA tripled the percentage of uptake of pure monolaurin into the cells of *Escherichia coli*.

Therefore, distilled monoglycerides in general and glycerol monolaurate in particular give the formulator a unique opportunity to explore the full benefits of a multifunctional additive. It is rare to find a chemical that has such diverse properties, i.e., preservative, emulsifier, emollient, thickening agent, deodorant, substantivity, and transdermal effects. It is rarer to use a chemical which has no toxicity. Monoglycerides are classified by the FDA as generally regard as safe (GRAS). This designation is given to food additives which are considered nontoxic. It remains for future formulators to investigate the role of these unique and safe lipids. Distilled monoglycerides may be the new "active" in cosmetic preparations and given more press than aloe, vitamin E, collagen, etc. In these cases, performance will replace hype.

X. ADDENDUM

Classical cosmetics formulae are presented using pure monoester as an emulsifying/preservative agent. An anionic emulsion made with ordinary glyceryl monolaurate (monester content <60%) is less stable, colored, has an objectionable odor compared with the same emulsion made with lauricidin (monoester content >90%) (see Figure 12.6 and 12.7).

Cosmetic emulsions made with pure monoesters are superior products both from a manufacturing and use point of view. The anionic emulsion even with mineral oil is not sticky/greasy on the skin.

Readers are also referred to recent review (59).

RESEARCH PHOTOMICROGRAPH

DATE 3-1-87   NEG. NO. ———

FILM 107C   MAG. 400X
                          (Ref. 104-14B)
EXP. 2 Sec.   LIGHT

NOTEBOOK REF. 104-15-1

SAMPLE DESCRIPTION: 104-10A

CONTROL ANIONIC

EMULSION (No Lauricidin)

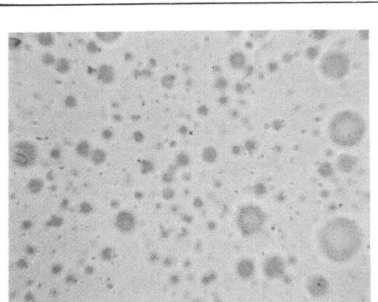

Figure 12.6 Anionic emulsion using standard commercial glyceryl monolaurate. [Microphotography (400×) courtesy of Larry Lundmark.]

RESEARCH PHOTOMICROGRAPH

DATE 3-1-87   NEG. NO. ———

FILM 107C   MAG. 400x
                    (Ref. 104-14B)

EXP. 2 Sec   LIGHT

NOTEBOOK REF. 104-15-3

SAMPLE DESCRIPTION: 104-10C
ANIONIC EMULSION BASE
2% LAURICIDIN

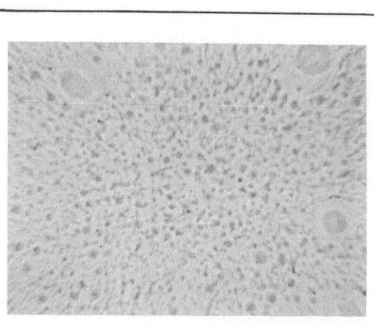

Figure 12.7   Anionic emulsion using lauricidin. [Microphotography (400×) courtesy of Larry Lundmark.]

Anionic Emulsion

Part A	%
Mineral oil	3.5
Stearic acid	3.0
Cetyl alcohol	1.0
Octyl palmitate	2.0
Glyceryl monolaurate (Lauricidin[a])	2.0
Part B	
Propylene glycol	3.0
Methyl parabens	0.1
Disodium EDTA	0.1
Triethanolamine (TEA)	1.0
Water	qs to 100

Procedure: Heat A and B separately to 60–65°C with agitation. Add A to B with agitation until thoroughly mixed. Cool with agitation to 40°C.

[a] Emulsion with standard commercial glyceryl monolaurate (Fig. 12.6) had particle sizes larger than emulsions made with lauricidin (Fig. 12.7). Generic monolaurin produces off color and odor to final product.

"Oil of Olay"[a] Type Lotion

A. Prepare laurin wax premix

Part I	%
Water	7.85
Potassium hydroxide	3.00

Part II	
Stearic acid	37.00
Glyceryl monolaurate	30.00
Cetyl alcohol	15.00
Octyl palmitate	7.00
Butyl paraben	0.15

Add part I to part II at 65°C. Mix for 15 min and cool to ambient temperature.

B. Preparation of water-phase premix

Water (1)	83.34
Carbomer 934	0.12
Methyl parabens	0.20
Propyl parabens	0.10
Water (2)	3.00
Sodium hydroxide	0.03

Add water (1) to vessel. Begin agitation and disperse Carbomer. Heat to 80°C add parabens. Add water (2) and sodium hydroxide solution. Maintain at 80°C.

C. Preparation of oil phase

Laurin wax	6.71
Soy sterol	1.00
Mineral oil	3.50

Add laurin wax to vessel. Heat to 80°C and form melt. Dissolve soy sterol. Add mineral oil and maintain at 80°C.

D. Preparation of emulsion

Add oil phase (C) to water phase B at 80°C. Maintain temperature for 15 min with agitation. Cool to 45°C with mixing.

---

[a] Oil of Olay is a registered trademark of Richardson-Vicks

Hand Lotion

	weight
A. Dynasan 114	6.0
Cremophor Asolid + Cremophor O = 3:2 (7)	5.0
Glyceryl monolaurate (Lauricidin)	6.0
Caprylic/capric/triglyceride	10.0
B. Glycerol	8.0
Preservatives	q.s.
Water	q.s. 100.0
C. Perfume	q.s.

Procedure: Heat A to 75 to 80°C and emulsify B into A at the same temperature. Perfuming is effected at a temperature below 40°C. It is recommended to homogenize the hand lotion prior to filling.

Eye Makeup Lotion

	%
A. Cetearyl alcohol PEG-40 castor oil (and)	
sodium stearate sulphate (3)	5.0
Caprylic/capric/triglyceride	3.0
Glyceryl monolaurate (Lauricedin)	3.0
Trilaureth-4 phosphate	1.0
B. Glycerol	3.0
Preservatives	q.s.
Water	100.0
C. Perfume	q.s.

Procedure: Heat A to a temperature of 75–80°C and emulsify B into A at the same temperature. Perfuming should be effected at about 40°C.

Orange Beauty Lotion

	%
A. Glyceryl monolaurate (Lauricidin)	8.0
Trilaureth-4 phosphate	5.0
Caprylic/capric/triglyceride	7.0
B. Lemon oil	0.3
C. Preservative	q.s.
Water	q.s. 100.0
Ascorbic acid	0.2
D. Perfume	q.s.

Procedure: Heat A to a temperature of 75–80°C and add B. Emulsify at the same temperature mixture C into the A and B compound. Perfuming should be effected at about 40°C. Additionally homogenization of the overall mixture is recommended.

Compact Makeup

	%
A. Glyceryl monolaurate (Lauricidin)	8.0
Wool wax	4.0
White beeswax	7.0
Hard paraffin	10.0
Trilaurin (Softisan 100)	10.0
Prime stearin	3.0
Propylene glycol dicaprylate/dicaprate	10.0
Karion F	5.0
Caprylic/capric/triglyceride	
Neutral oil	5.5
Dynacet 850	12.5
B. Perfume	1.0
C. Ferrous Oxide PC 1136	0.5
Pure Oxy Siena 3179	0.5
Talcum powder	8.0
Zinc oxide	8.0
Titanium dioxide	8.0

Procedure: Melt A and add to C. Heat this mixture and then cool with agitation; then add B. Rehomogenization is recommended.

REFERENCES

1. DeNavarre, M. G., Oils and fats, the historical cosmetics, J. Am Oil Chem. Soc., 55: 435–437, 1978.

2. Swern, D. (ed.). Bailey's Industrial Oil and Fat Products, Vol. 24th ed., Wiley Interscience, New York, 1982, p. 134.

3. Lauridsen J. B. Food emulsifiers: Surface activity, edibility, manufacture, composition and application. J. Am Oil Chem. Soc., 53: 400–407, 1976.

4. Sonntag, N. O. V. Glycerolysis of fats and methyl esters; status, review & critique, J. Am. Oil Chem Soc., 59: 795A–802A, 1982.

5. Official & Tentative Methods of the American Oil Chemist Society, Vols. I & II, 3rd ed., AOCS Champaign, IL, 1966, Method Cd 11–57.

6. Marzo, A., Ghirardi, P., Sardini, D., and Meroni, G. Simplified measurement of monoglycerides diglycerides & triglycerides & free fatty acids in biological samples. Clin Chem., 17: 145–147, 1971.

7. Kabara, J. J. and Chen, J. S., Microdetermination of lipid classes after thin layer chematography, Aval. Chem., 48: 814–817, 1976.

8. Emdur, L. I., Lyle C., and Kabara, J. J., Quantitation of lipid classes following thin-layer chromatography, Anal. Lett. 10: 21–27, 1977.

9. Itabashi, Y. and Takagi, T., Gas chromatographic separations of di- and monoacylglycerals based on the degree of unsaturation and positional isomers, Lipids, 15: 205–215, 1960.

10. Halvarson, H. and Qviset, O., A method to determine the monoglyceride content in fats and oils, J. Am. Oil Chem. Soc., 51: 162–165, 1974.

11. Rilsom, T. and Hoffmeyer, L., High performance liquid chromatography analyses of emulsifiers: Quantitative determination of mono- and diaglycerols of saturated fatty acids. J. Am. Oil Chem. Soc., 55: 649–652, 1978.

12. Payne-Wahl, K., Spencer, G. F., Plattner, R. D., and Butterfield, R. O., High-performance liquid chromatography method for quantitation of free acids, mono, di- and triglycerides using an infrared detector. J. Chromatology 209: 61–66, 1981.

13. Takano, S. and Kondoh, Y. Monoglyceride analysis with reversed phase HPLC, J. Am. Oil Chem. Soc., 64: 1001–1003, 1987.

14. Braun, W. Q. and Shrewsbury, C. L., Oil Soap 18: 249, 1941.

15. Hine, C. H., Anderson, H. H., Moon, H. D., Dunlap, M. K., and Morse, M. S., Comparative toxicities of synthetic and natural glycerol, Arch. Ind. Hyg. Occup. Med., 7: 282, 1953.

16. Li, R. C., Suh, P. P. T., and Anderson H. H., Acute toxicity of monoacetin, diacetin, and triacetin, Proc. Soc. Exp. Biol. Med., 46: 26, 1941.

17. Ambrose, A. M., and Robbins, D. J., Toxicity of acetoglycerides, J. Am. Pharm. Assoc. Sci. Ed., 45: 282, 1956.

18. Hertig, D. C. and Crain, R. C., Foreign-body type reaction in fat cells, Proc. Soc. Exp. Biol. Med., 98: 347, 1957.

19. Anonymous, Food Additive Petition, Number 771.

20. Wretlind, A., The toxicity of low molecular triglycerides, Acta Physiol. Scand., 40: 338, 1957.

21. Fitzhugh, O. G., Bourke, A. R., Nelson, A. A., and Frawley, J. P., Chronic oral toxicities of four stearic acid emulsifiers, J. Toxicol. Appl. Pharmacol., 1: 315, 1959.

22. Mattson, F. H., Baur, F. J., and Beck, L. W., The comparative nutritive values of mono-, di-, triglycerides, J. Am. Oil Chem. Soc., 28: 386, 1951.

23. Ames, S. R., O'Grady, M. P., Embree, N. D., and Harris, P. O., Molecularly distilled monoglycerides. III. Nutritional studies on monoglyceride derived from cottonseed oil, J. Am. Oil Chem. Soc., 28: 31, 1951.

24. Harris, R. S., and Sherman, H., Comparison of the nutritive values of mono-, di-, and triglycerides by a modified pair-feeding technique, Food Res., 19: 257, 1954.

25. Anonymous, Food Additive Petition Number 884.

26. Orten, J. M., and Dajani, R. N., A study of the effects of certain food emulsifiers in hamsters, Food Res., 22: 529, 1957.

27. Anonymous, GRAS (Generally recognized as safe) Food Ingredients—Glycerine & Glycerides Ordering No. PB-221227, National Technical Information Service, U.S. Dept. of Commerce, Springfield, VA, 1973.

28. Krog, N., Functions of emulsifiers in food systems. J. Am. Chem. Soc., 54; 124–131, 1977.

29. Mattson, F. H. and Volpenhein, R. A., J. Lipid Res., 3: 281, 1962.

30. Ljusberg-Wahren, H., Herslop, M., and Larsson, K., A comparison of the phase behavior of the monoolein isomers in excess water. Chem. Phys. Lipids, 33: 211–214, 1983.

31. Brokaw, G. Y. and Lyman, W. C., The behavior of distilled monoglycerides in the presence of water, J. Am. Oil Chem. Soc. 35: 49–52, 1958.
32. Lutton, E. S., Phase behavior of aqueous systems of monoglycerides, J. Am. Oil Chem. Soc., 42: 1068–1070, 1965.
33. Lawrence, A. S. C. and McDonald, M. P., The investigation of lipid-water system (Part 1) by classical and N.M.R. methods, Molec. Cryst. 1: 205–223, 1966.
34. Krog, N., and Larsson, K., Phase behavior and rheological properties of aqueous systems of industrial distilled monoglycerides, Chem. Phys. Lipids, 2: 129–143, 1968.
35. Krog, N., Interaction between water and surface active lipids in food systems, in Water Relations of Foods (R. B. Duckworth, ed.), Academic Press, New York, 1975, pp. 587–611.
36. Larsson, K., Ark Kemi, 23: 35, 1964.
37. Chapman, D., The Structure of Lipids, Methuen & Co., Ltd., London, 1965.
38. Larsson, K., Z. Phys. Chem., 56: 173, 1967.
39. Johnson, D. H., The use of fatty acid derivatives in cosmetics and toiletries, J. Am. Oil Chem. Soc., 55: 438–443, 1978.
40. Blank, I. H., J. Invest. Derm., 18: 433–440, 1952.
41. British Patent 1,174,672, Larsson, V.K. 17 1969.
42. U.S. Patent 3,379,535, Top-Scor Products, Inc., 1968.
43. Kabara, J. J., Conley, A. J., Swieczkowski, D. J., Ismail, I. A., Lie Ken Jie, M., and Gunstone, F. D., Unsaturation in fatty acids as a factor for antimicrobial action, J. Med. Chem., 16: 1, 1972.
44. Conley, A. J. and Kabara, J. J., Antimicrobial action of esters of polyhydric alcohols, Antimicrob. Agents Chemother., 4: 501, 1973.
45. Kabara, J. J., Vrable, R., and Lie Ken Jie, M., Antimicrobial lipids: natural and synthetic fatty acids and monoglycerides, Lipids, 9: 753, 1977.
46. Kabara, J. J., Food-grade chemicals for use in designing food perservative systems, J. Food Prot., 44: 633, 1981.
47. Kato, N. and Shibasaki, I., Comparison of antimicrobial activities of fatty acids and their esters, J. Ferment. Technol., 53: 793, 1975.
48. Kato, A. and Shibasaki, I., Combined effect of different drugs on the antibacterial activity of fatty acids and their esters, J. Antibacterial Antifung. Agents, 8: 355, 1975.
49. Shibasaki, I. Recent trends in the development of food preservatives, J. Food Safety, 4: 35, 1982.
50. U.S. Patent 4,002,775, Kabara, J. J., 1977.
51. U.S. Patent 4,067,997, Kabara, J. J., 1978.
52. Ueda, S. and Tokunaga, H., Antiseptic effect of capric monoglyceride on pellicle-forming yeast, Seasoning Sci., 13: 1, 1966.
53. Koga, T. and Watanabe, T., Antiseptic effect of volatile fatty acid monoglycerides, J. Food Sci. Technol., 15: 310, 1968.
54. Beuchat, L. R., Comparison of anti-vibrio activities of potassium sorbate, sodium benzoate and glycerol and sucrose esters of fatty acids, Appl. Environ. Microbiol. 39: 1178, 1980.

55. Kabara, J. J., Fatty acids and derivatives as antimicrobial agents—a review, in Pharmacological Effects of Lipids (J. J. Kabara, ed.), American Oil Chemists Society, Champaign, IL, 1979, p. 1.

56. Kato, N. and Shibasaki, I., Combined effect of citric and polyphosphoric acid on the anti-bacterial activity of monoglycerides, J. Antibacterial Antifung. Agents, 4: 254, 1976.

57. Kabara, J. J. and Wermette, C. M., Cosmetic formulas preserved with food-grade chemicals, Cosmet. Toiletries, 97: 77–84, 1982.

58. Shibasaki, I. and Kato, N., Combined effects on antibacterial activity of fatty acids and their esters against gram-negative bacteria, in Pharmacological Effect of Lipids (J. J. Kabara, ed.), American Oil Chemists' Society, Champaign, IL, 19XX.

59. Rieger, M. M. Glyceryl stearate: Chemistry and use, Cosm. & Toil., 105: 51, 1990.

13

Glycerine in Creams, Lotions, and Hair Care Products

Rolf Mast

Neutrogena Corporation, Los Angeles, California

I. INTRODUCTION AND GENERAL PROPERTIES

A. A Wide Range of Useful Properties

As a road map for the use of glycerine, several of its qualities are paramount. These are its beneficial compatibility with human skin, its water-attracting properties, as well as its inherent physical properties which are further discussed in Chapter 10. Thus glycerine provides a nonfreezing, nonvolatile medium, which has inherent waterlike properties, as well as some oil/water compatibility. When these properties are combined with an almost total lack of toxic effects, one can readily see why glycerine is a favorite constituent for the cosmetic chemist. Especially, the cosmetic chemist should consider the use of glycerine for its proven moisturizing and protective effects on the skin.

Certainly there is nothing new about the use of glycerine in cosmetics. As Wells wrote in 1957 [1], "In fact glycerol can in many ways be regarded as an ideal compliment to water in cosmetic practice in that it is completely water miscible, with consequent depression of freezing point and increase in viscosity, and has also the valuable property of retarding the evaporation of water, and thus preventing the premature drying-out of aqueous solutions and emulsions containing it. Other properties of glycerol that must be taken into account are its clarity, lack of color and odor, sweet taste, preservative action, useful solvent properties, emolliency, non-toxicity and plasticizing action."

Wells also mentions two instances where it is useful for improving the adherence and spreading of powders on the skin. Firstly in vanishing creams (see below), where glycerine improved adhesion of the talcum powder used in post-application to enhance skin feel and appearance. Secondly, in the old-style theatrical cosmetics (such as Wet White or Blanche de Perle) glycerine improved spreading and adhesion of the pigments.

Other reviews of the cosmetic uses of glycerine have been written [i.e., 2–4]. Moxey [2] discusses many of the beneficial properties noted above, including its help with pigments and perfumes: "In pigmented products glycerine enables the pigments to spread out, more uniformly—a property of value in liquid and cream makeup formulations. In many products, it is often difficult to maintain the same perfume note, due to evaporation or internal reactions. Glycerine possesses a fixing power for perfumes and its incorporation in a product ensures that the note is conserved until all the product is used up." Two other properties of glycerine which are not so often thought about, are its sweet taste—a big attribute in toothpaste—as well as its insolvency in hydrophobic materials, making it useful as a protective barrier to the skin for certain injurious agents.

An interesting comment made by Mannheim [5] is that in the leather industry, glycerine is used to increase material suppleness. Leather is fibrous protein which stiffens as it dries out. This is analogous to stratum corneum, which presages well for a beneficial effect on the skin.

Another point to make is that glycerine is frequently used in combination with other humectants, as for instance in the patent by Brun and Koulbanis [6]. Here, a humectant mixture, which is added to various skin cosmetics, contains 3–8% sodium lactate, 12–24% glycerine, 30–42% urea, and up to 0.5% soluble collagen. They claim this as a particularly favorable mixture for maintaining suppleness of the skin.

B. Improvements in Product Manufacture and Stability

Several properties of glycerine make it very useful for helping construct the product per se, as well as improve the performance of the product in use. In particular these are its ability to restrict water loss; to prevent the formation of surface crusts (the latter attribute due to it being liquid, but nonvolatile); its depression of aqueous freezing point (which often prevents freeze damage of the formula); and its cosolubilizing properties (which can help with dispersion of perfumes and polar constituents). Also, under the correct conditions, its effects on product rheology can improve viscosity and suspending power. In products containing high solids, by using glycerine instead of water, the manufacturer protects it against possible change from a free-flowing smooth paste into a stiff concretelike mass. Two examples that come to mind are toothpaste and buffing creams. Certainly the wetting-out of the particles by glycerine is an important factor here. It can also facilitate the periodic dispensing of product through aerosol or turret top spouts, where otherwise excessively dry product can plug up the orifices.

C. A Large Variety of Products are Affected

Further, there is nothing new about the wide range of products in which glycerine is used. Thus, Ralph G. Harry in his book *Cosmetic Materials* [7, p. 136] wrote "It improves the spreading properties of creams and helps to preserve their consistency in virtue of its humectant properties. It is employed in aftershave creams, all purpose creams, astringent creams and lotions, beauty milks, bleaching creams, cleansing pads, cosmetic stockings, cream rouge, cuticle softener, deodorants, hand lotions, liquid powders, lubricating creams, powder creams, shaving creams of the saponaceous and brushlesstypes, tooth-pastes and liquid dentifrice, vanishing creams etc." I do not want to fill out the "etc." with an exhaustive list here, but I think the point is made.

II. CREAMS AND LOTIONS FOR CARE OF DRY SKIN CARE

A. Introduction

This section deals largely with emulsion products intended for use on dry/flaky but otherwise normal skin. Thus this section deals with the everyday skin care cream and lotion. Where very similar formulations also claim therapeutic benefits, these are discussed also, for simplicity of overall organization.

By using virtually any of the commercially available products, nearly everyone has experienced the immediate soothing effect of an oil/water/humectant mixture on the skin. The duration of these soothing benefits which last for hours, even days, after discontinuation of product use, is a common perception well substantiated by objective measurement.

In the same manner that one can consider the essential components of a car to be the wheels, engine, chassis, and body, one can consider the essential elements of emulsion skin treatments to be water, oil, dispersant system, and humectant. Then, just as cars are differentiated through styling, horsepower, stereo components etc., so too are skin products through viscosity, concentration, and claim ingredients. So, looking at the fundamental construction of most creams and lotions one will see the following structure.

Water 50–90% . . . Oils 1–25% . . . Humectants . . . 1–25% . . . Emulsifiers 1–10%

Thickeners, fragrances, preservatives, skin-healing agents, slip agents, special claim ingredients, etc. can then be added to achieve whatever special effect is desired.

B. Glycerine Mucilages for Dry Skin and Closely Related Disorders

Having briefly described emulsions, the first type of skin care item to be discussed will not be an emulsion at all, but simpler glycerine–water mixtures. This is because, as a humectant with well-proven benefits, glycerine is fundamental, because it can alleviate dry skin without the necessity for other components. It might be added, that these glycerine mucilages have been known for many years and are often referred to as glycerine/rosewater systems. They are called this because rose-scented water was used as the traditional diluent for the glycerine, which is thus the sole ingredient that has any substantial benefit to repair of dry skin. Simple glycerine mucilages are not very popular today due to the stickiness and feel of the products. These problems can be alleviated by the addition of oils, which, if correctly formulated impart an elegant feel, as well as provide additional skin benefits.

Harry [8, p. 90–91] states that the original "Glycerine and Rosewater" preparations for skin healing were oil-free mucilages consisting originally of equal parts of water and glycerine. Later, dilution was made to about 20–25% of the humectant. In order to keep the mucilage nature of the product, but reduce stickiness even more, the glycerine can be reduced still further, and gum tragacanth added to restore viscosity. Thus the following formulation is typical of gum mucilage preparations which contain no oil phase, says Harry.

Material	% by wt
Gum tragacanth	2.0
Glycerine	10.0
Titanium dioxide	0.2
Water	87.8
Perfume, preservative, color	q.s.

Even with the most simple of formulas, however, the cosmetic chemist must recognize a wide range of possibilities and tastes, as evidenced by Flick listing a 50% glycerine formula that is further thickened by Carbopol 940:

Material	% by wt
Glycerine	49.5
Water	49.5
Carbopol 940*	0.5
Triethanolamine	0.5

*Polyacrylic resin, trademark B. F. Goodrich Chem.

In this context, Ward claimed a gel-based composition for the treatment of fissured/chapped skin, as well as for healing minor cuts [9]. The very simple composition used is as follows:

Material	% by wt
Glycerine	26.0 to 29.0
Water	64.0 to 71.0
Powdered tragacanth BP	1.6 to 2.0
Bergamot	1.1 to 2.0
Benzoic acid	0.03 to 0.2
Other preservatives, typically	0.07

As just stated, simple glycerine–water combinations precede this patent by many years, but in this case benzoic acid keeps the pH in the range of 3 to 5 which is claimed to aid the natural surface of the skin in controlling microorganisms. Certainly however, the skin benefits are due to the glycerine itself, a fact known from folk medicine as well as exact scientific study (i.e., 10) as discussed in Chapter 10.

There is nothing new about glycerine/rosewater, and there is nothing new about serum albumin being used as an antiwrinkle composition (as serum albumin dries on the skin it stretches it taught, giving a temporary reprieve from wrinkles). Even so, these two items appear together in a 1985 patent application, thus putting a new wrinkle on an old idea [11]. Basically the application teaches that rosewater with 1–2% glycerine is used as a carrier for human serum albumin (previous antiwrinkle potions had used the bovine variety).

Two Japanese patents are other examples of work that continue to closely emulate the traditional glycerine/rosewater treatment. A skin conditioner recipe from a 1986 patent [12] has the following composition:

Material	% by wt
Glycerine	68.0
Polyacrylic acid	2.0
Propylene glycol	1.0
NaOH (neutralizes the polymer)	1.4
Water	to 100%

In use, five parts of this concentrate was mixed with 95 parts of a hydroalcoholic diluent. Although the resulting formula would have a very low glycerine level, after evaporation from the skin, glycerine would be the major remaining ingredient.

In a remarkably simple 1985 patent, the following composition is claimed as effective for the treatment of frostbite and skin disorder [13]:

Material	Parts by volume
Glycerine	120
Water	100
Ethanol	100
Lemon extract	15

Again, after evaporation from the skin, glycerine would be the remaining ingredient, and presumably it is to that ingredient that the beneficial action can be attributed.

C. Astringent Glycerine Mucilages

Still in this context, but where glycerine is definitely not the only active ingredient, are the astringent mucilages detailed by Harry [7, pp. 97–98]. Astringent lotions have been used for pore closure prior to cleansing or makeup addition, as well as to minimize orange-peel skin caused by enlarged follicles. To this end, oil-free astringents such as the following have been used:

Material	Parts by wt
Potassium aluminum sulfate	1.0
Zinc sulfate	0.3
Glycerine	5.0
Rosewater	50.0
Water	43.7
Preservative and coloring matter	q.s.

The astringency, of course, is provided by the inorganic salts. Closely related to astringency are styptic compositions where the highly charged cation is used to agglomerate blood rather than tighten skin pores. Styptic formulas containing a humectant (such as glycerine at 10% by wt) and aluminum chlorate or perchlorate have been claimed [14].

D. Emulsion Formulas for Dry Skin

1. Introduction: Formula Complexity

The patent application of Dixon and Kelm [15] restricts the essential ingredients to a basic combination of water, oil, humectant, and emulsifier. Thus they claim a mixture of 5 to 30% water, 0.5 to 10% glycerine, volatile silicone oil, up to 1% of an alkoxylated silicone oil, as well as surfactant. Presumably, the skin conditioning of the glycerine is combined with an elegant skin feel due to the silicones. Fully formulated, this simple combination does get considerably more complicated, however, and provides a good object lesson on how the simple concept of glycerine skin treatment can lead to a complex final result:

Material	% by wt
Glycerine	20.00
Cyclomethicone	2.70
Alkoxylated silicone oil	0.30
Cetyl alcohol	3.00
PEG 100 monostearate	0.50
Isopropyl palmitate	1.50
Dimethicone	1.00
Stearic acid	0.50
Lanolin fatty acid	0.50
Emphos F27–85[a]	0.50
Carbopol 934[b]	0.10
Preservative	0.025
NaOH	0.11
Perfume	0.20
Titanium dioxide	0.20
EDTA	0.10
Water	to 100

[a] Glyceride phosphate ester, trademark Witco Chemical Co.
[b] Polyacrylic resin, trademark B. F. Goodrich Chem.

Quite apart from the patent mixture claimed (with presumably great skin feel as well as conditioning) the final product here utilizes a number of auxiliary components. These are: (1) a complex emulsifier/thickener combination (consisting of cetyl alcohol, PEG 100 monostearate, stearic acid, glyceride phosphate ester, Carbopol 934, and NaOH); (2) additional oils and emollients (consisting of isopropyl palmitate, dimethicone, and lanolin fatty acid); (3) preservatives; (4) TiO2 as a colorant; and (5) perfume.

Why does a simple idea like aqueous glycerine—if indeed it is so effective—need to get so complicated, and why are there the endless variety of aqueous glycerine formulations in existence? Apart from the arbitrary reasons of personal preference, marketing claim requirements, and differing availability of chemicals to perform similar tasks, is the fact that each formula has slightly different requirements, and there is an enormously complicated relationship between the performance requirements of the product, and the interactions of the chemicals that are used to assemble the formula in order to meet them. Some questions that might be asked are as follows. Is the product going to be a hand lotion suitable for daytime use? Alternatively, should it be a night cream, heavy, and containing specialized exotic ingredients? What about the perfume level? Is a fragrance necessary to cover the base (which often is the case for instance with protein additives), does one want to make a "fragrance-free" claim, or possibly incorporate an exotic recognizable scent? These types of questions, which need to be answered even before formulation work can begin, all have a pronounced effect on the ingredients that are to be used. Combine these factors with the almost limitless numbers of surfactants, oils, preservatives, perfumes, special ingredients, etc., and the result is an infinite variety of formulas.

It must be said that the need to produce a stable, microbiologically preserved formula, that meets the marketing objectives, are resolved to a large extent on a trial and error basis, so experience by the formulator is a very useful asset. As a for-instance, thickening can be accomplished by polymers, inorganic dispersions, or fatty monomers, each of which has different relationships to stability, skin feel, opacity, etc. as well as the other ingredients in the formula. It is not unknown, for instance, for a particular thickener to react with the preservative system and produce an unstable emulsion on storage at higher temperatures. Thus the thickener (or any other ingredient) must be carefully chosen and tested so as to ensure the correct overall balance of properties is maintained.

2. Creams and Lotions: No Fundamental Distinction

From a consumer point of view this is an important distinction. The usual assumption is that lotions are cheaper "workhorse products" typically used on a frequent basis for hand care. On the other hand, creams are perceived as more expensive, "heavy duty," and often for use on the face. Whether a skin care emulsion is defined as a cream or a lotion is a function of its thickness. In general, if the product flows under its own weight it is a lotion, otherwise a cream. In practice, however, a multitude of factors can affect viscosity with no real correlation to solids content, or to the constituents used in the product. Also, the terms "cream" and "lotion" are used very loosely throughout the industry. With these factors in mind there is no real point in contextually separating them in this chapter, except to mention that characteristic in passing. Certainly, for most products, the most minor (and varied) of changes can turn a cream into a lotion and vice versa.

3. Mineral Oil-Containing Emulsions

Possibly the two most widely used ingredients for skin conditioning are the humectant, glycerine, and the emollient, mineral oil. With this thought in mind, the following Penreco formula [16, p. 162] could be considered as a typical emollient cream.

Material	% by wt
Beeswax	3.00
Spermaceti substitute	3.00
Drakeol 7[a]	30.00
Glyceryl monostearate	12.00
Methylparaben	0.15
Propylparaben	0.15
Glycerine	8.00
Water	43.40
Perfume	0.30

[a] Mineral oil, trademark Penreco Co.

The spermaceti substitute and beeswax both help to emulsify, as well as add sheen and body to the emulsion.

Similarly, the following ICI Americas formulation is another example of a typical, high-activity mineral oil/glycerine skin emulsion [18, p. 165].

Material	% by wt
Mineral oil	16.00
Cetyl alcohol	4.00
Lanolin	2.00
Arlacel 165[a]	10.00
Tween 60[b]	1.00
Glycerine	5.00
Water	62.00

[a] Self-emulsifying glycerol monostearate, trademark ICI Americas.
[b] PEG (20) sorbitan monostearate, trademark ICI Americas.

Given that the cetyl alcohol is present for viscosity and sheen, in a sense the lanolin can be considered to be the only optional ingredient, in an otherwise very basic, and I am sure effective, formulation.

Of course, as stated above, we are dealing with an unlimited number of formula permutations and combinations, even when the range of ingredients themselves are kept fairly limited. Of necessity, therefore, the formulator must use his art to arrive at the desired end-effect. For instance, the rich formula above is considered suitable as a face cream and for night use. Simple adaptation of these materials, however, leads to a much lighter product, suitable for use as a hand cream [16, p. 18].

Material	% by wt
Glycerine	12.00
Propylparaben	0.15
Triethanolamine	0.17
Water	69.98
Cetyl alcohol	10.00
Lanolin	6.00
Mineral oil	2.00
Ethomeen C/25[a]	0.50
Carbopol 934[b]	0.20

[a] Ethoxylated tertiary amine, trademark Armak Chemical
[b] Polyacrylic resin, trademark B. F. Goodrich Chem.

The greatly reduced mineral oil, and much higher cetyl alcohol will provide the "drier" feel that is necessary for a good hand cream. At the same time, the glycerine level is raised so as to maintain good skin conditioning properties.

4. Petrolatum-Containing Emulsions

From work conducted in the study of dry skin products, this author feels very strongly that very effective and cosmetically elegant products for the alleviation of dry skin can be formulated using 5–12% petrolatum and 5–10% glycerine. With these types of formulations the persistence of the skin moisturizing effects (i.e., the time after cessation of treatment, before dry skin was again clinically observed) was even better for these types of formulations than for pure petrolatum. Also, with care in formulation, the cosmetic feel was much better than the highly greasy, undiluted, petrolatum. Some type of synergism between these two basic skin treatments seems very likely.

It is not surprising that the literature contains many examples of petrolatum–glycerine skin care emulsions. From a formulation point of view, petrolatum can be used very similarly to mineral oil. Typical is the following lotion [17].

Material	% by wt
Veegum[a]	1.50
Cellulose gum CMC-7MF[b]	0.50
Water	67.90
Glycerine	6.00
Allantion	0.10
Petrolatum	12.00
Acetulan[c]	3.00

Material	% by wt
Amerchol L101[d]	4.00
Arlacel 165[e]	5.00

[a] Magnesium aluminum silicate, trademark R. T. Vanderbilt Co.
[b] Trademark Hercules Chemical.
[c] Lanolin alcohol acetate, trademark Amerchol Co.
[d] Lanolin derivative, trademark Amerchol Co.
[e] Self-emulsifying glycerol monostearate, trademark ICI Americas Inc.

In addition to the normal type of petrolatum–glycerine emulsion just described, formulators have worked with very high levels of these ingredients to good effect. For example, a completely anhydrous composition (which, of course, is not an emulsion, but is included here for convenience) containing white petroleum jelly 40 parts, lanolin 40 parts, cocoa butter 10 parts, and glycerine 60 parts has been claimed to alleviate not only dry skin, but psoriasis and skin cracks after 4 to 7 days of 4 times daily use [18]. Similarly a "gravitationally and thermally stable emulsion" comprised of petrolatum 49 parts, glycerine 10 parts, glyceryl monooleate 1 part, and water 40 parts, was claimed to be particularly beneficial for skin conditioning [19]. Probably these two formulations are not the most elegant from the point of view of skin-feel, but the claims made for salutary skin conditioning, point to the fact that in spite of all the modern advances with special ingredients (like liposomes, elastins, biological extracts, and whatever) glycerine and petrolatum remain fundamental to the cosmetic moisturization of skin.

5. Stearic Acid-Containing Emulsions: Vanishing Creams

In addition to the glycerine and mineral oil type of basic skin cream/lotion is another basic recipe called the "vanishing" cream product. As the name implies, these formulations, which contain stearic acid as a major ingredient, are designed to disappear rapidly into the skin. Vanishing creams have been defined as a suspension of stearic acid in a gel of stearic soap [20]. Not surprisingly, they are designed to be used primarily as a hand care item. Glycerine (or other humectants) is also commonly incorporated into vanishing creams. Suzuki has done a very detailed rheological study on simple vanishing cream systems consisting of 14% partially neutralized stearic acid with 1% glyceryl monostearate as an auxiliary emulsifier, with and without the addition of 30% humectant [21]. Glycerine and the other humectants improved the viscosity and smoothness of the creams, preventing the crystallization of the larger stearic acid particles.

In practice most vanishing creams also contain a large amount of emollient oil, as the following typical example, from Amerchol Corp. demonstrates [16, p. 208].

Material	% by wt
Amerchol 400[a]	2.0
Solulan PB-10[b]	3.0
Stearic acid	20.0
Isopropyl palmitate	12.0
Glycerine	4.0
Triethanolamine	1.2
Water	57.8

[a] Lanolin alcohols, trademark Amerchol Co.
[b] PPG (10) lanolin alcohols, trademark Amerchol Co.

In the above formula, the mixture of the stearic acid and isopropyl palmitate provides a skin-conditioning action due to a very high active level, while at the same time maintaining a very pronounced dry feel because of the large amount of stearic acid. This is good for hands which need protection without a lot of oiliness. A vanishing cream type, which takes this attribute of dry feel even further, is taken from a Hercules Inc. formula [16, p. 208].

Material	% by wt
Stearic acid	15.00
Span 60[a]	1.50
Zinc stearate	5.00
Glycerine	6.00
Cellulose gum CMC-7HF	
(2% in water)	37.50
Water	33.00
Preservatives	q.s.

[a] Sorbitan monostearate, trademark ICI Americas Inc.
[b] PEG (20) sorbitan monostearate, trademark ICI Americas Inc.

In this particular formula, the zinc stearate adds an extra level of protection, making this a "protective" hand cream. The stearic acid/zinc stearate combination gives a very dry skin feel, so the glycerine aids not only in adjusting to a good feel and spreading action for these formulations, but also in improving the overall skin conditioning after application.

6. Facial, Eye, and Baby Creams and Lotions

Most of the formulations discussed so far are for hand, leg, and overall body use. There is nothing inherently different in emulsions intended for the face, eye, or infant care. Some practical differences, however, are that the formulator need not be worried about residual greasiness, as with a hand product, but should bear in mind the delicate nature of the facial and baby skin as well as the possibility of eye contamination. Thus these products will be rich, protective, and very nonirritating. Fortunately, with the types of formulations already described this is easily achievable. In particular, harsh surfactants (such as sodium lauryl sulfate), fragrances, and any material which has not been thoroughly tested and approved for eye safety, is to be avoided. With these thoughts in mind, again glycerine is an ideal constituent for these especially innocuous products.

Example formulas, which embody these principles are now given.

First a baby lotion [16, p. 37].

Material	% by wt
Drakeol 7[a]	26.00
Penreco Snow[b]	12.00
Beeswax	3.00
Arlacel 60[c]	3.00
Tween 60[d]	4.00
Propylparaben	0.15
Water	43.60
Methylparaben	0.15
Glycerine	8.00
Veegum[e]	0.10

[a] Mineral oil, trademark Penreco Co.
[b] Petrolatum, trademark Penreco Co.
[c] Sorbitan monostearate, trademark ICI Americas Inc.
[d] PEG (20) sorbitan monostearate, trademark ICI Americas Inc.
[e] Magnesium aluminum silicate, trademark R. T. Vanderbilt Co.

The basis for a pigmented glycerine-containing facial or foundation cream is given by Harry [7, p. 72]. As with nearly all other products, glycerine is not essential to make a facial or a foundation cream, however, in this particular semisold formula, it will help to maintain spreading and dispersion of the pigments on the skin. Thus, in addition to skin conditioning, covering of small blemishes can also be achieved.

Material	% by wt
Cetyl/stearyl alcohols	8.0
Stearic acid	8.0
Water	64.0
Preservative	q.s.
Glycerine	10.0
Powder base[a]	10.0
Color	q.s.
Perfume	q.s.

[a] Powder base contains mixtures of solids such as titanium dioxide, talc, kaolin, magnesium stearate, and zinc oxide.

Barnett emphasizes the need for eye creams to be nonirritating, nonpenetrating, and unperfumed. Barnett further states that nonionic emulsifiers should be used. Soap or detergent type anionic emulsifiers, and especially cationic emulsifiers are to be avoided with eye-care creams [22, pp. 27–104]. Of the formulations recommended by Barnett for eye care that contain glycerine, an oil-in-water [22, p. 55], and a water-in-oil, [22, p. 57] example are given, respectively:

Material	% by wt
Beeswax	3.00
Spermaceti	3.00
Mineral Oil (65/75 Saybolt)	30.00
Glyceryl monostearate, pure	12.00
Propylparaben	0.15
Methylparaben	0.15
Glycerine	8.00
Water	43.70

Material	% by wt
Peanut oil	8.00
Protegin X[a]	26.00
Spermaceti	5.00
Lanolin	5.00
Cetyl alcohol	2.00
Glyceryl monostearate, pure	1.00
Mineral Oil (65/75 Saybolt)	5.00
Propylparaben	0.15
Antioxidant	0.15
Methylparaben	0.15
Glycerine	3.00
Water	44.55

[a] Nonionic emollient blend, trademark Goldschmidt Chem. Corp.

Barnett also gives several examples of creams and lotions for the eye area that do not have glycerine, but certainly it is a very appropriate ingredient for use in that delicate skin region.

III. CLEANSING CREAMS

Cleansing creams have been broken down into two general categories: emulsion creams and lotions (with which we are interested in here), and the "liquefying" type which are usually anhydrous in character consisting of a mixture of hydrocarbon oils and waxes [23]. Of course, excellent cleansing products can be made from mild surfactant materials, but they are not the subject of this chapter. Designed to remove oily makeup from the face, cleansing creams need a high oil content in order to act as a good makeup solvent. It is important that formulations to be used on the face be nonirritating. Glycerine can enhance cleansing cream properties as a result of its cosolubilization and pigment-dispersing properties. Also, because the product is used in the face and eye area, glycerine is beneficial for its skin-conditioning effects. Further, if the cleanser is merely wiped off, as is sometimes recommended, glycerine and other skin conditioners will be left as residue. A typical formulation in this category [16, p. 150] is as follows:

Material	% by wt
Mineral oil	20.00
Cetyl alcohol	4.00
Lanolin	2.00
Arlacel 165[a]	10.00
Arlatone B[b]	6.00
Glycerine	6.00
Water	52.00

[a] Self-emulsifying glycerol monostearate, trademark ICI Americas Inc.
[b] Nonionic emulsifier mixture, trademark ICI Americas Inc.

A 1983 patent [24] stresses the beneficial effect of a correctly formulated cleansing product on the skin. The "improved hypoallergenic facial skin activator . . . and cleanser" contains 50–80 parts water, 8–35 parts unsaturated vegetable oil, 5–10 parts glycerine, as well as smaller amounts of emulsifiers, preservatives, etc. Soma states that the glycerine is an important aid in moisturizing the skin.

Another patent, which is interesting because the emulsion is packaged as an aerosolized mousse, rather than the typical cream or lotion, is given by Synder [25]. Unlike the oil formulas discussed, this product relies mainly on mild surfactants for its cleansing action, however, it is still an emulsion product. Synder claims that the product can be removed without aqueous rinsing, thus some conditioning agents, including the high glycerine level, are left behind to condition the skin. The emulsion is packaged in a pressurized aerosol container, with nitric oxide as the propellant. Of course, more typically, mild surfactant formulations should be well rinsed.

IV. OTHER GLYCERINE-CONTAINING SKIN CARE EMULSIONS

As already discussed, the possibilities here are endless, so no attempt will be made to cover the potential subject matter of this section. Rather, two illustrative patents will be given, and the subject will be dropped. The point to be made with these two examples is that the subject of glycerine in skin care emulsions is an open-ended one.

In the first example, an elastin hydrozylate is mixed into a standard, but concentrated, oil-in-water emulsion [26]. A typical formula from this work is 15% elastin hydrolyzate (70,000 M Wt) 10 parts, glycerine 10 parts, sorbitol 2 parts, liquid paraffin 20 parts, water 35 parts, and 1% neutralized Carbopol (trademark B. F. Goodrich Co.) 22 parts. These emulsions are stable and hold water for a long period when applied to the skin. One interpretation of this patent is that a high glycerine emulsion is used as an effective vehicle to allow the use of the type of sophisticated cosmetic ingredient that promotes consumer interest and high prices. Considering this formulation in general, and without comment as to the benefits of such glamorous items like elastin, liposomes, plant extracts or whatever, it certainly makes sense to add the reliable standbys like glycerine and oils to make a product as effective as possible.

A second formula of interest, because of the high titanium dioxide level, is the following, which is used for coloring as well as conditioning facial skin [27].

Material	% by wt
Glycerine	30
Potassium hydroxide	0.2
Tri-Na EDTA	0.05
Stearic acid	3.0
Squalane	15.0
Cetostearyl alcohol	3.0
Polyoxyethylene oleyl alc. ether	1.5
Glyceryl monostearate	1.5
Cyclic polydimethylsiloxane	9.0
Titanium dioxide	7.0
Perfume	0.2
Water	to 100%

This formula is claimed to be a good skin conditioner, which one would expect with such a high level of glycerine. One can strongly surmise, however,

that in this case the glycerine is also wetting and anchoring the titanium dioxide powder to the skin as well as giving good dispersion and spreadability of
the powder during manufacture and application.

V. USE OF GLYCERINE IN CLEAR GELS AND STICKS

A. Microemulsions/Gels

The qualities of glycerine which make it useful in the formulation of clear gels
and sticks are, of course, related to its complete water miscibility, cosolubilizing properties, and viscosity. Clear gels, in which aqueous glycerine is thickened with a water-soluble polymer, have already been discussed. Of interest in
this section, are the types of clear gels that contain hydrophobic materials in
the form of colloidal particles so tiny as to be invisible. Although glycerine is
not essential to these types of products, the following is a good example of a
clear, glycerine-containing microemulsion [taken from [2]:

Material	Parts by wt
Crodafos N3 Neutral[a]	4
Volpo 10[b]	4
White oil	8
Glycerine	22
Water	18

[a] Phosphated oleyl ether, trademark
Croda Inc.
[b] Ethoxylated oleyl alcohol, trademark
Croda Inc.

The ingredients are mixed at 90°C. Viscosity can be adjusted from liquid to
gel by changing the ratio of glycerine to water.

Another Croda gel formula with a large amount of glycerine uses Polychol
15 as the emulsifier [2]. It was formulated as a general purpose base.

Material	% by wt
Polychol 15[a]	20
Glycerine	20
Isopropyl myristate	17
Super Hartolan[b]	3
Water	40

[a] Polyoxyethylene (15) lanolin ether,
trademark Croda Inc.
[b] Lanolin alcohol, trademark Croda Inc.

A more recent microemulsion, formulated specifically as a skin moisturizer, is shown in the following example [28].

Material	% by wt
Cyclomethicone	20.0
Squalane	14.7
Ritachol 1000[a]	2.0
Isopropyl myristate	3.0
Jojoba oil	10.0
Water	20.8
Glycerine	4.0
Methylparaben	0.15
Propylparaben	0.15
Germall 115	0.2
Polysorbate 21	25.0

[a] Ethoxylated fatty alcohol blend, trademark R.I.T.A. Corp.

The essential ingredients claimed in the skin-moisturizing microemulsions from ref. 38 are: a moisturizer, a microemulsion-forming surfactant, a polysiloxane, and a skin humectant, which in the example given here is glycerine.

Although it is outside the scope of this chapter, it is interesting to note that glycerine is also a common constituent of self-foaming soap gels such as those used in the gel shaving creams. The well-known solubilizing effect of glycerine on soap is a key help in this context. For instance, in a patent by Su, a self-foaming shave gel is made by mixing 100 g of glycerine at 49°C, with a previously melted mixture consisting of 100 g stearic acid, 30 g of polyglycerol 3-diisostearate, and 30 g of cetyl alcohol. This glycerine induced microemulsion is mixed with a proprietary aqueous gel and pentane to give a semitransparent gel with good spreadability and foaming properties [29].

B. Sticks

Solidified sodium stearate sticks, as described, for instance, by Barker [30], are well known in cosmetics. Essentially they are a hydroalcoholic colloidal dispersion of sodium stearate, often plasticized by humectants such as glycerine. Major uses for these sticks have been in the application of deodorants and perfumes. For illustration, a high alcohol, translucent formula for an attractive looking cologne stick is as follows [16, p. 235]. The glycerine acts both as a humectant and a plasticizer to improve stability and application. Furthermore, glycerine (as well as alcohol, water and propylene glycol) acts as a solvent for soap in its own right and thus helps gel formation upon cooling.

Material	% by wt
Stearic acid (triple pressed)	5.2
Glycerine	2.10
SDA-40 alcohol	83.85
Sodium hydroxide (50% solution)	1.35
Water	4.5
Timica Sparkle[a]	1.00
Perfume	2.00

[a] Mica (and) titanium dioxide, trademark Mearl Corp.

A very similar fragrance stick can be made by using sodium stearate directly [31].

Material	% by wt
Ethyl alcohol	81
Carbitol	3
Glycerine	3
Water	5
Sodium stearate	6
Perfume	2

Kassam et al. have studied the effects of different humectants on sodium stearate-based sticks using the following formula [32]: stearic acid 5.22%, sodium hydroxide 0.78%, water 2.00%, humectant x%, and 95% alcohol to 100%: where x was, respectively, 10, 12.5, 15, 17.5, and 20%. The humectants used in this study were glycerine, propylene glycol, polyethylene glycol 400, and polyethylene glycol 600. Hardness, disintegration, weight loss, and rheology were extensively studied. All of the formulations were stable over 18 months, although synerisis did occur, especially at the higher humectant levels. The humectants have the pronounced attribute of reducing stick weight loss. There is a great deal of information on hardness and rheology in this paper, but some interesting conclusions made were: "Generally speaking, sticks made with polyethylene glycol 600 or propylene glycol tend to be harder than the corresponding ones containing glycerine or polyethylene glycol 400 . . . propylene glycol gives sticks with higher plastic viscosity; they indeed have a tough texture; glycerine or polyethylene glycol-600 based sticks have intermediate plastic viscosities and acceptable spreading characteristics, and polyethylene glycol 400 based sticks have the lowest plastic viscosities and best spreading." Further, this report has an interesting discussion concerning the effect of the humectants on the soap micelles. It is surmised that the higher yield values caused by glycerine and propylene glycol in the sticks may be due "to hydrogen bonding effects between the humectants and the soap molecules, stiffening the overall micellar structure."

Of course there are as many different formulations as there are formulators, as the following two literature examples illustrate. First, a stick which incorporates the three humectant types as studied by Kassem et al. as well as a higher water level, has been claimed to have high dimensional stability [33].

Material	Amount
Water	40 ml
Stearic acid	10 g
Sodium hydroxide (360 Be)	3.7 ml
Glycerine	15 ml
Propylene glycol	40 ml
Polyethylene glycol 4000	20 g
Ethanol with 0.2% trichlosan	30 ml

Secondly, it has been claimed that cosmetic gel sticks containing 6–70% polyhydric alcohol (examples given are with propylene glycol, although glycerine is claimed), 3–10% soap, 10–60% diisopropyl adipate, and deodorant, improve smoothness and reduce hair pull during application. These formulations also typically contain about 8% ethanol and 5–30% of a fatty alcohol [34]. Similarly, Luebbe and Davis advocate a clear deodorant stick with high levels of polyhydric alcohol (40–70%), water (10–40%), soap (3–10%), emollient (1–20%), and ethanol (<12.5%) [35]. The high levels of propylene glycol or glycerine are instrumental in forming a clear gel matrix. Luebbe and Davis utilized much higher levels of glycerine than in conventional sticks, so a lesson to be learned is that in addition to being aware of the fundamental advantages of an agent like glycerine, the formulator should be looking for radically different ways to utilize these advantages. One must be prepared to move away from the standard recipes given. This point will be elaborated in the section on unusual applications.

Finally, in this section is a very unusual recipe for a styptic pencil [36]:

Material	% by wt
Aluminum potassium sulfate dodecahydrate	90
Talc	5
Glycerine	5

Glycerine aids in fusion of the salt, and thus, stick formation. In related formulations that have less talc, it also makes the stick translucent.

VI.   GLYCERINE-CONTAINING CREAMS, LOTIONS, AND MUCILAGES WITH MEDICINAL CLAIMS

Of necessity this will be something of a potpourri category, but it is important to emphasize that glycerine is widely used in areas that are not just strictly cosmetic in nature.

A very basic formula in this category [37] is a topical antibacterial cream containing: an antibacterial agent (preferably povidone iodine), 15 to 40% of a hydrocarbon (mineral oil/petrolatum), and 5 to 15% of a polyol moisturizing component (preferably glycerine), as well as other surfactants to make an aqueous emulsion. An example given from this work is:

Material	% by wt
Povidone iodine	5
Stearyl alcohol	2
Cetyl alcohol	2
White petrolatum	12
Liquid petrolatum	10
Glycerine	8
Polysorbate 60	2
PEG monostearate	1
Sodium hydroxide	0.1
Water	57.9

The application contains quantitative data from 25 patients, showing improved wound healing and antibacterial activity. Thus the following general statement is made: "Clinical trials using a povidone iodine cream according to the present invention have shown that the cream affords an improved penetration of the povidone iodine into wounds and that the cream has a superior therapeutic profile when compared with PVPI products (especially ointments) of the prior art which contain a low content of hydrocarbon and moisturizing components." One can conclude that, amongst other things, the glycerine has a definite positive influence on the rate of skin healing.

The wounds just discussed were in fact caused by burns. This relates to other work now described, in which glycerine-containing formulas were studied for burn treatment [38]. Here glycerine, propylene glycol, and polyethylene-glycol were studied (in an acrylic mucilage) for the treatment of purulent wounds and burns. The presence of an optimum osmotic level, which in turn is related to the humectant activity, was demonstrated for burn treatment. Consequently, the concentration of the material used is a factor to consider in the analysis of glycerine's beneficial effect in burn-treating ointments.

Humectant materials are good for skin disease states other than wounds and burns, as a 1987 patent teaches [39]. Skin diseases which are treatable by the disclosed items are: hyperkeratotic skin diseases, seborrheic eczema, dermatomycosis, and onychomycosis. Thus, urea 15%, lactic acid 10%, propylene glycol 69.50%, glycerine 5%, and thickening agent 0.5%, cured tinea man (fungal infection) after 1 month treatment. Although glycerine is not an essential ingredient in this particular patent, it is interesting to speculate on the beneficial effects of humectants in general for conditions other than "dry skin." Further, while the formulations do seem uniquely effective, I do not think it too farfetched to relate the observations of this work with the beneficial effects of glycerine (and other humectants) to the general improvements to skin condition which they occasion.

A high glycerine containing formula which is claimed effective for a number of skin conditions including acne, insect bites and wrinkles contains essence of lavender and gum benzoin [40]. A formula from this patent is as follows: glycerine 40%, essence of lavender 1.88%, ethyl alcohol, 40.75%, gum benzoin 2.55%, and water to 100%. It is difficult to say how effective these products are, but certainly the patent is of possible interest as a point of further study.

Another, older therapeutic formula with a glycerine and water base is the following antipruritic [41]:

Material	Amount
Menthol	0.2 g
Hexachlorophene[a]	0.1 g
Glycerine	10.0 cc
Isopropyl alcohol	35.0 cc
Distilled water	to 100 cc

[a] Previously widely used, this antiseptic is not longer permitted in cosmetics.

VII. UNUSUAL FORMULATIONS AND APPLICATIONS

A. Films and Sheets

Glycerine-containing films and sheets for skin treatment have appeared in the literature in several versions. These are, preformed films, films that dry on the skin, and treatments placed on a separate support film.

Unsupported gelled films, suitable for general skin treatments and wound dressings, have been formulated by Doi [42], and Cornell [43], respectively. Doi classifies glycerine (or ethylene glycol, etc.) as a hydrophilic adhesive-forming agent and N-hydroxyimide ester as the crosslinking agent in the following sheet-forming composition: gelatin 12, calcium chloride 12, N-hydroxyimide ester 2.5, glycerine 22, ascorbic acid 1.5, 4-hydroxybenzoic acid 0.2, and water 49.8 parts by weight. The prominence of glycerine in forming these coherent sheets is not surprising when one considers its strong interaction with proteins and water (see Chap. 10). Without wishing to encourage anyone to violate any patent, it is interesting to speculate that modifications of this formula might form good artificial substrates for skin in laboratory testing. Cornell's gelled wound dressing is similar in many ways. Here a representative formula contains: Gelamide 250 (polyacrylamide, trademark American Cyanamide Co.) 1 part, glycerine 20 parts, crosslinked polyacrylic acid 8 parts, water 72 parts. In this case one can speculate that the glycerine is beneficial not only in physically structuring the film, but also in helping with subsequent wound healing.

Of course glycerine-containing films can also be formed in situ on the skin. For instance, in a Shiseido patent [44], rubber latex, polyvinyl alcohol, and a polyol such as glycerine or propylene glycol are mixed in water to form a peelable cosmetic pack (face mask). With facemasks, as with everything else, there are as many different ways to put them together as there are formulators who work on it. For example, the formula below [45] is completely different from the Shiseido one just mentioned.

Material	% by wt
Pullulan (MW 150,000)	20.00
Cellulose gum	5.00
Glycerine	2.00
Ethanol	5.00
Water	68.00

One can speculate that as the water dries, the residual glycerine aids in plasticizing the film, and in maintaining suppleness and integrity as the film is subsequently peeled away.

A third type of film or sheet is where the cosmetic treatment is applied to a preformed substrate such as fabric or paper. Given the existence of this preformed structure, glycerine is not necessary for film integrity as in the examples just discussed, however, it certainly can still aid in product functionality. Thus, Smith and Reilly teach nonwoven fabrics impregnated with glycerine-containing emulsions [46]. Glycerine is not an essential part of the invention.

Another example in this area is fabric or paper treated with polyethylene glycol, glycerine, and/or sorbitol, and a finely divided solid to make a nondrying cleansing substrate for the skin [47].

B. Multiphased Formulations

Much effort in cosmetics is devoted to making stable, uniform materials, be they emulsions, films, gels, or whatever. Sometimes, however, it is productive to engage in a little lateral thinking, as for instance in two patents by Mori which teach a three-layered facial oil [48, 49]. Mori claims that the lotions have aesthetic value because they produce a beautiful suspension when shaken. A representative formula contains: sodium lactate (50% solution) 10, glycerine 10, polyoxyethylene glyceryl monooleate 20, and liquid paraffin 50% by weight. I do not know whether glycerine is instrumental in forming the three layers. Nevertheless, the idea of nonhomogeneous formulations does open up possibilities for putting together products that may be particularly beneficial and attractive. For instance, some readers may remember the two-phase conditioning shampoo that was on the market several years ago. With an oil phase floating on top of the surfactant, the idea was to provide a pronounced cleansing and conditioning in one application (which can be achieved by cationic polymers in a homogeneous system also). The idea did not catch on then, but there is always a time and place for everything.

A similar two-layered skin lotion has also been claimed as follows: polyglycerine nonylphenyl ether 10, squalane 15, glycerine 5, sodium lactate 1, ethanol 15, perfume 0.1, preservative 0.1, and water 62.8% by weight [50].

C. Internal Phase with Very High Glycerine Levels

A contrast to the standard way of using glycerine in emulsions, workers at Shiseido have made emulsions with an external oily phase and very high levels of glycerine in the internal phase. In one embodiment of this approach, 49 parts of glycerine were dispersed into 50 parts of paraffin oil using one part of an organoclay [51]. Obviously, the insolubility of glycerine in paraffin is instrumental in this composition, which is claimed as a stable base for cosmetics and ointments. It certainly makes me curious about what can be done with this as a starting point for different items, and again emphasizes the importance of not always being narrowly led by the prior art. Two very similar patents, also from Shiseido, both contain about 35% water and 35% glycerine as the inner phase with silicone oil outer phase [52, 53]. The products are claimed for general cosmetic and pharmaceutical use. The fact that water repellency is one of the claims, suggests that the internal glycerine structure persists after application, and this may further stimulate the "creative juices" of the interested reader.

D. Highly Concentrated Glycerine Gels

As already well covered, there is nothing new about thickening glycerine and water combinations with polymeric thickeners, however, the formulation of anhydrous glycerine gels does deserve comment here. Thus one glycerine based paste, which can be used in cosmetics or dentifrices, consists of nothing more than glycerine thickened with about 1% of sodium carboxymethyl cellulose [54]. A very similar patent uses hydroxyethyl cellulose rather than CMC [55].

Heated to 120°C, carageenan is another material that can form a gel with glycerine [56]. Methacrylates and diethyl sebacate were added to the glycerine/carageenan gel to form an ointment base.

Finally, getting back to the standard glycerine mucilages (often referred to as glycerine/rosewater), Hiuga et al. teach a concentrated gel thickened with polyacrylic acid [12]. For example, a conditioner was prepared consisting of crosslinking type polyacrylic acid 2.0, glycerine 68.0, propylene glycol 1.0, sodium hydroxide 1.4, and water to 100% by weight. The patent teaches that this concentrate is diluted with other products, such as an emulsion lotion, prior to use. Consequently, one would end up with a fairly standard glycerine lotion of the type discussed above. Nevertheless, this does help to emphasize that workers continue to recognize the benefits of the material and find different ways of utilizing it.

E. Liposomes

Liposomes, those cosmetically fashionable phospholipid-based colloids, have also been manufactured with high glycerine levels [57]. The trusted old workhorse in a fancy new harness. As an example, soybean lecithin 2.0, and glycerine 5.0 parts, were heated at 70°C followed by the addition of 0.5 part purified water. Stirring, sonication, and then mixing with 0.2 part methyl paraben, 0.1 part perfume, and 10.0 parts ethanol were employed to obtain a cosmetic liposome lotion.

Obviously, as with the other areas mentioned, the possibilities here are endless. This example merely serves to remind formulators that although glycerine is an old standby it does not have to be restricted to traditional uses. Utilization in items from sheet wound dressings, to liposome cosmetics for the hair and skin, and just about everything else in between is possible.

F. Human Body Dust

Innovation can take two basic forms, find a new way to deal with an old need, or invent a new need that we were not previously aware we had. I am not sure where a cosmetic that controls the release of dust and other undesirable substances from the human body in clean room environments falls, but indeed it is available [58]. The cosmetic contains one or more compounds selected from glycerine, sorbitol, urea, etc., and also a film-forming agent selected from sodium alginate, ethyl cellulose, polyvinyl alcohol, etc. Presumably the combination of humectant and film-forming agent does make a coherent, lasting, and substantive film on the skin, a factor of more general interest than just clean room preparations. For instance, what about a flea and parasite controller for underdeveloped areas?

G. Calcium Lactate-Glycerine Adduct

Just discussed were formulations relating to definitive and newly perceived end-use applications. By contrast the calcium lactate-glycerine adduct may be a chemical just waiting for the end-use inspiration to materialize [59]. To prepare the adduct, 30.8 g of calcium lactate are mixed with 92.0 g of glycerine and refluxed in methanol. Cooling and mixing with acetone yields the crystals which are filtered off. The starting calcium lactate has five water molecules which are obviously replaced by the glycerine. Presumably the final compound has special miscibility and penetration properties originating from glycerine's carbon backbone. The special interface that glycerine occupies between water and organic materials is again evident in this work. Formulation work in hair, skin and nails would all be very interesting with this adduct, which presumably has astringent as well as highly moisturizing properties. (Readers must of course recognize its proprietary nature and work under license if applicable.)

VIII. GLYCERINE PRODUCTS FOR EYE, NAIL, AND FOOT CARE

There is not a great deal of special literature of these areas, but for the sake of completeness a few formulations designated for these special body regions will be mentioned.

A. Fingernail Treatments

Two very different recipes are included in this section. The first, a nail hardener, is closely related to the astringents and styptic compositions previously mentioned (from ref. 2, p. 29).

Material	% by wt
Aluminum acetate	2
Glycerine	5
Formaldehyde	0.1
Perfume	qs
Water	to 100

Presumably, both the small aluminum cation and formaldehyde, crosslink and harden the nail keratin. Glycerine helps to achieve an even coating and enhance the penetration, says Moxey.

A quite different cosmetic cream for fingernails is as follows [60].

Material	% by wt
Water	38.52
Dipropylene glycol	10.00
Glycerine	20.00
Keratin hydrolyzate	0.07
Tri-sodium EDTA	0.01
Xanthane gum	0.05
Squalane	15.00
Petrolatum	10.00
Cetostearyl alcohol	2.00
Polyoxyethylene oleyl alc. ether	2.00
Glyceryl stearate	2.00
Ethylparaben	0.10
Butyl paraben	0.10
Perfume	0.10
Vitamin E acetate	0.05%

This is a fairly conventional cream, where the moisturizing/conditioning effect of the humectant–oil combination is intended for the nail keratin rather than the stratum corneum keratin of the skin care products.

Simpler yet is a nail lotion containing glycerine and a protein hydrolyzate as the active materials [61]. In this work a nail lotion contained glycerine 3–5.5%, geranium oil 0.1–3%, ethanol 18%, methyl cellulose 0.3–0.6%, protein hydrolyzate of Chlorella 1–3%, and water to 100%.

B. Eye Care

Because it is so innocuous, can be used for isotonicity, and has good co-solubilization properties, glycerine makes an excellent constituent of eye drops and lotions. Two illustrative examples are included here.

An isotonic antiinflammatory eye drop was formulated using 2.75 g glycerine, 0.05 g hyaluronate, 0.05 g dexamethasone disodium phosphate, and water to 100 g [62].

In a second example glycerine is used to solubilize an indomethacin eye treatment in a composition containing 10 g indomethacin, 3 g glycerine, and 87 g of water [63].

C. Foot Care

Normally feet do not require moisturizing treatment against dry skin. In fact the opposite is often the case and overmoist feet require treatment for unwanted fungal growth, such as the often referred to "athlete's foot." Glycerine can be used even in this environment, however, as the following funjistatic, deodorant, foot cleansing composition demonstrates [64].

Material	% by wt
Water	59.30
Igepal CO-630[a]	16.5
Ninox L[b]	5.00
Soap	8.25
Pine oil	2.50
O-benzyl-p-chlorophenol	2.25
Tetra sodium EDTA	0.45
Glycerine	5.00
Isopropanol	0.75

[a] PEG-9 nonyl phenyl ether, trademark GAF Corp.
[b] Lauryl dimethyl amine oxide, trademark Stepan Co.

Possibly the glycerine helps with cosolubilization of the ingredients and/or with activating the fungistat on the skin.

IX. GLYCERINE AND HAIR CARE

Although more limited, glycerine finds its place in hair care cosmetics both because of its useful effects on the formula itself as described in the sections above, as well as its direct "moisturizing" effects on hair.

It should be noted that damage in hair stems from various causes including sun exposure, bleaching, coloring, and permanent waving treatments as well as repetitive shampooing. All this has the effect of stripping the cuticle and causing the hair shaft to split, resulting in hair that feels dull brittle and lifeless. As with the keratin in stratum corneum, glycerine and other humectants can partly reverse these effects by assisting in the formation of a compatible aqueous environment for the keratin fibrils.

A good example of the use of glycerine on restoring "moisture," or structural integrity, to damaged hair is found in the work of Newell in a series of patents concerning a method for restoring normal moisture level to hair with moisture deficiency [65–67]. The patents involve various regimens for using a moisture stabilizing shampoo, an intensive conditioner and a deep heat treatment. Example of the formulations employed in his work are as follows:

Moisture Stabilizing Shampoo

Material	% by wt
Sodium DL-2-pyrrolidone-5-carboxylate	
(50% aqueous solution)	0.100
Glycerine	0.100
Protein	0.100
Dynol SAM[a]	42.000
Ninol 2012[b]	1.000
Lauryl dimethyl amine oxide	2.000
Deriphat 160C[c]	0.100
Water	to 100
Methylparaben	0.150
EDTA	0.100
Citric acid	0.190
Monomethylol dimethyl hydantoin	0.100
Perfume	0.300
Coloring agent	0.015
Ammonium chloride	0.600

[a] Sodium lauryl sulfate, trademark Richardson Co.
[b] Coconut diethanolamide, trademark Stepan Chemical Co.
[c] Sodium N-lauryl-iminodipropionic acid, trademark General Mills

Supplemental Gel Conditioner

Material	% by wt
Sodium DL-2-pyrrolidone-5-carboxylate	
(50% aqueous solution)	2.00
Glycerine	1.50
Protein	0.50
Carbopol 940[a]	0.35
SD Alcohol 40	25.00
Nonionic surfactant	0.500
Perfume	0.10
Water	to 100

[a] Polyacrylic thickener, trademark B. F. Goodrich Chem.

Deep Heat Treatment Conditioner

Material	% by wt
Sodium DL-2-pyrrolidone-5-carboxylate (50% aqueous solution)	5.0
Glycerine	5.0
Protein	2.5
Barquat CT-429[a]	4.3
DL-pantothenyl alcohol	0.1
Acid stabilized glycerol monostearate	1.0
Promulgen D[b]	1.0
Mineral oil	2.0
Isopropyl Myristate	2.0
Cetyl alcohol	3.5
Ethylene glycol monostearate	2.0
Perfume	0.4
Coloring	0.2
Water	to 100

[a] Cetyltrimethyl ammonium chloride, trademark Lonza Inc.
[b] Cetyl alcohol—polyethyleneglycol complex, trademark Robinson Wagner Co.

Newell teaches that the more severe the dryness of the hair, the more one needs to use the conditioning treatments, especially those with high humectant levels such as the 5% glycerine-containing deep heat treatment conditioner above. In essence, the hair moisture-restoring complex taught by Newell comprises a mixture of sodium DL-2-pyrrolidone-5-carboxylate (50% aqueous solution), glycerine, and protein which can be added to a variety of hair treatment formulations.

A very simple glycerine-containing hair treatment consists of 5–20 parts urea, 3–5 parts glycerine, and 100 parts water. This recipe is claimed to increase the volume of the hair, as well as act as a treatment for seborrhea and other scalp diseases [68]. Effective or not for seborrhea, it should certainly provide a strong moisturizing treatment. In another treatment example, although one would have to view the claims with a certain degree of reserve, glycerine has been claimed as an essential ingredient in a hair growth stimulant [69]. In particular the formulations consist of, waxes 5–10, fats 35–40, oils 20–25, mineral oil 5–10, glycerine 5–10, citrus fruit juice 6–9, fungus juice 6–9, and vitamin E, 3–5% by volume. Another formula uses 25% glycerine in a gel for hair growth [70]. In this case the mixture contains carboxyvinyl polymer–sodium alginate 2, triethanolamine 1.5, glycerine 25, ethanol 3, thistle root extract 2, perfume 1, and water 64.5%. Continuing in the treatment vein, and to demonstrate the variety of uses to which glycerine can be put, one patent claims its use for hair loss prevention in cancer chemotherapy [71]. In this work a cap is made that is prepared from polyvinyl alcohol and glycerine. This chilled mixture is applied to the patients head during chemotherapy with the intent of preventing hair loss by lowering body temperature in the head region.

I am not sure exactly what relevance it has to the protective effects of glycerine on hair in general circumstances, but it, and related materials have been shown to reduce the damage caused by thioglycolates [72]. Thus, the reaction of hair with thioglycolic acid was enhanced by increasing pH, by adding guanidine salts, and by oxidative pretreatment. On the other hand thioglycolic reduction was slowed by the addition of propanol, triacetin, glycerine, and propylene glycol. In this work the method of measurement used was single fiber stress decay.

Another use of glycerine (and other humectants) in hair care over the years has been in the so-called hair color-restoring complexes. These are the metal salt (usually lead acetate) and sulfur mixtures which are used to gradually darken hair. Two examples of formulations in this context are as follows. First a hair-coloring recipe from Holland [73].

Material	% by wt
Sodium thiosulfate	0.2–5
Bismuth nitrilotriacetic acid complex	0.1–5
Diethanolamine	(to pH 8–11)
Glycerine	5–70
Water	q.s.

A similar hair-coloring recipe from Barrucco [74], uses glycerine in an aqueous-alcohol solution of lead acetate, sulfur, sodium thiosulfate, and vegetable oil.

Finally, in a recipe which is probably made obsolete by the ozone issue, a gelled foaming shampoo was made which had the following composition [75]. This interesting formula has the following composition:

Material	% by wt
Glycerine	35
Freon 11	32
Perfume	1
Monamid 779[a]	25
Monamid 150 IS[b]	2
Triethanolamine methyl cocoyl taurate	5

[a] Coconut oil diethanolamide neutralized with lauryl ether sulfate, trademark Mona Chemical Corp.
[b] Isostearic acid diethanolamide, trademark Mona Chemical Corp.

Although not in commercial use, this recipe does illustrate the use of glycerine in making nonaqueous compositions.

In conclusion, both for hair and skin, glycerine is an important part of health and beauty.

REFERENCES

1. Wells, F. V., Glycerin in cosmetics, Soap Perfumery Cosmet., February: 194–196, 218, 1957.
2. Moxey, P., The role of glycerine in cosmetic and toiletry products, Mfg. Chem. Aerosol News, November: 27–30, 1967.
3. Glycerine Producers Association, Uses of glycerine, available through The Soap and Detergent Association, New York, pp. 8–9.
4. Newman, A. A. (with additional chapters by L. V. Cocks), Glycerol, CRC Press, Cleveland, 1968 p. 176–179.
5. Mannheim, P., Structure modifiers in cosmetics, Soap Perfumery and Cosmet., July 1959, 713, 720.
6. French Patent 81 01971, Brun, A. and Koulbanis, C., New humectant compositions encorporating sodium lactate, glycerine, urea and collagen, 1982.
7. Harry, R. G., The Principles and Practice of Modern Cosmetics, Volume Two, Cosmetic Materials, Leonard Hill Ltd., London, 1950, p. 136.
8. Harry, R. G., Harry's Cosmeticology, 6th ed., Vol. 1 (J. B. Wilkinson et al., eds.), Chemical Publishing Co., Inc., 1973, pp. 90–91.
9. British Patent 1,402,282, Ward, P. E., Preparation for skin treatment, 1975.
10. Bissett, D. L. and McBride, J. F., Skin conditioning with glycerol, J. Soc. Cosmet. Chem., 35, 345–350, 1984.
11. European Patent Application, 85114086.3, Miller, D. G., Antiwrinkle cosmetic application, 1985.
12. Japanese Patent 86218508, Huiga, T., Sendai, H., and Kodoma, K., Skin conditioners containing poly(acrylic acid), glycerine, propylene glycol, and an alkaline compound, 1986.
13. Japanese Patent 85246309, Yoshizawa, O., Skin lotion containing ethanol, glycerine, a lemon fragrance and water, 1985.
14. British Patent 1,416,296, Blass, J. M., Styptic preparations for use on human skin, 1975.
15. European Patent Application 0 076 146, Dixon, T. J. and Kelm, G. R., Skin conditioning composition, 1982.
16. Flick, E. W., Cosmetic and Toiletry Formulations, Noyes Publications, 1984, p. 162.
17. Modified from R. T. Vanderbilt Co. Inc., Cosmetics Formulary No. 319.
18. U.S. Patent 4,454,118, Johnson, Z. M., Treating psoriasis with cosmetic compositions, 1984.
19. European Patent Application EP 103,910 A1, Bratton Howard, N., Emollient containing skin conditioning composition, 1984.
20. Ristic, N., Arh. Farm., 19: 1, 28, 1969.
21. Suzuki, K., Rheological study of vanishing cream, Cosmet. Toiletries, 91: 23–31, 1976.

22. Barnett, G., Cosmetics Science and Technology, Vol. 1, 2nd ed., (M. S. Balsam and E. Sagarin, eds.), Wiley-Interscience, New York, pp. 27–104.

23. Masters, E. J., Cosmetics Science and Technology, Vol. 1, 2nd ed., (M. S. Balsam and E. Sagarin, eds.), Wiley-Interscience, New York, p. 5.

24. U.S. Patent 4,375,480, Soma, W. D., Facial skin activator emulsion and method of skin moisturizing and cleansing, 1983.

25. European Patent Application EP 213827 A2, Synder, W. Earl, Pressurized aerosol nonfoaming skin cleansing and conditioning mousse, 1987.

26. Japanese Patent 84231007 A2, Pola Chemical Industries Inc., Cosmetic emulsions containing elastin hydrolyzates and polyhydric alcohols, 1984.

27. Japanese Patent 86271206 A2, Shiseido Co., Ltd., Cosmetics containing polyhydric alcohols, silicones, and powders, 1986.

28. European Patent Application 226337 A1, Linn, E. E. C., West, M. P., and York, T. O., Skin protecting microemulsions, 1987.

29. German Patent 3,445,749, Su, D. T. T., Stable gel-like shaving cream, 1985.

30. Barker, G., Solidified sodium stearate based sticks, Cosmet. Toiletries 92: 73–75, 1977.

31. Schuler, Robert F., Cosmetics Science and Technology, Volume III, 2nd ed., (M. S. Balsam and E. Sagarin, eds.), Wiley-Interscience, New York, p. 652.

32. Kassem, A. A., Mattha, A. G., and El-Khatib, G. K., Influence of some humectants on the physical characteristics of solidified sodium stearate based sticks, Int. J. Cosmet. Sci., 6: 13–31, 1984.

33. Hungarian Patent 38830 A2, Ban, M., Ban, Hotya, L., Tisoczky, I., and Torok, I., Gel base with high dimensional stability for cosmetics and household pesticides, 1986.

34. U.S. Patent 4,617,185, DiPietro, D. M., Improved deodorant stick, 1986.

35. European Patent Application 107330 A2, Luebbe, J. P. and Davis, J. A., 1984.

36. Bell, S. A., Cosmetics Science and Technology, Volume III, 2nd ed., (M. S. Balsam and E. Sagarin, eds.), Wiley-Interscience, New York, p. 35.

37. European Patent Application 0 185 490 A2, Holtshousen, P. D., Antibacterial cream, 1985.

38. Alekseev, K. V. and Bondarenko, O. L., Study of the osmotic activity of slightly crosslinked acrylic copolymer-based gels, Farmatsiya, 38(1): 22–5, 1989.

39. PCT International Patent, WO 8704617 A1, Moberg, S., Pharmaceutical compositions containing propylene glycol and/or polyethylene glycol and urea as active main components and use thereof in treating skin disorders, 1987.

40. French Patent 73 10085, Levrier, A., Base for cosmetics and pharmaceuticals, 1974.

41. Frazier, C. H. and Blank, I. H., A Formulary for External Therapy of the Skin, pub. 1954, as quoted in Ref. 12.

42. Japanese Patent 8868510, Doi, Sheet-type cosmetic packs containing proteins, gelation retardants, adhesives, and hydroxyimide ester crosslinking agents, 1988.

43. European Patent Application 282316 A2, Cornell, J., Gel composition in sheet or paste form for dressing skin wounds, 1988.

44. Japanese Patent 8462512 A2, Shiseido Co. Ltd., Cosmetic packs containing rubber, polyvinyl alcohol, and polyols, 1984.

45. Wiggers de Vies, J. V. W., Cosmet. Toiletries, 91: 8, 1976.
46. U.S. Patent 4,559,157, Smith, J. A. and Reilly, J. E., Cosmetic applicator useful for skin moisturization, 1985.
47. German Patent 3,447,499, Schwartzkopff, U., Ohl, K., Heilemann, T., and Brecht, G., Nondrying cleansing cloth, 1986.
48. Japanese Patent 8879811, Mori, K., Multilayered cosmetic lotions containing sodium lactate, glycerine and surfactants, 1988.
49. Japanese Patent 8883011, Mori, K., Multilayered cosmetic lotions containing sodium pyrrolidonecarboxylate, glycerine and surfactants, 1988.
50. Japanese Patent 8827409, Shiseido Co. Ltd., Two-phase cosmetics, 1988.
51. Japanese Patent 86209035, Shiseido Co. Ltd., Polyhydric alcohol in oil type emulsions for cosmetics, 1986.
52. Japanese Patent 87215510, Shiseido Co. Ltd., Emulsions containing silicone oil and polyhydric alcohols for cosmetics and pharmaceuticals, 1987.
53. Japanese Patent 87216635, Shiseido Co. Ltd., Cosmetic emulsions containing water and polyhydric alcohol and siloxanes, 1987.
54. Japanese Patent 7435411, Aimoto, K., Ohta, M., and Hashimoto, S., Water soluble nonaqueous paste composition, 1974.
55. Japanese Patent 8195937, Lion Corp., Anhydrous paste preparations for cosmetics, 1981.
56. Japanese Patent 87123112, Hasegawa, K., Nakajima, K., Eguchi, T., and Oota, M., Pharmaceutical and cosmetic bases containing hydrogels and methacrylate polymers, 1987.
57. Japanese Patent 85153938, Takenochi, M., Preparation of liposome-containing cosmetics, 1985.
58. Japanese Patent 88112506, Sato, H., Cosmetics controlling dust release from human body, 1988.
59. German Patent 3,710,177, Reul, B. and Petri, W., Preparation of a calcium lactate-glycerin adduct as a calcium source for pharmaceutical, cosmetic, and food use, 1988.
60. Japanese Patent 86/263907, Shimada, T. and Nakamura, A., Cosmetic creams for fingernails, 1986.
61. Koroleva, N. B., Aleshinkova, T. N., Mayatskaya, T. V., and Timofeeva, I. V., Nail Lotion, Otkrytiya Izobret, 12, 14, 1987.
62. British Patent Application 2196255 A1, Dikstein, S., Isotonic humectant eyedrops, 1988.
63. Japanese Patent 8627917, Hara, K. and Kamya, T., Aqueous eye lotion containing indomethacin and glycerols, 1986.
64. U.S. Patent 4,668,419, Moseman, R. E., Liquid foot treatment composition, 1987.
65. U.S. Patent 4,220,166, Newell, G. P., Method of restoring normal moisture level to hair with severe moisture deficiency, 1980.
66. U.S. Patent 4,220,167, Newell, G. P., Method of restoring normal moisture level to hair with slight to moderate moisture deficiency, 1980.
67. U.S. Patent 4,220,168, Newell, G. P., Method of restoring normal moisture level to hair with normal moisture deficiency, 1980.

68. German Patent 3,514,087, Szekely, L., Hair preparations, 1986.
69. German Patent 3,246,265, Rudder, V., Hair growth stimulants, 1984.
70. Romanian Patent 66642, Comanescu, B., Dediu, V. I., Rautia, P., and Schiopu, E., Gel for hair growth and care, 1979.
71. Japanese Patent 59115379, Nippon Oil Co., Gels for the prevention of hair loss in cancer chemotherapy, 1984.
72. Wickett, R. R. and Mermelstein, R., Single-fiber stress decay studies of hair reduction and depilation, J. Soc. Cosmet. Chem., 37: 461–73, 1986.
73. U.S. Patent 4,310,329, Holland, D. O., Aqueous compositions for darkening keratinous materials, 1982.
74. Canadian Patent 1,041,429, Barrucco, A., Hair rinse, 1978.
75. Canadian Patent 1,029,306, Mackles, L., Self-foaming essentially nonaqueous gel shampoo composition, 1978.

14

Glycerine in Oral Care Products

Morton Pader

Consumer Products Development Resources, Inc., Teaneck, New Jersey

I. INTRODUCTION

Glycerine has played a role in the formulation of oral hygiene products virtually from the earliest development of sophisticated versions of such products, for example, toothpastes as we know them today. Its functions in oral care product formulation have not changed, although the absolute amount used for that purpose around the world has been affected by the advent of substitutes. The major competitive products for glycerine in oral care products today are sorbitol and sorbitol-based materials. Sorbitol products have successfully replaced glycerine, either partially or totally, in many major over the counter (OTC) oral product applications. Recent developments have led to interest in the use of xylitol in oral products. Readers should keep in mind as they follow the discussion below that the properties of glycerine in toothpaste and similar oral hygiene aids can be matched, at least partially, by other available materials.

The choice between sorbitol and glycerine in oral care formulations is largely a matter of cost effectiveness. Glycerine has properties distinct from pure sorbitol in those formulations. In recent years, however, the introduction of modified sorbitol preparations, such as noncrystallizing sorbitols, have endowed sorbitol preparations with properties which were previously exclusive to glycerine in oral care applications. The preference for glycerine over a sorbitol preparation, and vice versa, has thus become a matter of economics in some oral care formulations, and is determined by such factors as cost and availability.

These, in turn, are determined by both market and political factors, such as whether or not crops from which sorbitol derives are supported by government subsidies, the price of petroleum from which glycerine can be derived, the demand for vegetable oil sources, etc.

Glycerine can be used exclusively in most oral care products. Sorbitol and appropriate derivatives can also be used exclusively, but usually with some compromise in the organoleptic properties of the final product. A mixture of glycerine and sorbitol is generally preferred in the United States, and is to be recommended as a general rule in toothpaste manufacture.

Glycerine is about 0.8 times as sweet as sugar, sorbitol about 0.6 times as sweet. This difference in sweetness level can be compensated for by the addition of synthetic sweeteners. Even all-glycerine products are formulated with synthetic sweeteners in those instances where glycerine alone does not provide adequate sweetness.

Glycerine is employed as a formulation aid in both OTC and prescription items for care of the oral cavity. Primary applications are in dentifrices, oral rinses, and dental gels [1]. A dentifrice has been defined as a substance used with a toothbrush to clean the accessible surfaces of the teeth. Current dentifrices also play the important role of acting as vehicles for delivery to the oral cavity of therapeutic and cosmetic agents, such as fluoride for anticaries activity and pyrophosphates for antitartar effect. An oral rinse, or mouthwash, has historically been defined as a solution containing breath-sweetening, astringent, demulcent, detergent, or germicidal agents which is used for freshening or cleaning the mouth or for gargling. Today's oral rinses are frequently the vehicles for the delivery of active agents, cosmetic or therapeutic, to the oral mucosa or dental hard tissues. The terms "oral rinse" and "dental rinse" are preferred to "mouthwash" by some according to their respective areas for use, viz. oral mucosa or the teeth, respectively. A new type of dental rinse was introduced to the market a few years ago, a prebrushing dental rinse, for rinsing prior to toothbrushing, to remove and loosen some dental plaque initially and leave it in condition for ready removal by the toothbrush subsequently. The term "dental gel" is usually used to distinguish a dosage form for the delivery of an anticaries agent to aid in the prevention of tooth decay. The dental gel does not contain dental abrasives. It is usually applied by the dentist or dental hygienist.

II. ROLE OF THE HUMECTANT

The "humectant" in oral care products must serve many functions and possess properties beyond simple humectancy. A dictionary definition of humectant is "a substance which promotes the retention of moisture." That, of course, is still the primary purpose of humectants in oral care products. The product formulator, however, must be concerned with functions and characteristics other than water retention. The choice of "humectant," in the broader sense in which the term is used by the formulator of oral care products, is decided by multiple, interrelated factors. These are considered here with emphasis on glycerine; alternatives to glycerine are discussed elsewhere in this volume.

Glycerine's humectancy, in conjunction with its low volatility, makes possible its most important functions—to prevent drying out of product and/or package during exposure to the atmosphere during use of the product by the consumer and during manufacture and packaging of the product, and prevention of "cap locking" after the closure has been removed from the product container, be it tube, pump, or bottle, and product is present at the interface of the closure (as in the cap threads) and the bulk container.

Beyond that basic, conventional function, glycerine plays many "non-humectant" roles in oral care products. These will be discussed in this chapter. They may or may not be important in all oral care products. Strongest demand is placed on the humectant's properties in toothpaste. Many of the functions of the humectant in other oral care products are strongly related to toothpaste requirements or are specific to special dental products.

III. GLYCERINE IN DENTIFRICES

A. Functions

The humectant in toothpaste can be considered in two categories: (1) its role in determining bulk toothpaste properties and (2) its role in determining the characteristics of the properties of the dentifrice in the mouth. Whether the humectant is glycerine, a sorbitol preparation, or a combination of the two, its functions in determining bulk properties are:

1. To provide a vehicle into which oral care agents can be incorporated with the production of a smooth, homogeneous mixture. The agents can be therapeutic (fluoride) or cosmetic (abrasive, flavor, synthetic sweetener).
2. To provide resistance of the formulation to "drying out" during exposure of the product to the atmosphere, either as a ribbon or at the container's closure.
3. To prevent microbial deterioration during storage of packed product, during distribution, storage, and use at home.
4. To develop a product with a "short," cohesive, thick paste structure, in conjunction with a gum binder/thickener; i.e., to structure the rheology.
5. To provide a system wherein the solid and liquid phases are stable, i.e., where those phases do not separate during distribution and storage.
6. To provide the means whereby a transparent or translucent toothpaste can be formulated.

The humectant can fulfil the following functions of a toothpaste perceivable to the consumer during the use of the product:

1. To impart sweetness
2. To rapidly release the flavor, as well as other additives; i.e., to provide a paste which disperses very quickly in the mouth
3. To contribute to regulation of the amount and properties of the foam
4. To help make the overall toothbrushing experience pleasant, i.e., to leave the mouth with a subjective feeling of cleanliness and freshness

B. Toothpaste Formulations

Dominant toothpaste formulations in the United States and elsewhere are of two types, those containing a high percentage of humectant and a low percentage of abrasive, and those where the abrasive dominates the humectant. These two types of toothpaste are outlined in Table 14.1 [1]. Specific examples of these two types of toothpastes, drawn from the literature, are given in Table 14.2. Table 14.3 presents formulas for silica abrasive toothpastes containing two different sources of fluoride [2].

Table 14.1 Dominant Dentifrice Formulations in the United States

	Low abrasive type	High abrasive type
Abrasive level	10–25%	40–50%
Abrasive	Silica xerogel	Dicalcium phosphate
	Silica precipitates	Dicalcium phosphate dihydrate
		Alumina trihydrate
		Calcium pyrophosphate
		Calcium carbonate
		Insoluble sodium metaphosphate
Humectants	Sorbitols	Sorbitols
	Glycerine	Glycerine
	Polyethylene glycols	
Thickeners	Silica aerogels	Carboxymethylcellulose
	Pyrogenic silicas	Carrageenan
	Silica precipitates	
	Carboxymethylcellulose	
	Carboxyvinyl polymers	
	Xanthan gum	
Surfactant	Sodium lauryl sulfate	Sodium lauryl sulfate
Flavor, preservative, water, etc.		

Table 14.2 Examples of Dentifrices with Low and
High Abrasive Levels

Low abrasive level[a]	
Silica xerogel	14.0%
Silica aerogel	7.50
Sodium carboxymethylcellulose	0.60
Saccharin	0.20
Sorbitol solution (70%) or	
glycerine, or a mixture	67.02
Dye solution (red)	0.47
Flavor	2.00
21% sodium lauryl sulfate-79%	
glycerine mixture	7.00
Sodium hydroxide (30% solution)	0.31
High abrasive level[b]	
Alpha alumina trihydrate	55.0%
Glycerine	20.0
Sodium carboxymethylcellulose	1.0
Sodium lauryl sulfate	1.5
Flavor	0.9
Water	to 100%

[a] Adapted from M. Pader et al., U.S. Patent 3,538,230
(1970)
[b] From B. R. Pugh et al., British Patent 1,188,353
(1970)

Translucent (transparent) dentifrices require that the refractive index of the aqueous phase is roughly the same as that of the silica abrasive, about 1.47. The refractive index of anhydrous glycerine is 1.43 at 25°. The refractive index of 70% aqueous sorbitol is 1.458 at 25°. Combinations of the three materials will yield clear gels, upon which base translucent or transparent products can be structured.

Toothpaste humectant systems are known as "lean" systems. They contain relatively little water. Nonetheless, they must accommodate and be compatible with substantial amounts of organic binders and thickeners as well as inorganic and organic salts. The thickener/binder usually is a water-soluble gum or resin, such as carboxymethyl cellulose, xanthan gum, carrageenam, or a naturally occurring gum. It helps provide "body" and viscosity to the paste. The viscosity imparted will be influenced by the ratio of glycerine to water. The gums are not soluble in glycerine, but a given weight percentage of gum will thicken the glycerine-water mixture in accordance with the amount of gum

Table 14.3   Examples of Dentifrices with Special Additives

Sodium monofluorophosphate dentifrice

	%
Silica xerogel[a]	14.00
Silica thickener	8.00
Sodium monofluorophosphate	0.78
Sorbitol, 70% solution	46.72
Glycerine, 96%	20.90
Carboxymethylcellulose, grade 9MX[b]	0.30
Polyethylene glycol 1450	5.00
Sodium lauryl sulfate, dentifrice grade	1.50
Flavor	2.00
Sodium benzoate	0.10
Sodium saccharin	0.20
Color solution	0.50

[a] W. R. Grace & Co.
[b] Hercules.

Sodium flouride dentifrice

Precipitated silica abrasive[c]	20.00
Sorbitol, 70% solution	50.20
Glycerine	18.00
Carbomer 940 (Carbopol 940)[d]	0.25
Xanthan gum	0.50
Sodium lauryl sulfate, dentifrice grade	1.20
Sodium fluoride	0.24
Flavor	0.90
Sodium saccharin	0.10
$Na_3PO_4 \cdot 12H_2O$	1.50
$Na_2HPO_4 \cdot H_2O$	0.60
Titanium dioxide	0.70
Water plus color	5.81

[c] J. M. Huber Corp.
[d] B. F. Goodrich.

solubilized. Obviously, it is preferable to use an amount of gum that will not exceed its solubility in the water-glycerine.

The importance of the humectant liquid in toothpaste rheology is readily appreciated by reference to Table 14.4, where it is emphasized that in building a toothpaste structure the continuous liquid phase is the fundamental unit determining viscosity. Addition of glycerine to water will increase the viscosity of the water. That increase in viscosity will still be apparent on addition of an organic thickener/binder; in other words, the viscosity of a water-glycerine-

Table 14.4 Building a Model of Toothpaste Rheology

1. The continuous liquid phase (exclusive of surfactants) is the fundamental unit determining viscosity.
2. The polymer increases the viscosity of the continuous liquid phase. It determines shear-dependent properties and behavior with time (shear thinning, thixotropy).
3. The abrasive particles increase viscosity markedly. The increase depends on particle amount, size, size distribution, and shape.
4. The surfactant has a variable effect on rheological properties. It usually reduces viscosity.
5. Other toothpaste ingredients exert minimal effects on toothpaste structure.

organic thickener system will be greater than that of the corresponding water-organic thickener system. The increase in viscosity of a water-organic thickener system on replacement of some water with glycerine is predictable.

C. Preservative Function of Glycerine

Use of glycerine and/or other polyols in toothpaste was, until the introduction of clear "gel" toothpastes, a practice based in tradition. There is no practical need for many of today's toothpastes to use humectants at the very high levels at which they are used; opaque toothpastes, for example, need use only that amount of humectant necessary to structure the paste appropriately, provided that other humectant functions can be accommodated and alternative means of bacteriological preservation are available. Glycerine at appropriately high levels in water will prevent the support of bacterial growth. That traditionally represented a major reason for its use in toothpaste—the product was self-preserving. Today, however, one should consider the advances in preservatives and gums and thickeners, and improvements in processing and packaging facilities over the past decades as part of creative formulation of toothpastes. In short, the preservative aspect of humectants in toothpaste, including glycerine, may be rethought productively.

The classical view of glycerine in toothpaste was based on the observation that glycerine at a sufficiently high concentration in water will lower the water activity below that necessary for growth and metabolism of most bacteria.

The water activity (a_w) of a system subject to microbiological contamination determines the ability of that system to resist a microbial challenge in the absence of a chemical preservative [3]. Bacteria require high concentrations of water in the immediate environment if they are to grow and multiply. They are, in this respect, more demanding than several other kinds of organisms; they can even be considered aquatic organisms. Water is needed for metabolic and nutritive purposes to maintain the structural integrity of the bacteria within the cell wall. The availability of water to the bacteria is governed largely by the solute content of the medium as well as by the absolute amount of water therein. This feature is the basis for the control of microbial contamination in dentifrices without reliance on chemical preservatives active at low concentrations.

The water in a humectant system is not entirely free and available to the microorganisms; many of the water molecules are associated with the solute molecules. This results in a lower vapor pressure for the solution than for pure water, and the measurement of "free" water concentration or thermodynamically active water can be expressed by a_w, or;

$$a_w = \frac{\text{Vapor pressure of solution}}{\text{Vapor pressure of pure water}}$$

The value for a_w is determined experimentally by measurement, under equilibrium conditions, of the humidity of the atmosphere of a closed container containing the medium at incubator temperature (i.e., a_w = relative humidity/100). The relative humidity (RH) is taken as the percent moisture in the air of the vessel compared with the percent moisture at the saturation value of air over pure water at the same temperature.

The optimum a_w for bacterial growth and survival depends on many factors, including the characteristics of the bacteria and the environmental conditions other than the concentration of solute. There are no hard rules for predicting the optimum a_w for survival and multiplication of specific bacteria, and each product formulation must be tested under conditions of interest with specific microbial challenges. Generally, a_w values for bacterial growth usually are in the range of 0.90–0.999, and only a few microorganisms will grow below a_w = 0.965.

The lowest a_w compatible with growth has been found to be about 0.75 for some halophilic bacteria. Such deviations are quite rare. They are exhibited by certain halophilic bacteria and a few other xerophilic bacteria, but the types of potentially pathogenic bacteria of concern in dentifrice, pharmaceutical, and cosmetic manufacture and in home use, e.g., enterococci, pseudomonads, staphylococci, and *Escherichia*, will generally require a relatively high a_w to survive and grow. At sufficiently low a_w, readily achieved in dentifrice systems by the use of glycerine and other humectants, complete inhibition of bacterial growth can be achieved. If the a_w is borderline, growth generally is characterized by increased lag, decreased rate of bacterial fission, and smaller bacterial populations.

Osmotic pressure can be an extremely important factor in the ability of a microorganism to live in a particular medium [4]. The cell wall acts as a semipermeable membrane. Again, rules to predict the role of osmotic effects on specific bacterial populations can be found only in very broad outline. The size of the solute molecule, the concentration of the solute, the composition of the bacterial protoplasm, all will exert an influence. Only a small number of bacterial types can tolerate the osmotic pressures of commonly used humectant systems.

The value of a_w is a colligative property. A 30% solution of glycerine generally will not support microbiological growth. A 40% solution of sorbitol is necessary to achieve that effect [3].

The optimal water/glycerine ratio is determined by several factors, each of which is formulation-dependent. These are dictated by proliferation of the functions which toothpastes serve today, and the incorporation into the aqueous phase of some toothpastes of large amounts (several percent) of salts such as potassium nitrate and soluble pyrophosphates. If reduction of the glycerine content is economically desirable or feasible on grounds of other commercial considerations (e.g., declaration of package content weight), the formulator is advised to establish that ratio as part of the final formulation.

IV. MICROBIOLOGICAL ASPECTS OF GLYCERINE

Glycerine is accepted virtually without reservation as a component of oral care products which will interact little or not at all with oral microorganisms responsible for oral diseases, such as dental caries. This situation would appear to be based on historical factors rather than on published studies on the metabolism of glycerine by, for example, *Streptococcus mutans*, a common resident of dental plaque. Recent dental literature, reporting studies on the metabolism of glycerine by organisms involved in caries and periodontal disease, is scarce. This contrasts with the relatively extensive literature on sorbitol and related polysaccharides which has been generated to support use of these materials in the food supply.

Glycerine in toothpaste or mouth rinse dosages contacts the oral cavity dental structures for less than a few seconds (in toothbrushing) to perhaps half a minute or less (from oral rinses). Hence, its uptake and utilization from toothpaste and oral rinses by oral microorganisms must be limited.

The utilization by microorganisms of polyhydric alcohols such as sorbitol and xylitol has been investigated extensively. For the most part, the studies were concerned with the potential of these substances to promote or inhibit dental caries when they were introduced into the diet. Long-term studies on the presence of glycerine or sorbitol in oral care products would be necessary to define the ability of those materials to promote dental caries when applied topically from oral care products. Historical usage, however, provides sufficient justification for the incorporation of polyhydric alcohols such as glycerine and sorbitol(s) into properly formulated oral care products. This justification is especially true today, when additives with antimicrobial properties are introduced into products for their specific attributes in oral care.

Polyhydric alcohols have been tested for their ability to affect the growth and metabolism of microorganisms in dental plaque [5]. Suspensions of dental plaque and of salivary sediment were incubated in the presence of a number of polyhydric alcohols, including ethylene glycol, glycerine, erythritol, D-arabitol, ribitol, xylitol, sorbitol, and mannitol. The reaction mixtures were assayed for pH, lactate, pyruvate, and total keto acids. Glycerine did not produce detectable amounts of acid, while sorbitol and mannitol systems showed decreased pH as the result of the formation of small amounts of acid. It was also found that radio labeled sorbitol was bound to the plaque/salivary sediment [5].

Glycerine can be bound by *Streptococcus mutans*, an important resident of dental plaque [6]. Various species and strains of oral streptococci synthesize glycerol teichoic acids (lipoteichoic acids). These glycerol phosphate polymers function in membrane stabilization. It was found that [^{3}H]glycerine was taken up by strains of *S. mutans*. This and other studies indicate that exogenous [^{3}H]glycerine can be incorporated into cell-associated and extracellular lipoteichoic acids by oral streptococci. The evidence suggested that glycerine can enter the bacterial cell wall by facilitated diffusion. Sodium fluoride inhibited the uptake of [^{3}H]glycerine.

Lack of production of acid from glycerine in a salivary sediment–dental plaque system suggests that binding of the glycerine is not of practical consequence in the development of dental caries where it is employed in oral care products, particularly because they almost universally are formulated with fluoride or another agent with antimicrobial properties.

In general, however, it must be recognized that glycerine is a substance that can support bacterial function. It can serve as a bacterial nutrient and enter metabolic processes of a limited number of bacteria capable of acid production [3]. It can support microbial growth when present in concentrations under roughly 30%. It can also penetrate the bacterial cell wall faster than sorbitol; for example, the relative rates of penetration into *Azotobacter vinelandii* have been found to be ethylene glycol > glycerine > erythritol > sorbitol > mannitol [7]. Organisms able to metabolize glycerine appear to do so through pathways involving dihydroxyacetone phosphate. As a component of fats, glycerine can be metabolized by many organisms via its connection to carbohydrate metabolism [4]. The original popularity of glycerine as a dentifrice humectant probably can be attributed to its indisputable safety as an additive subject to ingestion along with relatively low fermentability, at a time when starch-derived polyol alternatives were of comparatively little commercial importance.

V. GLYCERINE IN ORAL RINSES

A mouthwash or gargle historically has been a liquid product used to clean the oral cavity and freshen the breath. Some mouthwashes claim ability to soothe sore throat. The mouthwash is designed to be used posttoothbrushing or at special times during the day (e.g., prior to a social engagement). Recently, prebrushing dental rinses were introduced into the market (i.e., liquid products which when used just prior to toothbrushing can remove a small amount of dental plaque directly, and then leave the residual plaque more readily removable by the toothbrush). These formulations contain major amounts of glycerine. Mouthwashes have been used in the more recent past as delivery systems for not only breath fresheners but also as the vehicle for fluoride, antimicrobial agents, astringent agents, and anticalculus (antitartar) agents. Glycerine plays essentially the same role in all of these products—providing sweetness, preventing "cap-locking," providing desirable organoleptic attributes of sweetness and mouthfeel, and functioning as a formulating aid. Glycerine is a component of most oral rinses marketed in the United States; rarely has it been supplanted by sorbitol.

A typical base formula for an antimicrobial mouthwash consists of an aqueous solution of ethyl alcohol (12–30%), flavor stabilizer such as polyoxyethylene-20 sorbitan monolaurate, an antimicrobial agent (0.05–0.1%), flavor oils, buffers, and colorants. Antitartar and fluoride actives can be added to this base directly without major changes because of the large quantity of water present and the lack of influence of glycerine on the solubility of active agents at the glycerine concentrations used.

The formulation of a representative antimicrobial mouthwash is given in Table 14.5, that of a fluoride oral rinse in Table 14.6. Bite, or flavor impact, is determined mainly by the contents of ethyl alcohol and flavor. Glycerine content is generally 10–20%, not sufficiently high to prevent solubilization of active agents.

VI. GLYCERINE IN SPECIAL TOPICAL PRODUCTS

Topical fluoride gels to be administered by the dental professional are available. Where the fluoride compound is unstable in water (e.g., stannous fluoride), it is formulated in a glycerine base. That base is glycerine (essentially anhydrous) which contains flavoring and thickening agents. The Food and Drug Administration has defined a "treatment gel" as a gel containing 0.4% stannous fluoride in anhydrous glycerine containing suitable thickening agents to adjust viscosity [8]. The Food and Drug Administration has also tentatively accepted stannous fluoride in a stable base as a concentrate, for dilution with water just prior to use as an aqueous rinse.

VII. GLYCERINE AS A MANUFACTURING AID

The ability to obtain glycerine commercially either as 96% material or 99+% is a feature of glycerine which endows it with a very important capability in oral care product manufacture. A gum thickener/binder is a critical component of all toothpastes, and xanthan gum has been used in the manufacture of prebrushing dental rinse. Incorporation of the types of binders used in toothpastes,

Table 14.5 Formulation of a Medicinal-Type Mouthwash

	% by wt
Ethyl alcohol	15.0
Flavor (menthol 40%, eucalyptol 8%, cinnamic aldehyde 5%, anethole 5%, eugenol 2%, thymol 40%)	0.2
Glycerine	10.0
Polyethylene glycol (20) sorbitan isosterate	0.5
Saccharin, sodium	0.03
Caramel color (to desired shade) and water	to 100
Adjusted to pH 6	

Table 14.6 Formulation of a Fluoride Oral Rinse

	% by wt
Alcohol	5.0
Flavor (spearmint oil 45%, peppermint oil 10%, methyl salicylate 25%, menthol 20%)	0.25
Glycerine	15.00
Polyoxyethylene/poloxypropylene block polymer (Poloxamer 338)	1.0
Saccharin, sodium	0.05
Sodium fluoride	0.05
Sodium benzoate/benzoic acid	0.1
FD&C dyes to desired color, and water	to 100
pH 6	

usually modified cellulosics or of natural origin, is frequently difficult because the gum "balls up" on addition to the mix or solution; small balls of hydrated material surrounding nonhydrated material form when the gum is added to an aqueous system. Several techniques have been proposed to overcome this problem, including mechanical devices to evenly distribute the gum during its addition. A popular and effective technique is based on predispersion of the gum in anhydrous glycerine. The gum is added to the glycerine to form a smooth suspension of fine, unhydrated gum particles. This suspension can then be added to the aqueous phase of the toothpaste or oral rinse without the formation of large gum balls. The gum can then hydrate quickly, evenly, and thoroughly.

The technique is used advantageously even in mixed glycerine–sorbitol systems; all or part of the glycerine is set aside to make a gum slurry.

A similar technique has been used to introduce sodium lauryl sulfate into dentifrice. This detergent as a commercial powder frequently is hard to work with, primarily because of its potential to be irritating to mucous membranes. Dispersion of sodium lauryl sulfate in the glycerine yields a liquid product which is readily handled and metered.

REFERENCES

1. Pader, M., Oral Hygiene Products and Practice, Marcel Dekker, New York, 1988, Chap. 9.
2. Pader, M., Cosmet. Toiletries, 102:81, 1987.
3. Roquette Freres, Technical Bull. (not dated)
4. Lamanna, C. and Malletts, M. F., Basic Bacteriology, Williams & Wilkins, Baltimore, 1965, Chap. XIII.
5. Mäkinen, K. K., in Microbial Aspects of Dental Caries (H. M. Stiles, W. J. Loesche and T. C. O'Brien, eds.), Proceedings Spec. Suppl. to Microbiology Abstracts, 1976, Vol. II, p. 521.
6. Ciardi, J. E., Reilly, J. A., and Bowen, W. H., Caries Res., 14:24, 1980.
7. Lamanna, C. and Mallette, M. F., Basic Bacteriology, Williams & Wilkins, Baltimore, 1965, Chap. XI.
8. Food and Drug Administration, Anticaries Drug Products for Over-the-Counter Human Use; Tentative Final Monograph, Fed. Reg., 50:39853, 1985.

15

Glycerine in Bar Soaps

Eric Jungermann

Jungermann Associates, Inc., Phoenix, Arizona

Beth Lynch

Neutrogena Corporation, Los Angeles, California

I. INTRODUCTION

Glycerine is a commonly used additive in bar soaps. Additionally, it often finds its way, either by design or accidentally, into soaps as a byproduct of the historical saponification process.

Soap is one of the oldest chemical substances known to man. Saponification, the most elementary way for manufacturing soap, was known as early as 2500 B.C. when Sumerians reported the preparation of soaplike substances by heating oils and an alkali ash rich in potassium carbonate. Soap was found in the writings of the Egyptians and Greeks primarily because of its medicinal properties. In the Middle Ages the art of soapmaking was carried on in Italy, France, and Germany; these early soap makers leached sodium and potassium carbonate from wood ashes and treated it with slaked lime as shown in Equation (1). They used the resulting caustic soda to

$$Na_2CO_3 + Ca(OH)_2 \rightarrow 2NaOH + CaCO_3 \tag{1}$$

saponify a variety of animal and plant fats. An important point in the history of soapmaking occurred early in the nineteenth century with the development of the LeBlanc process for the large-scale manufacture of caustic soda. This transformed soapmaking from a cottage industry to an important industrial process.

The basic saponification reaction, shown in Eq. (2), consists of reacting fats and oils, which are mixtures of tryiglycerides, with an alkali to produce soap and glycerine.

$$C_3H_5(OCR)_3 + 3 \ NaOH \rightarrow 3RCOONa + C_3H_5(OH)_3 \qquad (2)$$
$$\underset{O}{\overset{\|}{}}$$

Fats   Alkali   Soap   Glycerine

The fats and oils commonly used in the manufacture of soap include tallow, coconut oil, palm oil, palm kernel oil, and in the case of certain specialty soaps, castor oil. Fats and oils are triesters of fatty acids and glycerine; the fatty acids most prevalent are saturated acids, such as lauric, myristic, palmitic, and stearic acids, and unsaturated acids such as oleic and linoleic acid.

Early soapmaking methods did not recover the glycerine, leaving as much as 9% in the product. Addition of salt resulted in graining out the soap, producing a harder soap mass and also separating out most of the glycerine formed in the reaction. This became the basis for the kettle soap process which today still flourishes in many parts of the world. In these processes, not more than 0.5% of glycerine is left in the soap. Economic recovery of the glycerine is important because of the higher value of the glycerine than the finished soap. After World War II, soapmaking took another leap forward with the introduction of continuous processes for the manufacture of soap, reducing production time to a matter of hours; this was a major improvement over the kettle soap processes which often requires a 5 to 7 day production cycle [1].

Another important commercial process in the manufacture of toilet soap is the neutralization reaction between fatty acids and sodium hydroxide shown in Eq. (3). In this reaction, obviously no glycerine is produced or left in the finished soap.

$$RCOOH + NaOH \rightarrow RCOONa + H_2O \qquad (3)$$

Fatty Acid   Lye   Soap

These processes can be carried out batchwise or continuously. When glycerine is needed in the formulation, it must be added later in the manufacturing process. The various saponification reactions and the type of glycerine removal or addition are summarized in Table 15.1

II. FUNCTIONS OF GLYCERINE IN BAR SOAPS

The presence of glycerine in bar soap formulations can have a number of functions and effects:

1. To provide a transparent or translucent appearance
2. For skin-conditioning effects

Table 15.1 Soap Processes

Process	Ingredients	Glycerine treatment	Products
Saponification (batch)	Fats	Leave in soap	Early soaps
Saponification (batch)	Fats	Leave in and add more	Specialty, transparent bars
Saponification (batch or continuous)	Fats	Remove as much as possible	Standard opaque bars
Saponification (batch or continuous)	Fats	Remove all or most and add back some	Superfatted, specialty, moisturizing bars
Neutralization (batch or continuous)	Fatty Acids	No glycerine present	Standard opaque bars
Neutralization (batch or continuous)	Fatty Acids	Add glycerine	Superfatted, special-ty, moisturizing

3. As a processing aid
4. To enhance product appearance

The use of glycerine in the manufacture of transparent and translucent soaps, usually classified as specialty soaps, is probably the most important example in the above list. Transparent soaps have often been referred to as "glycerine soaps." In the mid 80s, manufacturers added glycerine to some mass-marketed opaque bars to endow them with skin care and skin feel properties, as well as to improve the foam characteristics of the bar. Typical examples in this latter category are Camay and Monchel.

A. Transparent and Translucent Soap

The oldest known example of a transparent soap is Pear's Transparent Soap. First sold in Great Britain in 1789, it is still an item of commerce today, though somewhat improved since olden days. It claims to have won "25 highest awards at exhibitions between 1851 and 1939." Pear's is an example of a soap made with resin, sugar, and alcohol, three ingredients often found in transparent bars. It is made by first saponifying a blend of tallow, palm oil, and resin, drying the resulting soap, dissolving that soap in alcohol along with glycerine and sugar, then distilling off the alcohol. The molten transparent soap is then cast.

The crystal structure of ordinary bar soap has been studied by Maclennan [2] and Ferguson et al. [3]. The key to transparency lies not in the presence of these crystals, but rather in their size. Soaps are transparent when their crystals are too small to provide optical discontinuities to ordinary light. McBain

Figure 15.1

and Ross refer to the crystals as ultramicroscopic [4]. The crystalline nature of soap explains the vulnerability of transparent soap to seeding with impurities or soap crystals or sugar crystals. Physical strain or shock from improper stamping can also lead to crystal growth and loss of transparency.

The generally accepted definition for transparent soap is a bar that permits reading 14-point boldface type through a 1/4 inch soap layer [5]. Other test methods have been developed including the Colgate-Joshi Translucency Test [6] and the Translucency Voltage Test [6].

Transparent soaps can be prepared by the semiboiled or by the cold process. Fats are reacted with 25–35° Be alkali using the amount needed for complete saponification; the reaction is carried out in a soap kettle or crutcher with mechanical agitation. The glycerine produced is retained in the soap. When the soap is heated to near 100° C, the reaction can be completed in about one hour. Additives are added, after which the soap is poured into frames (fig. 15.1). In the cold process, 35–45° Be alkali is used, and the mixing occurs at room temperature. The ingredients are emulsified at that stage, but the saponification reaction actually proceeds for several days after the soap is run into frames. Details of the semiboiled and cold processes are given by Davidsohn et al. [7].

Table 15.2 Composition of Transparent Soap Bars

	A	B	C	D	E	F
Sodium soap	27	54	60	46	50	78
Glycerine	11	10	10	5	20	2
Water	9	9	26	29	18	17
Fatty acid	21	0	0	0	0	2
Triethanolamine	29	0	0	0	0	0
Propylene glycol	0	15	0	0	0	0
Anionic surfactant	0	11	0	0	0	0
Rosin soap	0	0	3	0	12	0
Alcohol	0	0	1	5	0	0
Sugar	0	0	0	15	0	0
Sorbitol	0	0	0	0	0	0
Other	3	1	0	0	0	1

Since the semiboiled and cold processes do not allow for washing impurities out of the soap, very pure raw materials must be used. The quality of raw materials is important with transparent soaps because of color and clarity considerations. Transparent soaps contain additives that interfere with the formation of large crystals. Alcohol, sugar, glycerine, sorbitol, castor oil, and other materials with hydroxyl groups are recommended for retarding crystal growth. Choosing fats that are less saturated or have extra hydroxyl groups, such as castor oil, also aid transparency. However, unsaturated fats make soap softer and stickier and decrease its lather, so they must be combined with saturated fats which are less favorable to transparency. The alkali used usually is sodium hydroxide, although potassium hydroxide and triethanolamine are also used, and tend to give better transparency.

Examples of soap formulas that use glycerine to enhance transparency are given in Table 15.2 Example A is a triethanolamine-based formula that can be made by the semiboiled process. Its clarity is a result of the high levels of triethanolamine and glycerine. Example B contains an anionic surfactant in addition to soap, as well as glycerine and propylene glycol. Soaps C and D are typical of the transparent soaps made using an alcohol as solvent (about 25% of kettle charge) from which most of the alcohol is recovered in a condenser. The bars are aged, so very little alcohol remains in the final product. Example E is a semiboiled rosin and glycerine soap. The formula given in example F is also semiboiled, but has been spray dried, milled, plodded, and stamped in the manner described in a Lever patent which covers specified proportions of free fatty acid, polyhydric alcohols, salt, and water and certain drying and milling conditions [8]. Formulas A through E are typical because they are formed into bars by casting the molten soap into molds, frames, or slabs.

Formula F instead of being cast, is dried and worked and extruded. Other patents exist for making transparent soap with conventional extruding equipment, but these processes are not widely used [9–11].

Because of their alkalinity and the presence of alcohol, some transparent glycerine soaps have a reputation of relative harshness and causing skin dryness. A patent by Fromont describes neutral transparent soaps that are blends of sodium and triethanolamine soaps [12]. The commercial soap made under this patent was found to be exceedingly mild, probably due to the absence of alcohol, its high glycerine and triethanolamine levels, and excellent rinsability properties. The pH of this soap is also lower than that of standard soaps. The inventor referred to this property as "neutrogenous" and the soap is marketed worldwide under the name "Neutrogena." A series of clinical studies and other tests conducted in recent years have demonstrated the unique properties of this particular transparent soap. It was shown to have superior rinseability compared with eighteen common soap and syndet bars and combars [13]. In irritation tests that include rinsing, it was found to be less irritating than the average bar [14]. Some typical chamber irritation test results for bar soaps follow [15].

Soap Chamber Test Results

Camay	6.4
Dial	2.4
Dove	2.0
Irish Spring	3.8
Ivory	2.8
Neutrogena	2.0

A recent advance in transparent soapmaking is in the area of continuous processing. A patent by Jungermann et al. describes a process for making triethanolamine-based soap by continuous mixing and casting into molds. It claims improved quality and lower cost [16]. A process for making transparent soap by a batch process then continuously casting it into a ribbon that is cut into bars has also been reported [17]. Another patent describes the use of some additive systems that improve the transparency of triethanolamine based soaps [18].

Glycerine has a minimal effect on the processing of transparent soaps since most are not dried or refined. In molten transparent soaps, glycerine is formed in situ from the saponification reaction of fats and oils. In some cases, additional glycerine is added. Glycerine can be added prior to the reaction with alkali if a dilution of the acid portion is desired. It can be added after the reaction occurs to help lower the temperature. It can also be added concurrently with heat-sensitive additives just prior to casting. Glycerine has a similar viscosity profile to that of molten soap, so it has little effect on mixing and pumping, whether the process is continuous or batch. It has a lower specific heat than soap, so it will speed the heating and cooling rate slightly. The difference is negligible in soaps that are cast into frames and allowed to cool overnight, but it will affect throughput on continuously cast soaps. High glycerine soaps will increase throughput by about 5%.

Translucent soaps transmit light, but are not transparent; they often contain glycerine to enhance their translucency. An example of this function of glycerin is a translucent formula with 6% glycerine that can be processed on Mazzoni finishing equipment [19].

Soap:	79
Moisture and volatiles:	14
Glycerine:	6
Other:	1

A patent [6] for incorporating germicides into translucent soaps by dissolving them first in the superfatting agent is another example of a translucent soap formula with glycerine.

Soap:	61
Water:	17
Super fat:	11
Glycerine:	6
Germicide, etc.:	5

This formula has been spray dried from 27% to 17% moisture then amalgamated, milled five times, plodded, and stamped.

B. Skin Conditioning

Another desirable property of glycerine in bar soap is its skin-conditioning effect. High levels are necessary to achieve consumer-perceivable skin conditioning. Glycerine has long been known for its skin care properties, but only recently has this benefit been demonstrated in bar soaps. Glycerine was shown to impart skin care benefits to bar soap when present at 10% level. It was found to reduce skin roughness when measured clinically; in consumer tests, the users perceived a definite feeling of softness and moisturization.

Dahlgren et al. [20], compared 50:50 tallow:coconut soap containing 10% glycerine versus the same base without glycerine. They found them to be significantly different on some clinical and consumer-perceived attributes. In one test, the two bars left the skin the same as measured by expert tactile and visual evaluation, electrical impedance, transepidermal water loss (TEWL), and sonic transmission. However, in a second test, the 10% glycerine soap bar left a smoother skin condition as rated by an expert tactile evaluation, although expert visual evaluations of dryness and redness were not significantly different. More evidence of the benefit of glycerine to skin came from visual analysis of photographs of skin replicates which revealed fewer features associated with roughness from the glycerine bar. Consumers who used the two bars at home for two weeks rated the glycerine bar higher for leaving the skin feeling soft and smooth and feeling moisturized. This 10% glycerine soap was found to increase the amount of glycerine in the stratum corneum, a possible explanation for how glycerine improved skin smoothness without reducing dryness measured clinically. Many transparent soaps contain enough glycerine to impart a skin-conditioning benefit. Opaque and translucent soaps may also contain therapeutic levels of glycerine, although this requires special process considerations.

Opaque soaps with high levels of glycerine often use higher levels of coconut soap; rich lather is achieved even in the presence of the additive load. A typical composition of a glycerine skin-conditioning specialty soap bar is shown below.

Sodium soap: 50/50 tallow/coconut:	70%
Free fatty acid:	7
Glycerine:	10
Water:	10
Other:	3

In spite of the perceivable smoothing or moisturizing effect glycerine can have, it is not presently used at therapeutic levels in mass-marketed soaps, probably because of cost considerations. Higher molecular weight polyols are more commonly used because, at lower levels, they leave a smooth and soft afterfeel on the skin that connotes moisturization. Moderate levels of glycerine have a positive effect on lather texture which provides a signal of skin care benefit. A well-known soap formula with enough glycerine to connote moisturization is shown below:

Sodium soap: 50/50 tallow/coconut:	75%
Free fatty acid:	6
Glycerine:	4
Water:	12
Other:	3

The next example is a complexion bar in which glycerine modifies the lather to signal a skin benefit, which is useful when the actual skin care ingredients do not affect lather.

Sodium soap: 75/25 tallow/coconut: 78
Free fatty acid: 4
Glycerine: 2
Water: 13
Skin conditioner: 3

In the large-scale production of opaque or translucent soaps, the saponification reaction yields "neat soap," a highly viscous mass containing about 30% water. It will contain small amounts of glycerine if the soap is made from triglycerides; additional glycerine can be added to the neat soap prior to drying; alternatively, glycerin can be added to the dried soap pellets during the finishing process. The best point of addition depends on the amount of glycerine to be added and whether formulas with different glycerine levels are to be made from the same soap pellet.

A uniform mix is easily achieved when the glycerine is added to the liquid neat soap which can then be dried on standard soap drying equipment, such as a vacuum spray dryer. A slight reduction in throughput can be expected because glycerine reduces the vapor pressure and heat capacity of soap. The effect of reduced vapor pressure is only partially offset by the effect of lower heat capacity. A low moisture content is desirable with heavy additive loads to keep soap bars firm. This is where glycerine's impact on drying is most noticeable.

Adding glycerine to the dried soap during refining along with color and perfume works well for low levels of glycerine. For high levels, the soap base must be very dry; this makes the refining process very important because dry soap can be unhomogeneous in moisture resulting in gritty soap. Nyguist et al. patented a process that prerefines the dry, 8–12% moisture, soap prior to the addition of glycerine and subsequent refining stages. Prerefining the dry soap breaks up the overdry particles before glycerine's lubricating effect can impede their break up [21]. The resulting product is free from grit. The prerefining step is necessary when a high level of glycerine is blended into soap pellets.

C. Processing Aid

At levels lower than the therapeutic levels studied by Dalhgren et al., glycerine imparts important properties to bar soaps. One such property is to improve the processability of extruded soaps. At low levels, glycerine can aid in processing soap by making it harder and more plastic hence easier to compact and extrude and remove from a die. Examples of formulas that use low levels of glycerine for this purpose are discussed below.

A translucent soap bar patented by Wood-Rethwill et al. [22] is based on sodium and potassium soaps, and a high-molecular weight polyol skin-feel additive; glycerine is used to harden the bar so it is more easily processed on standard refining, extruding, and stamping equipment.

Sodium soap: 73
Potassium soap: 4
Polyol (PEG-12): 4
Water: 14
Glycerine: 2
Other: 3

In another specific example, an 80/20 blend of tallow/coconut fatty acids was saponified with a 90/10 blend of NaOH/KOH along with 0.6% salt and 0.8% glycerine. The neat soap was dried to 15% moisture and formed into pellets. The pellets were mixed with an additional 2.0% glycerine and 3.8% PEG-12 (MW 600) in the amalgamator, and the soap mass was then extruded, cut, and stamped into bars. The resultant bars had an excellent translucency index value. The additional 2% glycerine was found to significantly increase both translucency and bar firmness when compared to similar formulations and processing without the glycerine.

Plain or "pure" soaps can contain glycerine without violating their purity claims since glycerine is naturally present in soap. Such soaps will benefit from improved processing because the glycerine will plasticize the otherwise brittle soap and improve its compaction and gloss.

Sodium soap: 84
Glycerine: 1
Water: 14
Other: 1

Like plain soaps, some deodorant soaps contain very low levels of additives and no superfat. Their appearance and stamping can be improved by the addition of a small amount of glycerine.

Sodium soap: 84
Glycerine: 2
Water: 12
Germicide: 1
Other: 1

D. Enhanced Appearance

Another function of glycerine at low levels is to enhance the appearance of extruded and stamped bars. It can give bars gloss and reduce the appearance of scuffs and marks. To make soap bars mar resistant, petrolatum and glycerine are added to prevent whitening and scuffing. Decorative figurine soaps are especially benefitted because their intricate shapes are easily marred, especially when left unwrapped and allowed to dry out. A typical formula for mar-resistant soap is as follows:

Sodium soap: tallow/coconut: 76
Petrolatum: 5
Glycerine: 4
Water: 12
Other: 3

III. CONCLUSION

Enhanced transparency and translucency, skin conditioning, lather improvement, improved appearance and processing are some of the important functions of glycerine in bar soap formulas. Glycerine is widely used in many kinds of bar soaps, often at relatively high levels, particularly in transparent soaps. Whether its presence is the result of the saponification of triglycerides or it is intentionally added to achieve a desired effect, glycerine is naturally associated with soap as a desirable additive in the minds of consumers and formulators alike.

REFERENCES

1. Jungermann, E., Bailey's Industrial Oil and Fat Products, Vol. 1, 4th Ed. John Wiley and Sons, New York, 1979, Chapter 8.
2. Maclennan, K., J. Soc. Chem. Ind., 42: 393–404, 1923.
3. Ferguson, R. H., Vold, R. D., Rosevear, F. B., Oil Soap, 16: 48–51. 1939.
4. McBain, J. W. and Ross, S., The structure of soap, Oil Soap, 21:97–98, 1944.
5. Wells, F. V., Soap Chem. Spec., 31: 39, 1955; 31: 43, 1955.
6. U.S. Patent No. 4,490,280, Joshi, D. P. and Divene, A.: (assigned to Colgate), 1984.
7. Davidsohn, J., Better, E. J., and Davidsohn, A., Soap Manufacture, Vol. 1, Interscience, New York, 1953.
8. U.S. Patent No. 3,155,624, Kelly, W. A., (assigned to Lever Bros.), 1964.
9. U.S. Patent No. 3,969,259, Kelly, W. A. (assigned to Lever Bros.), 1976.
10. U.S. Patent No. 2,970,116, Kelly, W. A. (assigned to Lever Bros.), 1961.
11. Ooms, J., Retureau, B., and Imbaud, A. (assigned to Procter and Gamble), 1982.
12. U.S. Patent No. 2,820,768, Fromont, L., 1950.
13. Wortzman, M. S., Scott, R. A., and Wong, J., J. Soc. Cosmet. Chem., 37: 89, 1986.
14. Frosch, P. J., Irritancy of soap & detergent bars, in Principles of Cosmetics for the Dermatologist (Frost, P., and Horowitz, S. N. eds.), C. V. Mosby Co., St Louis, 1982.

15. Jungermann, E., Cosmet. Toiletries, 97: 77, 1982.
16. U.S. Patent No. 4,758,370, Jungermann, E., Hassapis, T., Scott, R. A., and Wortzman, M., (assigned to Neutrogena Corporation), 1988.
17. Brittania Soap Machinery Company, private communication.
18. U.S. Patent No. 4,408,338, Lindbergh, G. A., (assigned to Purex Corporation), 1984.
19. Barnhart, Jr., J. W., Cosmet. Perfumery, 88: 105, 1973.
20. Dahlgren, R. M., Lukaconic, M. F., Michaels, S. E., and Visscher, M. O., Proceedings of Second World Conference on Detergents, 1988, pp. 127–134.
21. U.S. Patent No. 4,405,492, Nyguist, J. D., Kwasniewski, G. K., Thornton, A. W., Vest, P. E., and Ducklo, K. E., (assigned to P&G), 1983.
22. U.S. Patent No. 4,879,063, Wood-Rethwill, J. C., Jaworski, R. J., Myers, E. G., and Marshall, M. L. (assigned to Dial Corporation), 1989.

16

Alternatives to Glycerine in Cosmetics

Dale H. Johnson

Helene Curtis, Inc., Chicago, Illinois

I. INTRODUCTION

The principal function of glycerine in cosmetic products is its action as a humectant. Humectants are hygroscopic materials that promote the retention of water by either the product or the skin. They generally absorb water vapor from air under humid conditions, and retard its loss under dry conditions. In cosmetic products, humectants function both in the container and on the skin or hair. Since they influence the moisture exchange between product and hair, they are added to emulsions, especially oil-in-water types, to reduce drying out upon exposure to air. When a cream or lotion is applied to the skin, the proper choice of humectant will enhance the feel and texture of the product, and contribute to the cosmetic elegance so critical to consumer acceptance.

The "ideal" humectant doesn't exist for every case. Although glycerine is widely used for its humectant properties, as this monograph attests, a variety of other materials are utilized under varying conditions, to obtain either superior performance or more cost-effective benefits than glycerine. It is my intent to provide information about virtually all of the significant glycerine replacements suitable for cosmetic products. This will provide a basis for the formulator to make more informed choices as he or she devises an experimental development plan.

The material in this chapter should provide the formulating scientist with information necessary to determine the materials most appropriate to evaluate

Table 16.1   Properties of the Ideal Humectant

Hygroscopicity	The product must absorb moisture from the atmosphere and retain it under normal humidity conditions
Humectant range	Within the normal range of relative humidity, the change of water content should be small in relation to humidity changes
Viscosity	A low-viscosity humectant is readily mixed into a product, while a high viscosity can help stabilize emulsions or suspensions
Viscosity Index	The viscosity-temperature curve should be relatively flat
Compatibility	The humectant should be compatible with a wide range of raw materials; solvent or solubilizing properties are desirable
Color, odor, taste	Good color, odor, and taste are essential
Toxicity	The humectant should be nontoxic and nonirritant
Corrosion	The humectant should be noncorrosive to normal packaging materials
Stability	The humectant should be nonvolatile and should not solidify nor deposit crystals under normal temperature conditions
Reaction	The humectant should preferably be neutral in reaction
Availability	The humectant should be freely available and should be as inexpensive as possible

in a particular situation. The advantages and limitations of several alternative materials can be reviewed, and a rational experimental design can be developed so that one can determine the optimum materials and use levels to obtain the desired performance characteristics.

This chapter is organized to provide basic information about the various raw materials utilized as glycerine replacements, indicating general areas of application along with a summary of the relevant safety information.

II. HUMECTANTS

A. Properties of the Ideal Humectant

Although the "ideal humectant" does not exist, it is useful to examine the criteria for one, so that various candidate materials can be evaluated against the standard. In 1952, Griffin, et al. [1] summarized the properties of such an ideal humectant (Table 16.1).

By far the largest group of glycerine replacements can be classified as polyols. Since glycerine itself is a polyol, this is no surprise. Propylene glycol and sorbitol are the most widely used humectants after glycerine. These will be treated next, followed by other polyols in approximate order of frequency of their use.

III. PROPYLENE GLYCOL

C.A.S. Number 57-55-6
Other Names 1,3-Propanediol

A. Description

A clear, colorless, water-white viscous liquid available in USP, FCC, or technical grades. USP or food grades are most suitable for cosmetic applications. Propylene glycol has a bitter taste which restricts its use in oral products and cosmetics subject to incidental ingestion such as lip glosses. A summary of basic physical properties of propylene glycol can be found in Table 16.2.

B. Safety

This material is used in foods and is considered to be relatively harmless (LD_{50} rats, oral-30 g/kg) [2].

Although commonly regarded as nonirritating to skin, there is conflicting evidence. It has been reported to be a primary irritant and a cause of delayed contact hypersensitivity [3] in a study of dermatological patients. Conversely, in an extensive study by Drill and Lazar[4], propylene glycol was reported to have no toxicologic or other adverse properties.

Fox has recently discussed the apparently conflicting safety data on propylene glycol [5] and concluded that the test conditions affect the result. A Japanese study of propylene glycol and other humectants concludes that it has only a very low irritation potential [6].

C. Cosmetic Properties

Propylene glycol is more volatile and less viscous than glycerine. It is not as hygroscopic as glycerine, having one less hydroxyl group, but exhibits greater solvent powers for the same reason.

Table 16.2 Properties of Important Polyol Humectants

Property	Glycerine	Propylene Glycol	Sorbitol	1,3-Butylene Glycol
Freezing point (°C)	18.6	Supercools	NA	−50
Boiling point (°C)	290	187	295[a]	207.5
Specific gravity (25°C)	1.261	1.036	1.280	1.005
Flash point (°C)	160	107	None	109
Ref. Index (20°C)	1.4746	1.4320	1.4600	1.4412
Viscosity (cP) (25°C)	954	44	110	103.9
Molecular wt	92	76	182	90

[a] @ 3.5 mm.

Propylene glycol does not support the growth of microorganisms; in fact, it exhibits bactericidal and bacteriostatic properties. A major consideration, often leading to the choice of propylene glycol over glycerine in a formulation, is one of microbiological integrity. Propylene glycol acts as a solvent for many commonly used preservatives such as the parabens, and also reduces the oil:water partition coefficient to a greater degree than glycerine. Thus a higher concentration of the preservative is achieved in the aqueous phase where it is needed for antimicrobial efficacy.

Propylene glycol has a low freezing point and acts as a freeze-thaw stabilizer in many systems. Both glycerine and propylene glycol in concentrations from 5–10% reduce the coefficient of expansion of emulsions at freezing temperatures. Thus the bursting of bottles can often be avoided [7].

Propylene glycol usually reduces the viscosity of systems to which it is added. Due to its greater solvent power when compared with glycerine, propylene glycol should be seriously considered when clarity is a prime consideration. It is also recommended for use in shampoos to reduce the cloud point, as well as solubilize water-insoluble components such as colors, henna, botanical extracts, parabens, BHA, and fragrance oils. In the case of flavor and fragrances, 1,3–butylene glycol has shown particular efficacy as discussed subsequently in this chapter.

Propylene glycol is often used to lower and control viscosity in shampoo preparations. Its presence in a shampoo formulation can reduce the tendency of some systems to form undesired gels, thus maintaining smooth flow characteristics and batch to batch consistency.

IV. 1,3-BUTYLENE GLYCOL

C.A.S. Number 107-88-0

Butylene glycol is a clear, practically colorless, viscous hygroscopic liquid. It is odorless and has a slightly sweet, characteristic taste. It is also known as 1,3-butanediol. A compilation of key physical properties is included in Table 16.2. Butylene glycol is produced by the catalytic hydrogenation of acetaldehyde with metallic catalysts, with subsequent purification by distillation. It is available 99.5% pure and is classified as Food Grade by the Food Chemicals Codex.

A. Safety

This product is practically nontoxic with an acute oral LD_{50} of 23 g/kg in rats [8] and 11 g/kg in guinea pigs [9]. The material has been reviewed by the CIR [10] and classified as "safe as presently used in cosmetics."

Butylene glycol caused no more than minimal skin irritation when tested under occlusion on the skin of rabbits for 24 hours or daily for four consecutive days [11]. Some apparently conflicting data exist regarding eye irritation.

According to NIOSH [12], 505 mg of undiluted butylene glycol applied to the rabbit eye was an irritant, but no irritation was observed in another study with only 0.1 ml of the undiluted material, or when a 40% aqueous solution was used [13]. The manufacturer classifies the product as essentially nonirritating to the eye in the Material Safety Data Sheet [14].

B. Cosmetic Properties

Butylene glycol functions similarly to propylene glycol in many respects, but has shown superiority in its ability to retard fragrance loss, as well as enhance the preservation of cosmetics against microbial attack. These benefits are probably due to the increased solvent properties associated with the additional methylene group compared to propylene glycol.

Butylene glycol finds particular use in hair sprays and setting lotions as a humectant. In a compilation of data from the voluntary filing of product formulations with the U.S. FDA in 1981 [10] p. 227, butylene glycol was contained in 165 formulations at concentrations from <0.1 to >50%.

In 1968, Osipow and co-workers [15] discussed work done to compare glycerine, propylene glycol, and butylene glycol with respect to preservation enhancement and retention of flavors and fragrances. They found that both propylene and butylene glycols enhance the water solubility of methyl paraben compared with glycerine. Against *Escherichia coli*, somewhat less butylene glycol was required to inhibit its growth (in the absence of parabens) compared with propylene glycol, but against other test organisms the two glycols acted similarly. In experiments with methyl paraben, butylene glycol appears to enhance its efficacy against *Staphylococcus aureus* to a greater degree than propylene glycol.

In the same article, Osipow et al. reported on comparisons of the two glycols with glycerine as humectants, when formulated into a simple potassium stearate vanishing cream. The creams were inoculated with a mixed culture of bacteria, yeasts, and molds. The formulations containing butylene glycol were most effective at inhibiting microbial growth, with the propylene glycol-containing cream also controlling the contamination. The system containing glycerine, however, exhibited proliferation and overgrowth of the microbes. The authors postulate three mechanisms by which the glycols enhance the preservation of cosmetic emulsions.

1. Glycols decrease the oil:water partition coefficient of the preservative.
2. They lower the minimum concentration of preservative required to inhibit the growth of microbes.
3. They may provide an unfavorable environment for growth of microbes when used at conventional or somewhat higher levels in cosmetic emulsions. They may even preclude the need for additional preservative ingredients.

In 1967 Harb first introduced the idea of butylene glycol as a replacement for glycerine in suppositories [16], some skin preparations, and a hydrous USP ointment [17]. In 1977, Harb presented the results of work done evaluating butylene glycol as a replacement for glycerine in shave lathers [18]. Benefits attributed to the butylene glycol included a softer, more pearlescent cream with better rub-in qualities and a more emollient feel on the skin. The lather was denser and firmer and its wetting power was improved over the glycerine containing control. It was noted, however, that lather dried faster with butylene glycol than with glycerine. Both materials, however, greatly increased the time that lather would stay moist on the face compared with when no humectant was used. Harb also reported that the glycol-containing shave creams retained their scents more effectively than those with glycerine.

Improved retention of fragrances and flavors was shown by Osipow et al. [16] in studies of a variety of flavor and fragrance compounds. In a comparison with propylene glycol, 1% solutions of an essential oil in each glycol were left uncovered in evaporating dishes for 24 hours at room temperature. Analysis of the residue showed a 2-3-fold increase in retention of the key volatile components when butylene glycol was compared with propylene glycol. This effect was also seen in a 50/50 mixture of the glycol with ethanol, in which case only 2 hours was required for the test.

This improved retention of fragrance and flavor components manifests itself in two ways in the finished product. First, long-term stability is enhanced in the package, as the tendency for such components to migrate through plastic packaging is significantly reduced. Second, the duration of fragrance would be expected to be extended after application to the skin. In addition to the obvious applications in colognes and perfumes, extended masking of a malodorous component in a product would be beneficial. This property has been taken advantage of for many years since perfumers and flavorists widely use butylene glycol as a solvent for essential oils, other fragrance and flavor components.

V. ETHOXYLATED GLYCERINE

C.A.S. Number 31694-55-0 (glycerol ethoxylates)

Ethoxylation is a time-honored process for modifying the properties of a material to make it more suitable for particular applications. Thus lanolin and lanolin fractions have been subjected to ethoxylation to varying degrees to obtain a wide variety of functional products retaining desirable properties of the starting material while overcoming particular drawbacks. Similarly, glycerine is available with varying levels of ethylene oxide added. Lipo supplies Liponic EG-1 (CTFA: Glycereth-26) [20] and Liponic EG-7 (Glycereth-7) [21]. Ethoxylation reportedly reduces the tackiness of glycerine while maintaining or even enhancing its humectant properties. Heterene Chemical and PPG/Mazer also supply ethoxylated glycerine.

A. Safety

Glycereth-26 is 100% nonirritating to unwashed rabbit eyes and rabbit skin. The acute oral LD_{50}(rats) is greater than 5 g/kg.

B. Cosmetic Properties

Lipo reports that Glycereth-26 has a nongreasy, smooth luxurious feel with particular usefulness in pre-electric shaves, aftershaves, and skin fresheners. They emphasize an exceptional lubricious feel which it leaves on the skin. It is also recommended as an effective humectant for creams and lotions with a particular suggestion for its consideration in shave creams for humectant, lubricant, and foam modification properties. It also eliminates the tackiness sometimes associated with sorbitol.

The lower ethoxylate, Glycereth-7, would be expected to behave more like glycerine than the higher ethoxylates. Both materials should be considered for ethnic hair formulations as partial glycerine replacements to reduce tackiness.

VI. SORBITOL

C.A.S. Number 50-70-4

Sorbitol is a hexahydric alcohol with the empirical formula $C_6H_{14}O_6$ which occurs naturally in fruits, but is prepared synthetically from glucose by high-pressure hydrogenation. The structure is shown in Figure 16.1.

A. Safety

Sorbitol is nontoxic, being classified as GRAS (Generally Recognized As Safe) by the FDA as a food additive. It is nonirritating to both skin and eyes.

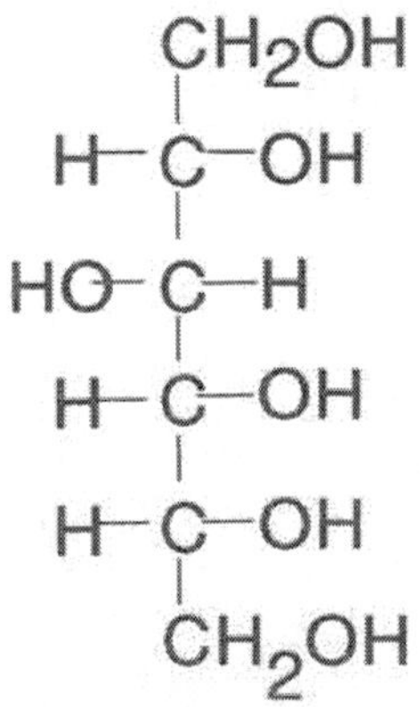

$$
\begin{array}{c}
CH_2OH \\
| \\
H-C-OH \\
| \\
HO-C-H \\
| \\
H-C-OH \\
| \\
H-C-OH \\
| \\
CH_2OH
\end{array}
$$

Figure 16.1 Structure of Sorbitol.

Since sorbitol is widely occurring in foods that are frequently ingested in significant quantities, its relatively free usage as a food additive is not surprising.

Sorbitol is most commonly available for cosmetic and drug use as a 70% aqueous solution, usually meeting the criteria of the USP or the Food Chemicals Codex (FCC). The 70% solution is a clear colorless, water-white, viscous liquid.

Sorbitol is also available in crystalline form (100% active) where its principal use is as the major component of sugarless gums. It is also utilized in pharmaceutical tablets and compressed mints and hard candy when non-cariogenic properties are desired. Since sorbitol does not provoke the same insulin response as sucrose in the human body, it is the key bulk sweetener used in diabetic foods such as candy.

Sorbitol is widely used as a partial or total replacement for glycerine in creams and lotions. Simple substitution on an equal weight basis for glycerine is seldom possible. However, the changes in product consistency, viscosity, and feel may often be readily compensated for by minor formula modifications. Sorbitol is often more cost effective than glycerine, and more stable in price. Sorbitol, as well as other polyols, provides smooth application properties as shown by good spreading action. It also facilitates gradual release of water from emulsions, as well as lubricity and emollience in these systems.

Sorbitol retains a high level of water at various humidities and exhibits a low rate of moisture gain or loss. It is less hygroscopic than other polyols and is claimed to be more likely to give a smooth dry effect when the cosmetic product is spread on the skin. This is contrasted to the damp, slippery feel that can be imparted by more hygroscopic humectants. In stearic acid soap type hand creams, sorbitol is more effective at retarding moisture loss than glycerine or propylene glycol [1].

Sorbitol is widely used in shave creams; both in the tube and aerosol variety. It provides protection against premature drying out of the foam on the face, and softens the cream in the tube to help prevent it from drying out. Typical use levels in aerosol shave creams are in the range from 3–10%.

B. Oral Products

Sorbitol is widely used in toothpastes, alone or blended with glycerine. Economic considerations are a major factor in the tendency to maximize the sorbitol while minimizing the glycerine levels in these products. The appearance of clear gel dentifrices in the early 1970s [23] was based on the use of synthetic, amorphous porous silica xerogels as the abrasive agent. The vehicle is a blend of glycerine and sorbitol having matching refractive indices of about 1.4. These formulations can contain 65–85% humectant, resulting in considerable cost savings by allowing manufacturers to use lower cost substitutes for glycerine.

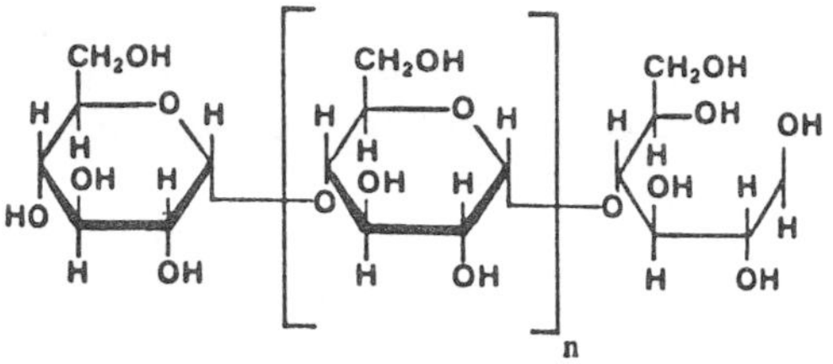

Figure 16.2 Structure of hydrogenated starch hydrolyzates.

An excellent treatment of humectant options in dentifrice formulation can be found in Pader's recent monograph [24]. In November 1970, U.S. Patent 3,538,230 was issued to Morton Pader and Wilfried Weisner for a clear gel toothpaste, which was the basis for Close-Up marketed by Lever Brothers.

VII. HYDROGENATED STARCH HYDROLYZATES

C.A.S. Number 68425-17-2

The general category of hydrogenated starch hydrolyzates (HSH) includes sorbitol in the sense that sorbitol results from the hydrogenation of a fully hydrolyzed starch, namely glucose or dextrose. Several manufacturers offer HSH with varying compositions and physical properties. Unlike glycerine, sorbitol, and propylene glycol, HSH represents a class of materials about which only some broad generalizations will be included. The reader is referred to the suppliers of these materials [21].

The general formula for the family of compounds referred to a hydrogenated starch hydrolyzates is shown in Figure 16.2.

The source of starch for HSH is generally determined by the economics and availability of a particular feedstock. For example, in the United States, the preferred starting material is corn syrup formed by the partial enzymatic hydrolysis of cornstarch. In Europe, potato starch is a common starting material.

A. Safety

The safety of the HSH is comparable to that of sorbitol. Roquette's Lycasin brand of hydrogenated starch hydrolyzate is the subject of a FDA GRAS Petition [26].

The physicochemical properties of HSH are largely influenced by the value of n, a measure of the degree of depolymerization achieved during hydrolysis. The degree of hydrolysis is generally characterized by the "dextrose equivalent" (DE). A high DE corresponds to a high degree of hydrolysis yielding dextrose which, upon hydrogenation, becomes sorbitol.

Since hydrolysis of a starch does not yield a single polysaccharide species, a mixture of polyols results when the hydrolyzate is hydrogenated. The mixture is generally characterized by specifying its sorbitol content, the hydrogenated disaccharide content (maltitol), the percentage of hydrogenated tri- to hexasaccharides, and the fraction of hydrogenated saccharides higher than hexa. Different HSH have been recommended and/or patented for dentifrice use based on different maltitol, sorbitol, and higher polyol content.

Roquette Corp. produces Lycasin brand of HSH, which is a proprietary composition covered by U.S. Patent No. 4,279,931. This material has 6–8% sorbitol, 50–55% disaccharides, 20–25% tri to hexasaccharides, with 18–23% of the higher saccharides. Greater detail about applications of these materials in dentifrice formulation can be found in Pader [24].

A significant application for HSHs as partial replacements for glycerine has arisen in the ethnic hair care field. The use of high levels of glycerine in conditioners has come about due to significant damage being done to blacks' hair from extensive styling treatments. Style-conscious black consumers who follow rapidly changing fashion trends often subject their hair to rapid, successive style changes; the result is overtreated, dry, fragile and otherwise damaged hair. Curl conditioners often contain 30–50% glycerine, which is heavy and sticky. Lipo [26] recommends its Polyol NC as a glycerine extender, substituted at 50–65% which will not adversely effect the feel or performance of the product.

The Liponic NC is derived from a mixture of mono-, di-, and trisaccharides which have been hydrogenated to polyols. The NC designation is widely used by polyol suppliers to indicate the material is a "noncrystallizing" grade. Sorbitol would not be suitable for such hair conditioner applications since crystal formation on the hair would be highly undesirable.

One of the main functions of humectant is to maintain the moisture levels in products, therefore, the comparison of various humectants' performance in that function can aid the formulator in choosing suitable candidate ingredients for a particular application.

Lonza has published [27] results of a moisture retention study for a prototype hand and body lotion containing 4% of various humectants. Glycerine, propylene glycol, and three HSHs from Lonza are used as humectants along with a control containing no humectant. The percentage moisture retained at 40° C as a function of time was tracked for 30 hours. Contrary to what one would expect from the work of Griffin and others [1], propylene glycol performed worse than all other humectants studied—even the control with no humectant. The three hydrolyzed starch hydrolyzates performed marginally better than sorbitol, and significantly better than glycerine. The three Lonza polyols used in the study are characterized by the manufacturer [28] with respect to the fraction of each material at varying degrees of polymerization as shown in Table 16.3.

Table 16.3   Degree of Polymerization of Hystar Hydrogenated
Starch Hydrolyzates courtesy of Lonza, Inc.

	Hystar 3375	Hystar 7000	Hystar HM-75
DP-1	14	45	13
DP-2	18	25	50
DP-3	10	8	12
DP-4	58	22	25

Polyol 7000 which has the highest fraction of sorbitol and lower oligosaccharides showed characteristics closer to that of pure sorbitol, while the polyols in the higher molecular weight range showed marginally higher water retention characteristics.

The results of this study cannot be compared directly with the earlier work of Griffin and others [1] which was carried out at 30° C. The results also cannot be generalized to other types of emulsions, as Griffin and others have reported that nonionic systems showed much less difference in moisture-holding ability when comparing glycerine, propylene glycol, and sorbitol.

VIII.   OTHER POLYOLS

There are several polyols which are used in cosmetics which have humectant properties, and also serve other functions such as solvents, plasticizers, and emulsifiers. Their humectant properties should be considered when choices are being made regarding a particular formulation. Hexylene glycol, ethoxydiglycol, and dipropylene glycol have been used extensively in cosmetic products, and have all been cleared by the Cosmetic Ingredient Review (CIR) [10] as being safe as currently used. DeNavarre summarized much of the data on a variety of polyols for their humectant properties. Table 16.4 reproduced from DeNavarre's book [7, p. 167] shows the equilibrium water concentration of 14 polyols at 50% relative humidity, 70–80° F. The formulator is referred to this book for additional details regarding calculation of hygroscopic properties of these materials and mixtures thereof.

IX.   ALKOXYLATED GLUCOSE

Several sugar derivatives have been used in cosmetic applications. Esters such as sucrose monopalmitate and sugar dipalmitate were first developed by Sugar Research Foundation in the United States as nonionic surfactants. More recently, Amerchol (Division of Union Carbide) has offered alkoxylated methyl glucosides—treated with ethylene or propylene oxide. They are commercially

Table 16.4

Name	Equilibrium hygroscopicity, % solids vs. RH (%)		
	30	50	70
Diethylene glycol	90	82	60
Polyglycol 400	95−	89	79+
Polyglycol 600	96−	90	80−
"Carbowax" 1000	98−	92	. . .
"Carbowax" 4000	99+	99−	96
Propylene glycol	91	82	68
Dipropylene glycol	96	89	77
Glycerol	89+	80	65+
Polyoxyethylene glycerol	94	87+	76+
Xylitol	92	84−	71−
Sorbitol(Arlex)[a] (85% soln.)	96+	87+	75+
Sorbitol(Sorbo)[a] (70% Soln.)	X	X	75−
Glucose	94	88+	79
Sodium lactate	84−	68−	50−

[a] Trade Mark ICI Americas, Inc.

known as Glucam E (ethoxylates) and Glucam P (propoxylates), exhibiting light color and low odor. They are promoted as "naturally derived" humectants and emollients. Typical properties of the 10 and 20 mole alkoxylates of methyl glucose are shown in Table 16.5 [29].

A. Safety

The methyl glucose ethers are very mild and safe ingredients. Repeated insult patch tests (human) have demonstrated they are nonsensitizers as well as not being primary irritants. They have also been shown to be well tolerated by skin over extended time periods [30]. The 20 mole propoxylate has shown potential for anti-irritant activity in various emulsion systems [31].

Several patents have been granted claiming special benefits for the glucose ethers. For example, in a Charles of the Ritz Co. patent [32] PPG-10 methyl glycoside ether is claimed to have imparted "staying power" to the skin-conditioning ingredients (propylene glycol dicaprylate/dicaprate) in an afterbath splash. Somewhat surprisingly, propoxylation of methyl glucosides improves oil solubility without affecting water solubility [33].

Goldemberg [34] found that the ethoxylated methyl glycosides reduced skin stinging of alcoholic products such as afterbath rubs, skin fresheners, and aftershaves. Guillot and others [35] showed that addition of only 2% of the 20 mole propoxylated methyl glycoside significantly decreased the primary irritation of two creams containing a known irritant.

Table 16.5   Typical Properties of Glucam Alkoxylated Glucose Ethers

Glucam	E-10	E-20	P-10	P-20
Appearance	Pale yellow medium viscosity syrup	Pale yellow thin syrup	Pale yellow viscous syrup	Pale yellow medium viscosity syrup
Odor	Practically odorless	Practically odorless	Practically odorless	Practically odorless
Acid value	1.5 max.	1 max.	1 max.	1 max.
Saponification value	1.5 max.	1 max.	1 max.	1 max.
Hydroxyl value	350–370	205–225	285–305	160–180
Iodine value	1 max.	1 max.	1 max.	1 max.
Moisture	1% max.	1% max.	1% max.	1% max.
Av. moles EO/PO	10 E.O.	20 E.O.	10 P.O.	20 P.O.
CTFA Name	Methyl Gluceth-10	Methyl Gluceth-20	PPG-10 Methyl Glucose Ether	PPG-20 Methyl Glucose Ether

These glucose ethers, especially the 20 mole ethoxylate and the 10 mole propoxylate are recommended as bar soap additives to reduce cracking, improve the lather, and impart an emollient afterfeel [29]. In this application, these materials act as both humectants and plasticizers.

Another application, cited in two patents [36, 37] is for the 20 mole propoxylate as a perfume fixative; it increases the intensity and lasting quality of perfumes without altering their aromatic characteristics. It is also claimed to reduce the pungency of alcohol contained in perfumes and colognes.

The strong water retention properties of the glucose ethers make them good choices as humectants which also provide significant emollient effects. They can form the basis for "oil-free" products, which provide a significant marketing benefit. Amerchol suggests an oil-free make-up remover formulation (Fig. 16.3) making use of the propoxylated methyl glycoside (Glucam P-10) for water solubility with minimal irritation around the eye. The PEG-20 methyl glucose sesquistearate serves as a mild surfactant together with the amphoteric to gently dissolve makeup.

Esters can be formed from methyl glucose which are then further reacted with ethylene oxide. This forms a water-soluble ester ether. Methyl Gluceth-10 is especially effective as a freezing point depressant.

X.   GLUCOSE GLUTAMATE

Glucose glutamate complex is the result of a reaction between glucose and glutamic acid. It is very hygroscopic and is claimed to impart humectant properties to a wide variety of cometic formulations. It is supplied by CasChem

Description: Crystal-clear solution designed to remove eye makeup without leaving an oily residue. Glucamate® SSE-20 serves as a mild surfactant that is used along with the amphoteric to gently dissolve makeup. Glucam® P-10 offers water-soluble emolliency to delicate tissue around the eye. Suitable to use with pads.

Formula	%
Glucam P-10	2.7
Glucamate SSE-20 (Peg-20 Methyl glucose sesquistearate)	2.0
Boric acid	0.3
Cocoamphocarboxyglycinate (and) sodium trideceth sulfate (and) hexylene glycol	5.0
Deionized water	90.0
Preservatives	Q.S.

Procedure: Dissolve Glucamate® SSE-20 and boric acid in water by gentle heating to 60°C. Add the rest of the ingredients. Mix until clear.

Figure 16.3   Oil-Free Eye Makeup Remover

Inc. under the trade name Wickenol 545. The material is supplied as a 51% solids aqueous liquid which is moderately viscous and opaque tan in color. The formula thought to represent the active part of this complex of a primary reducing sugar with an amino acid is shown in Figure 16.4.

Table 16.6 lists typical properties of this material

A.   Safety

The mean lethal dose (LD_{50}) (rats, oral) greater than 20.05 ml/kg. Nontoxic; primary skin irritation index is less than 5, indicating it is practically nonirritating in rabbit skin patch tests. It is also nonsensitizing via the guinea pig sensitization test. To the eyes (rabbit), it is a mild irritant when undiluted. No signs of acute dermal toxicity were seen in rabbits at the 2.0 g/kg level.

B.   Cosmetic Properties

The use of glucose glutamate in cosmetic-type applications is covered by U.S. Patent 3,231,472, licensed to Wicken Products. The rights to this material were acquired by CasChem, Inc., the current supplier [38]. The material is

$$CH_3O(CH_2O)_3CCH_2NHCH(CH_2)_2COOH$$

Figure 16.4   Glucose glutamate structure.

Table 16.6 Properties of Glucose Glutamate

Form:	Opaque, moderately viscous liquid
Color:	Off-white to tan
Odor:	Sugary note
pH (5% aqueous)	3.75–4.25
Total nitrogen (%):	1.2–2.2
Total sugars (%):	23.5–26.0
Solids (%):	50–55
Solubility:	Partially water soluble; V. slightly alcohol soluble

claimed to be substantive to hair and skin, leaving a residual film which produces a luxurious, smooth lustrous effect [39].

Substantivity on hair has been shown from a shampoo containing glucose glutamate compared with a control shampoo without the glutamate. Hair samples treated with shampoo containing glucose glutamate showed increased retention of glucose 2–5 times greater than a similar hair swatch treated with the control shampoo [40].

It is claimed that the presence of small residual amounts of glucose glutamate on the skin materially aids in restoring and maintaining the moisture balance necessary for a healthy, soft, and moist skin complexion [42]. For skin care products, the manufacturer recommends levels of 0.5–2.0% for normal to very dry skin.

XI. ORGANIC ACIDS AND SALTS

Helping skin maintain its optimum moisture level is probably the single most important goal of the majority of cosmetic treatments. Even sunscreens aid in this effort by helping prevent damage to the skin's moisture-holding systems. Thus moisture control in both products and skin has relied to a large extent on materials that will aid the skin in retaining its moisture under otherwise desiccating conditions. The hygroscopic and humectant properties of glycerine and its potential replacements are thus a major factor influencing a formulator developing a new product.

In an effort to understand how skin holds and releases moisture, many workers have demonstrated that there is a natural moisturizing factor (NMF) [41] in the skin which can be removed by means of water, other polar solvents, and detergent solutions. This material has been shown to have an amino-lipid nature [62]. Two major components of the NMF with pronounced hygroscopic properties have been shown to be the sodium salt of 2-pyrrolidone-5-carboxylic acid

Table 16.7   Typical Properties of Sodium Lactate–75%

Density (d_4^{25})	1.4075
Viscosity (poise at 25°C)	20
Freezing point	Below 10°C
pH (10% active)	7.0
Refractive index (n_D^{25})	1.4418
Equilibrium relative humidity (%)	32

(NaPCA) and sodium lactate [42]. These two materials constitute the most hygroscopic fraction of the water-soluble material in the stratum corneum. Laden patented sodium PCA as a humectant in cosmetics at levels of 2% or higher [52].

It is thought that the hygroscopic behavior of the NMF helps the corneum maintain its moisture content in drying atmospheric conditions. When this material is lost, not only does a reduction in water-holding capacity occur, but a very marked reduction in extensibility of isolated strips of stratum corneum is seen [44].

XII.   SODIUM LACTATE/LACTIC ACID

C.A.S. Number 72-17-3/50-21-5

Sodium lactate, a component of the natural moisturizing factor (NMF), is its most hygroscopic component, excepting sodium PCA. It works with the other components of the NMF to aid the stratum corneum maintain its moisture content under drying conditions. It is a better moisturizer than glycerine or sorbitol, but less effective than sodium PCA (Table 16.7).

Middleton [45] carried out three large double-blind consumer tests of hand lotions with lactic acid or sodium lactate at the 10% level, and compared the results after two weeks of dry winter weather, with a control and each other. Prior to the human studies, Middleton carried out laboratory tests on isolated guinea pig footpad corneum in which he compared glycerine, sorbitol, sodium lactate, and sodium PCA. He found none of these materials had effects on the corneum which persisted after rinsing. Investigation of the absorption of lactic acid as a function of pH showed that above pH 5, no detectable absorption occurred. Since only undissociated acid can be absorbed, the results are not surprising.

Although water-holding capacity increases seen after treatment of damaged guinea pig corneum with 10% lactate solutions did not persist after rinsing, an increased extensibility of the material treated with lactic acid did endure. This ability of lactic acid's influence on the extensibility of the corneum to be maintained after rinsing distinguishes it from other humectants under similar conditions. In tests with a hand lotion vehicle, similar results were obtained, showing only the lactic acid to have beneficial effects which lasted through rinsing. This is consistent with the requirement that absorption of the undissociated lactic acid is necessary for persistent effects.

In one consumer test with hand lotions containing 10% of either sodium lactate or lactic acid, both products showed significant reductions in hand skin dryness and flaking compared with a control. A numeric, but not statistically significant difference favoring lactic acid was seen. A year later under more severe cold weather conditions, both humectants showed superior performance to control, with the lactic acid version also significantly superior to the lotion with sodium lactate. A third consumer test compared lotions with 5 and 10% lactic acid and indicated little additional advantage for the higher level of humectant.

The superior performance of the humectant which can be absorbed into the stratum corneum has been demonstrated. Middleton [45] concludes that lactic acid may function by two mechanisms while sodium lactate has only one. He suggests that the temporary increase in water-holding capacity seen with sodium lactate causes the increase in extensibility of the stratum corneum. Lactic acid also functions in this capacity as a conventional humectant, but the persistent effect after rinsing must reflect an increase of extensibility due to absorbed lactic acid—not an increased water content. Thus both products functioned in the hand lotion test, but the more persistent lactic acid effect was reflected in superior performance in improving hand skin condition under severe dry, winter conditions.

Osipow [46] carried out a series of tests on sodium lactate/lactic acid systems and showed moisture loss to be comparable to glycerine in a variety of cosmetic emulsions. Only in antiperspirant creams and lotions was the humectant performance definitely inferior to glycerine. It is assumed this is due to the formation of aluminum lactate. The conclusions reached by the author include the following:

1. Sodium lactate is an effective humectant in cosmetic preparations. It is generally compatible with other cosmetic ingredients and it does not increase the difficulty of preparing stable emulsions.
2. Sodium lactate, in combination with lactic acid, has a further advantage in that it serves as a buffer, as well as a humectant.
3. Sodium lactate is an economical replacement for glycerine and other more conventional humectants.

In 1975, Clar and co-workers [47] reported on skin impedance measurements which demonstrated increased hydration of the skin from application of a 10% sodium lactate solution to the forearm of subjects. This method of evaluating moisturizers has been widely used since this pioneering work was published.

XIII.   SODIUM PCA

C.A.S. Numbers 28874-51-3 and 54571067-4

The sodium salt of pyrrolidonecarboxylic acid (NaPCA) is a naturally occurring humectant in skin with the structure as shown in Figure 16.5.

Chemical Name: DL-2-Pyrrolidone-5-carboxylic acid, sodium salt.
Other Names: Sodium pyroglutamate, sodium pyrrolidone carboxylate, glutamic acid lactam, sodium salt, or sodium salt of glutamic acid lactase.

A.   Safety

The material is practically nontoxic with an LD_{50} in mice of 10.4 g/kg(oral) [48]. At 50% concentration, it is slightly irritating to eyes and nonirritating to skin [32]. At 5% concentration it is nonirritating to eyes. It has been extensively tested and found not to be phototoxic or sensitizing to either humans or guinea pigs.

B.   Availability

Sodium PCA is available as a 50% aqueous solution from several suppliers. More recently Ajinomoto has offered the TEA salt also as a 50% aqueous solution under the trade name Ajidew T-50. The acid form is also available as a 100% powder which is nonhygroscopic and is useful as a neutralizing agent, at which time it becomes a humectant.

C.   Cosmetic Properties

Sodium PCA has been shown to constitute about 12% of the NMF (calculated as the acid [51]. It has very great water-binding properties—much more so than propylene glycol, glycerine, or sorbitol. The sodium or potassium salts of PCA absorb 60% of their weight in water at 65% R.H. This water-binding ability is 50% greater than glycerine, twice as much as propylene glycol, and six times that of sorbitol under the same conditions [48, p.11].

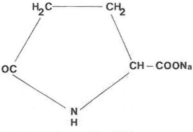

Figure 16.5   Sodium PCA.

Although sodium PCA is a very effective humectant, its high price has often led to its inclusion in products principally for label claim and marketing reasons. In such cases, other humectants, including sodium lactate, make excellent adjuvants to obtain a combination of efficacy and an appealing marketing story. The concept of using components of the NMF, or the NMF itself is not without its detractors [50].

In hair products, particularly after–shampoo rinses, sodium PCA has been the subject of a number of patents. In a series of patents assigned to Helene Curtis, Newell [51] reports on the use of NaPCA in combination with glycerine and protein hydrolysates in hair moisturizing products. In particular, it is claimed that when hair is treated with a formulation containing 0.01 to 1% NaPCA along with 0.05 to 5% each of glycerine and a protein derived from collagen, the normal moisture content is restored from initially moisture-deficient hair. This combination can be utilized in shampoos, conditioners, setting lotions, hair sprays, and other hair care products in such a manner as to help restore and maintain the proper moisture level in hair.

Formulated humectant blends are available which contain sodium PCA as a key component. An example is Prodew from Ajinomoto. It is available in two grades; Prodew 100 designed to be superior in moisture retention while Prodew 200 is superior in moisture absorption [54]. This material is formulated with components found in the NMF, and consists of L-proline, sodium PCA, sorbitol, sodium lactate, and hydrolyzed collagen.

XIV. HYALURONIC ACID

Hyaluronic acid is a member of a group of polysaccharides designated as a glycosaminoglycans (GAG) or mucopolysaccharides. It is a constituent of the intercellular ground substance of connective tissue. Though first isolated in 1934, its structure was not completely determined until 1958 [53]. The molecule consists of periodically repeating disaccharide units, with 1,000–10,000 linked in long, unbranched chains. Figure 16.6 shows that disaccharide units consist of one molecule each of N-acetylglucosamine and glucuronic acid linked by a beta-1,3-glycosidic bond. The disaccharide units are further linked 1,4-glycosidically to each other.

Hyaluronic acid (HA) or its sodium salt have the capability to reversibly absorb and release large amounts of water. The mechanism proposed to explain the water-binding behavior involves hydrogen binding between polar-OH groups and negative charges on the HA which can give up water when subjected to mechanical stress. A spongelike action has been proposed [54].

A. Safety

Hyaluronic acid is a physiological substance found in the vitreous body of the eye and in synovial fluid. It has a long history of medicinal use and is known

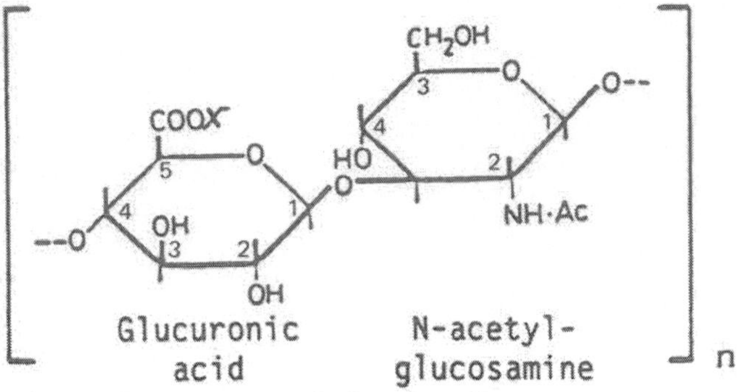

Figure 16.6 Structure of hyaluronic acid.

to be nontoxic. Immunological tests have shown that highly pure HA has no sensitizing properties. The material has an excellent safety profile as evidenced by its long-term medical use, such as replacing lost vitreous humor in the human eye.

B. Source

Hyaluronic acid is isolated from cock's combs or by a biofermentation process utilizing specially selected microorganisms that produce a relatively high-molecular weight fraction of the desired material. HA is very expensive, and only its functionality at very low concentrations allows its use in many products.

Hyaluronic acid is superior to almost all other natural and synthetic polymers for hydrophilic properties [55–57]. For example, a 2% aqueous solution binds the remaining 98% of the water to form a product with a gellike character—although it is not a true gel in the physical sense. Even a 0.2% solution retains a significant degree of structure and viscoelastic properties.

C. Cosmetic Properties

When applied to the skin, hyaluronic acid forms a viscoelastic colorless, transparent, and invisible film on the skin's surface. This film fixes moisture on the skin surface in the same way it does in the ground substance of connective tissue. It is claimed that the use of HA in topical skin products helps "maintain or restore the biophysical properties of skin, such as smoothness, elasticity and resilience, which gives it its youthfully fresh appearance" [54, p. 6].

The relationship between hyaluronic acid content, water content, and elastic and mechanical properties of the skin show age-dependent changes. The well-known loss of moisture with age is partially attributed to a decreasing content of HA in the skin [58]. Many claims are made for hyaluronic acid which will require further documentation as its use in cosmetics spreads.

The protective film, mentioned earlier, said to be left on the skin by hyaluronic acid treatment is presumed to protect the skin against dehydration by providing it with a very large amount of water of hydration. This would then enable the skin to maintain its soft, smooth, and elastic properties.

Hyaluronic acid, via its film-forming properties is claimed to enhance the defensive and protective mechanisms of the skin by reducing roughness and smoothing the skin profile. The special affinity of hyaluronic acid for the skin protein is given as a reason to expect the cosmetic effects of its application to persist for an extended period [54, p. 11].

XV. ACETAMIDE MEA

C.A.S. Number 142–26–7
Chemical Name N-acetyl ethanolamine
Formula $CH_3CONHCH_2CH_2OH$

A. Safety

Acetamide MEA is a clear liquid when supplied as 70% aqueous solution with zero skin and eye irritation and an oral LD_{50} (rats) of 24.8 g/kg. It has been shown to have counterirritant properties, especially with respect to sodium lauryl sulfate (SLS). Significant reductions in eye irritation scores are seen with equal active mixtures of acetamide MEA and SLS [59].

B. Cosmetic Properties

In shampoos, acetamide MEA is recommended to replace glycerine, propylene glycol, or hexylene glycol as a humectant or clarifying agent. It is reported to have little detrimental effect on either viscosity or foam stability. The material is also recommended for hair rinses and conditioners, as well as creams and lotions for its humectant properties combined with excellent nontacky skin feel.

The excellent pigment-dispersing properties of acetamide MEA allow it to function as a moisturizer in make-ups while requiring minimal homogenizing or milling during preparation.

The humectant properties of acetamide MEA are claimed to be superior to those of glycerine, when they are compared under the same conditions. For example, 70% acetamide MEA shows slightly lower weight loss than 70% glycerine when left standing in open air under "room temperature" conditions [59]. In another test, after being dried to constant weight, both acetamide MEA and glycerine were left at room temperature and the moisture absorption was measured for 3 hours. Glycerine, under those conditions absorbed only 60% of the amount of moisture take up by the acetamide MEA.

In a patent assigned to Alberto Culver Co., Dasher et al. [60] describe a preshampoo conditioner utilizing water-soluable quaternary compounds, polyethyleneimines, and acetamide MEA. In particular, suggested levels of acetamide MEA are from 3–18%, along with 0.5–5.0% of the water-soluble quaternary, and 0.5–5.0% polyethyleneimine. The balance of the composition consists of water, fragrance, and optional components. What is claimed is a composition which deposits cationic materials into cracks and crevices in the hair shaft; especially where it is damaged. This material complexes with anionic shampoo components which also penetrate the hair shaft, and are then said to remain "inside" the hair, while any excess on the surface is solubilized by the excess anionic shampoo.

XVI. ACYL LACTYLATES

C.A.S. Number 66988–04–3
Chemical Name Sodium isostearoyl lactylate

The reaction product of a fatty acid with lactic acid and subsequent neutralization to an alkaline salt are known as acyl lactylates [61]. C. J. Patterson Co. produces a series of this materials, originally used in the baking industry as dough conditioners, under the trade name of Pationic.

The isostearoyl derivative with the CTFA Adopted Name of Sodium Isostearoyl-2-lactylate, has been recommended for consideration as a hair and skin conditioner, as it is also utilized in its function as an anionic emulsifier [64].

A. Safety

These materials are essentially nonirritating at typical use concentrations (15%) both to skin and eyes. Their low degree of oral toxicity is expected from their wide use in food products. The oral LD_{50} is > 5.0 g/kg, which allows them to be classified as nontoxic by oral ingestion. [63]

B. Cosmetic Properties

The acyl lactylates, in particular, the isostearoyl derivative has been studied for effectiveness as a moisture retention agent. The supplier has published a study comparing moisture-holding abilities of the isostearoyl (ISL) derivative with propylene glycol and glycerine [64]. Testing under both high and low humidity conditions led to the following conclusions.

1. The ISL derivative holds onto its absorbed moisture to a higher degree than glycerine or propylene glycol. Extended heating of the material with 10% added water at 135°C is required to drive off significant levels of moisture.

2. Glycerine and propylene glycol absorbed higher percentages of moisture at 78% rh than the ISL, but then lost it more quickly at 25% humidity.

A subsequent study has been published [65] showing the effect on curl retention when isostearoyl lactylate was used as an additive to both shampoo and creme rinse formulations. In shampoo formulations where the Pationic ISL functioned as a conditioning additive, superior curl retention (15–26%) was seen after 3 h at 65% rh, compared with commercial conditioning shampoos. Performance was equal to Prell, a strong cleansing shampoo with no conditioning properties.

In a rinse/conditioner formulation, a 40% improvement in curl retention was seen after 3 h at 65% rh compared with three commercial rinse/conditioners [65]. The manufacturer concludes that the conditioning benefits of this substantive, anionic material do not necessarily include the negative, "limp hair" effect often seen with other conditioning additives.

The substantivity of the acyl lactylates was first observed in their use in baking when complexes with water-soluble wheat proteins were formed [68]. The conditioning effect of acyl lactylates on hair was patented by Osipow and Mana [67] in 1973. Substantivity and conditioning effects on skin have also been reported [68].

XVII. OTHER HUMECTANTS

Deshpande, and co-workers [69] studied 16 materials claimed to be good moisturizer/humectants as part of the work toward a Master's degree in cosmetic science. Many of the materials covered in this chapter were evaluated in this study to varying degrees—alone and in model formulations. The initial set of test materials was tested for moisture uptake and/or loss at 90% and 79% rh and ambient room temperature. The 8 best-performing materials were further evaluated with glycerine and sodium PCA acting as controls.

The 10 materials were then evaluated for moisture uptake at constant relative humidities of 20, 52, 79, and 90%. As expected, poorer performance was seen across the board in the lower humidity range. The authors reported "shock" at the poor performance of several materials at 52% humidity. For example, a 50% solution of sodium PCA lost 7.42% of its moisture after 1 day, and 20% after 5 days. This contrasts to gains of 21.31 and 28.2% moisture for glycerine.

Although one could expect a greater potential for weight loss from a 50% solution than from the 100% glycerine, weight loss from a 50% active humectant under ordinary humidity conditions is somewhat unexpected.

At 20% relative humidity conditions, several "humectants" showed significant weight loss; 36% for sodium PCA after 5 days, 19% for a 60% sodium lactate solution, also after 5 days. Acetamide MEA also showed weight loss at 52 and 20% rh, but since it is 75% active, it has proportionately less moisture to lose than the other solutions tested. Glycerine absorbed moisture, even at 20% rh; 9% in one day but only 6% after 5 days.

The interested formulator is referred to the excellent compilation by Deshpande for additional details. Also reported was work done using a model lotion formulation with each of the 10 additives used at the 5% active level. These lotions were stability tested at ambient and elevated temperatures, as well as refrigerator and freeze–thaw conditions. The function of these additives as humectants was not studied, as the containers were closed.

In general, Deshpande concluded that at 52–90% humidity, the following materials have excellent humectant properties: glycerine, methyl gluceth-20, methyl gluceth-10, sodium isostearoyl-2-lactylate, and sodium capryl lactylate. All collagen derivatives evaluated were "uniformly disappointing." Intermediate performance was seen with sodium PCA, acetulated PEG-10 lanolin alcohol, PPG-10 and PPG-20 methyl glucose ether, sodium lactate-60%, and acetamide MEA-75%. At the all important 20% relative humidity range, only glycerine and sodium capryl lactylate showed any appreciable activity.

XVIII. CONCLUSIONS

A wide variety of materials have been examined which are utilized in cosmetic and personal care products for many of the same purposes and function as glycerine would be considered. The major function of humectancy is the common property of all of the materials. The traditional glycols such as propylene, butylene, hexylene, and polyethylene glycols have many of the same properties as glycerine, but with important differences as summarized in this chapter.

In addition to these traditional compounds, other polyols such as sorbitol, hydrogenated starch hydrolyzates, and glucose derivatives have shown particular applications in which their performance was superior to glycerine. Organic acid salts such as sodium lactate, sodium PCA, and sodium hyaluronate have more recently gained more widespread usage as more sophisticated skin- and hair-moisturizing products are desired. Clearly such a review is limited in the amount of detailed information which can be included about a particular material or application. However, the references included should provide the formulator an entree to more extensive information as needed. In the past, glycerine's function as a vehicle, humectant, and solvent for cosmetic products was a major consideration in its choice by the formulator. Today, the function of the product on the user has led to development of more specialized materials which are often used in addition to glycerine, other polyols, and each other. Combinations of ingredients with particular properties for each appears to be the trend for the future.

REFERENCES

1. Humectants, in Harry's Cosmeticology, Chemical Publishing, New York, 1982 pp. 641–652.
2. Hunting, A. L. L., Encyclopedia of Shampoo Ingredients, Micelle Press, Cranford, NJ, 1983, p. 331.
3. Hannuksela, M., Allergic and toxic reactions caused by cream bases in dermatological patients, Int. J. Cosmet. Sci., 1: 257–263, 1979.
4. Drill, V. A. and Lazar, P., eds., Cutaneous Toxicity, Academic Press, New York, 1977.
5. Fox, C. Technically speaking, Cosmet. Toiletries, 100 (2): 17–18, 1985.
6. Motoyoshi, K. et al., The safety of propylene glycol and other humectants, Cosmet. Toiletries, 99 (10): 83–91, 1984.
7. deNavarre, M. G., ed., The Chemistry and Manufacture of Cosmetics, 2nd. ed., Vol. 2, Cosmetic Materials, D. Van Nostrand Co., Princeton, NJ, 1962, p. 165.
8. Windholz, M., ed., The Merck Index, 9th ed., Merck & Co., Ratway, NJ, 1976.
9. Fisher, A. A., Reactions to popular cosmetic humectants. Part III. Glycerin, propylene glycol, and butylene glycol, Cutis, 26 (3): 243, 269, 1980.
10. Final report on the safety assessment of butylene glycol, hexylene glycol, ethoxydiglycol, and dipropylene glycol, J. Am. Coll. Toxicol. 1, (5): 223–248, 1985.
11. CTFA submission of unpublished data, CIR Safety Data Test Summary, rabbit skin patch test on butylene glycol, CTFA Code 2017–75, March 9, 1979.
12. National Institute for Occupational Safety and Health (NIOSH), Toxic Substances List, U.S. Government Printing Office, Washington, DC, 1981.
13. CTFA submission of unpublished data, CIR Safety Data Test Summary, eye irritation test of butylene glycol, CTFA Code 2–17–74, Dec. 6, 1976.
14. Material Safety Data Sheet on 1,3-butylene glycol, Hoechst Celanese, Dallas, TX, May 18, 1988.
15. Osipow, L. I., Marra, D., and Resnansky, N., Drug Cosmet. Ind., 23:54, 1968.
16. Harb, N. A. et al., Use of 1:3 butylene glycol as a glycerol substitute in suppositories, and its pharmacological effects, J. Hosp. Pharm., 24(3): 114–116, 1967.
17. Harb, N. A., A study of some skin preparations containing 1:3-butylene glycol as a glycerol substitute, J. Hosp. Pharm. 24(9): 323–325, 1967.
18. Harb, N. A., 1:3 Butylene glycol as a substitute in shave lathers, Drug Cosmet. Ind., 121(4): 38, 1977.
20. Lipo Chemicals, Inc., Technical Bulletin #10–010 on Liponic EG-1, Paterson, NJ (undated).
21. Lipo Chemicals, Inc., Technical Bulletin #10–011 on Liponic EG-7, Patterson, NJ (undated).
22. Pader, M., Gel toothpastes: Genesis, Cosmet. Toiletries, 98: 71–76, 1983.
23. Pader, M., Oral Hygiene Products and Practices, Marcel Dekker, Inc., New York, 1988, pp. 266–279.
24. ICI Americas, Lonza, Lipo and Roquette are the major U.S. suppliers of sorbitol.
25. FDA GRAS affirmation petition #360286 filed December 1983.
26. Private communication, Steve Greenberg, Ph.D., Lipo Chemical Co., Paterson, NJ, December 1989.

27. Lonza, Inc., Technical Bulletin, Moisture Control and Sweetening Agents, Fair Lawn, NJ, October 1980.
28. Lonza Technical Bulletin, Liquid Carbohydrates, Fair Lawn, NJ, 1988.
29. Amerchol Technical Bulletin, Glucam Alkoxylated Glucose Derivatives, Edison, NJ (undated).
30. Guillot, M. et al., Safety evaluation of some humectants and moisturizers used in cosmetic formulations, Int. J. Cosmet. Sci., 4: 67–80, 1982.
31. Guillot, M. et al, Anti-irritant potential of cosmetic raw materials and formulations, Int. J. Cosmet. Sci., 5: 255–265, 1983.
32. U.S. Patent 4,482,573 assigned to Charles of the Ritz, 1984.
33. Alexander, P., Glucose derivatives in cosmetics, Mfg. Chem. 59–63, 1988.
34. Goldemberg, R. L., J. Soc. Cosmet. Chem., 30: 415, 1979.
35. Guillot, J. P. et al, 12th International Congress IFSCC, Paris, September 1982, p. 215.
36. Seldner, A., Polyol fragrance fixatives, U.S. Patent 4,264, 478 assigned to Americhol Corp., 1981.
37. Seldner, A., Polyol fragrance fixatives, U.S. Patent 4,324,703 assigned to Amerchol Corp., 1982.
38. CasChem Technical Bulletin, Wickenol 545, Cosmetic Humectant, Amino Acid Ester, #73431, CasChem, Inc., Bayonne, N.J (undated).
39. Wicken Products Technical Bulletin, WICKENOL 545, Glucose Glutamate Complex, Wicken Products, Inc., Huguenot, NY (undated).
40. Wicken Products Technical Bulletin, WICKENOL 545, Glucose Glutamate, No. 97060, Wicken Products, Inc., Huguenot, NY (undated).
41. Jacobi, O. K., About the mechanism of moisture regulation in the horny layer of skin, Proc. Sci. Sect. Toilet Goods Assoc., 31: 22, 1959.
42. Middleton, J. D., Sodium lactate as a moisturizer, Cosmet. Toiletries, 93: 85–86, March 1978.
43. Laden, K., Humectant, U.S. Patent 3,235,457, 1966.
44. Middleton, J. D., The mechanism of water binding in stratum corneum, Br. J. Dermatol., 80: 437, 1968.
45. Middleton, J. D., Development of a skin cream designed to reduce dry and flaky skin, J. Soc. Cosmet. Chem., 25: 519–534, 1974.
46. Osipow, L. I., A buffering humectant for cosmetics, Drug Cosmet. Ind., 88: 438, 1961.
47. Clar, E. J., Her, C. P., and Sturelle, C. G., Skin impedance and moisturization, J. Soc. Cosmet. Chem., 26: 337–353, 1975.
48. Ajidew Technical Bulletin, Ajinomoto Co., Inc., Tokyo, 1988.
49. Laden, K. and Spitzer, R., Identification of a natural moisturizing agent in skin, J. Soc. Cosmet. Chem., 18: 351, 1967.
50. Idson, B., 'Natural' moisturizers for cosmetics, Drug Cosmet. Ind., 136 (5): 24–26, 1985.
51. Newell, G. P., Hair moisturizing compositions, U.S. Patents 4,220,166, 4,220,167, 4,220,168, 1980; 4,374,125, 1983.
52. PRODEW Technical Bulletin, (PRD-8101), Ajinomoto Co., Tokyo, 1988.

53. Meyer, K., Chemical structure of hyaluronic acid, Fed. Proc., 17: 1075–1077, 1958.

54. Hyaluronic Acid Technical Bulletin, Pentapharm LTD., Basel, Switzerland (undated).

55. Yates, J. R., Mechanism of water uptake by skin, in Biophysical Properties of the Skin, N. R. Eden, ed., Wiley Interscience, New York, 1971, p. 485.

56. Pearce, R. H. and Grimmer, B. J., The Nature of the Ground Substance, in Advances in the Biology of Skin, Vol. II, W. Montagna, J. P. Bently, and R. L. Dobson, eds., Appleton Publishers, New York, 1970, p. 89.

57. Balazs, E. A. and Gibbs, D. A., The rheological properties and biological function of hyaluronic acid, in Chemistry and Molecular Biology of the Intercellular Matrix, Vol. 3, E. A. Balazs, ed., Academic Press, New York, 1970, pp. 1241–1254.

58. Fleishmajer, R., Perlish, J. S., and Bashey, R. I., Human dermal glycosaminoglycans and aging, Biochim. Biophys. Acta, 279: 265, 1972.

59. Shercomid AME-70 Technical Bulletin, Scher Chemicals, Inc., Clifton, NJ, July 1977.

60. Dasher, G. F., O'Cull, K. A., and Schamper, T. J., Quarternary ammonium compounds in pretreatment of hair before shampooing with an anionic shampoo, U.S. Patent 3,980,091, 1976.

61. Patco Products Technical Data Sheet, Pationic ISL developmental specifications, Bulletin No. 120, C. J. Patterson Co., Kansas City, MO, July 8, 1977.

62. Murphy, L. J., Sorption of acyl lactylates by hair and skin as documented by radio tracer studies, Cosmet. Toiletries, 94 (3): 43–47, 1979.

63. Patco Products Technical Data, Pationic acyl lactylates, physiological test data, Bulletin No. 100, C. J. Patterson Co., Kansas City, MO, November 2, 1977.

64. Patco Cosmetic Products Technical Data, Pationic ISL moisture retention study, PATCO Bulletin No. 183, C. J. Patterson Co., Kansas City, MO (undated).

65. Patco Cosmetic Products Technical Data, Pationic ISL curl retention additive in shampoo and conditioner rinse, PATCO Bulletin No. 204, C. J. Patterson Co., Kansas City, MO, January 20, 1980.

66. DeStefanis, V. A. and Ponfe, J. G., Cereal Chem., 54 (1): 13–24, 1977.

67. Osipow, L. and Mana, D., Fatty acid lactylates and glycolates for conditioned hair, U.S. Patent 3,728,447 assigned to C. J. Patterson, 1973.

68. Murphy, L. J. and Finney, The action of acyl lactylates on hair and skin, SCC Annual Scientific Seminar, Chicago, IL, May 17–18, 1978.

69. Deshpande, V. M., Ward, J. B., Kennon, L., and Cutie, A. J., Potential new skin-care humectants: An evaluation, Cosmet. Technol. 20–30, 1980.

17

Other Applications of Glycerine

Richard A. Reck

Consultant, Hinsdale, Illinois

Glycerine has a very wide range of applications and uses due to its unique combination of properties. Besides the applications in cosmetics and toiletries discussed in preceding chapters, approximately 69,000 MT of glycerine are used in other applications in the United States. A breakdown of these applications is shown in Table 17.1.

I. ALKYD RESINS

In the paint and coatings industry glycerine's usage has been a major factor. In the manufacture of basic alkyd resins three basic components are used: polyhydric alcohols, dibasic acids, and fatty acids. The dibasic acid is usually phthalic acid, the polyol is glycerine, and the fatty acid is polyunsaturated. Glycerine is the predominant polyhydric alcohol used because the alkyd reactions can be controlled using glycerine. The viscosity of the resins formed are low and find application in latex paints. The application of an alkyd utilizing glycerine is largely determined by its "alkyd ratio" or the proportion of glycerol phthalate in the resin. A high alkyd ratio ordinarily found in commercial products is 65. These resins would be high in viscosity and may require addition of aromatic solvents. The primary use for these alkyds are as hardeners for other alkyds of high oil content. When the alkyd ratio is 50, viscosity is still high and solvents may be required. These resins are used for baked coatings.

Table 17.1   Applications of Glycerine

Uses	Thousands of MT
Alkyd Resins	11
Wrappings	3
Tobacco	20
Explosives	2
Urethanes	15
Triacetin	8
Miscellaneous	10

Aliphatic solvents may be used when the alkyd ratio is reduced to about 40. Resins of this type may be applied by roller coating or spraying. Applications are in air-drying pigmented enamels, metal coating, auto and truck refinishing, general maintenance paints, and outdoor sign painting. Where the alkyl ratio is about 30, the viscosity is still lower and enamels made with these resins can be brushed. The main uses are architectural finishes, such as trim and trellis paints and where coatings of high flexibility are needed. A good example is in tubes used for holding flexible materials. During World War II, many marine finishes were formulated with this type of resin. Glyceryl phthalate resins frequently are formulated with other types of synthetic resins and film-forming materials. Alkyds are also used to improve the flexibility, adhesion, toughness, and other properties of urea-formaldyhyde resins and melamine formaldehyde resins. Color, mar resistance, hardness, and baking cycles are among the characteristics improved by such combinations. Paint technologists agree that no other film-forming materials complement the qualities of amino resins as do the alkyds.

Silicones have been combined with alkyds to produce new types of finishes that have superior quality and stability at high temperatures. Coating films formulated with silicone–alkyds have abrasion resistance, adhesion, hardness, and flexibility. also they have resistance to acids, alkalis, and solvents. A silicone–alkyd blend was developed for use in concrete waterproofing and can be applied similar to a standard alkyd and air-dried.

Fatty acids can also be styrenated and then converted to alkyds. Styrenation of fatty acids is followed by reaction of phthalic anhydride and glycerine. The resultant resins have excellent air-drying and baking characteristics. Styrenated castor oil acids are particularly useful in these products.

II. WRAPPINGS

Since glycerine is generally recognized as safe (GRAS), it is used in a wide variety of food packaging materials. Cellophane is plasticized by passing the film through a glycerine solution, followed by drying. The final film may contain 10–25% of its weight of glycerine which imparts flexibility, durability, and prevents shrinkage of the finished product. The softening action of the glycerine is not caused by an increase in moisture content by the presence of the glycerine, but by the direct effect of the glycerine molecules on the cellulose structure.

Glyceryl-plasticized cellophane may be modified and improved. Antisticking agents may be incorporated in the glycerine plasticizing bath to minimize the tendency of superimposed cellulose sheets to stick to each other. Moisture barriers may also be incorporated in the glycerol formulation. Flameproofing can also be accomplished by incorporating flame retardent materials into the glycerol formulation.

Glycerol formulations are also used in a wide variety of paper products. The main use is as a softener and plasticizer for glassine and parchment paper. Glyceryl formulations added to paper yields a more flexible sheet. Carbohydrates are added to the formulation to increase plasticization. Glycerol also is used as a softening agent for napkins, paper towels, and tissues.

III. TOBACCO

Glycerine is a very important component in the manufacture of tobacco products. Its use in this industry accounts for the major market for glycerine outside of foods and cosmetics.

Cigarette manufacturing is very carefully controlled, and moisture content is extremely important. Moisture content is rated very high among quality factors of cigarettes. The moisture content is the main quality factor in American-made cigarettes. Tobacco is capable of increasing its moisture content in high humidity conditions, but under normal atmospheric conditions most grades have a low moisture content and become brittle. Under these conditions, tobacco becomes useless and may be reduced to dust. Cigarette manufacturers combat this problem by treating the tobacco with solutions that contain hygroscopic or moisturing agents to produce a better, fresher-tasting product. Glycerine is the preferred reagent for this application.

Glycerine may be applied to tobacco as an aqueous solution or it may be sprayed on the blended or cut tobacco as a component of what is called "casing fluids." These solutions often contain small amounts of sugar and flavoring. The objective of this treatment is to keep the moisture at the optimum moisture level of 12%. The tobacco will contain 2–4% of glycerine to achieve this moisture level and the variance of the glycerine content will depend on seasonal and climatic conditions. Tobacco products other than cigarettes may also contain glycerine. Burly tobacco used in pipes and chewing mixtures may be dipped in glycerine solutions to prevent drying out of the product once it is removed from its wrapper.

In chewing tobacco, moisture retardation is only one reason for its use. Because the product is shipped and transported in all kinds of climates the glycerine will stabilize the product and keep the chewing tobacco palatable at even subzero temperatures.

IV. EXPLOSIVES

The use of glycerine in explosives is not an extremely large market for this product, but is highly recognized. The product produced is made by the reaction of glycerine and nitric acid and the resultant product is nitroglycerine or glyceryltrinitrate. The dynamite produced today is greatly modified. It is mainly an industrial explosive and is used for agriculture, quarrying, engineering, mining, harbor improvements, and other purposes. It is extremely sensitive and must be handled with care.

Dynamite formulations have extended to many mixtures, and all have nitroglycerine as a base. A common property of these many formulations is that they are not exploded by simple shock, inflammation, or moderate friction. The explosion is set off by the use of strong caps or detonators usually composed of mercury fulminate or a similar compound.

Straight nitroglycerine dynamites are made 15–60% active, have high detonation capability, and are used only where quick action or a shattering effect are needed. Because they resist water they may be used for underwater blasting, but specifically designed formulations may be better for this application. For example, there is underwater gelation dynamite which is a mixture of a colloidal solution of nitroglycerine and nitrocotton. The product also is mixed with sodium nitrate and wood meal. The result is a product that has plastic consistency, which may be extruded through preset molds in any shape desired.

Other formulations may be produced. The production of poisonous gases from dynamites has been reduced by improving the oxygen balance of the formulations. The use of these dynamites is regulated by the U.S. Bureau of Mines. However, lack of U.S. Bureau of Mines approval does not prevent local use of these formulations. These uses are regulated by each state.

Glycerine derivative explosives also are important in weapons manufacture. Pyropowder is the most common material used in military explosives. However, some types of weapons require the use of propellants, and these propellants have different formulations. These are called double-base powders and contain nitroglycerine and nitrocellulose as the main ingredients. The formulations may contain 60–80% nitrocellulose and 20–40% nitroglycerine.

Cordite is another explosive based on glycerine. The product was developed in the late 1800s by the French and contained nitroglycerine 40–50%, potassium nitrate, barium nitrate, and sodium nitrate. The product was called ballistite. The British government was not satisfied with ballistite and developed cordite. The name cordite was used because it was produced in form of cords and the main explosive was a mixture of nitrocellulose and nitroglycerine. Mineral jelly was added to facilitate lubrication of the gun barrels. The jelly was not a very good lubricant additive, but was important because it lowered the temperature of the explosion which reduced the erosion of the barrel, increased the volume of emitted gas, and maintained power. Cordite is still used by the British, but is not widely used in other countries.

V. URETHANE

The use of glycerine in urethanes started in the late 1950s and since then such applications has rapidly increased. Glycerine is the building block for polyethers which are produced from glycerine by the addition of ethylene oxide and/or propylene oxide to render trifunctional polymers in the 1000–4000 molecular weight range. Trifunctional polymers are reacted with diisocyanate to produce crosslinked flexible urethane foams. This reaction can be in the form of a one-shot process, where equivalent amounts of the polyether and diisocyanates are reacted or the production of prepolymers leaving an excess of diisocyanate for further reactions. The resulting foams are superior to foams produced by diols. Glycerine-based polyols are also used in the production of rigid foams and in urethane coating.

VI. TRIACETIN

Glycerine triacetate can be prepared by reacting glycerine with acetic acid under normal esterification methods, but the glycerine usually is reacted with acetic anhydride to produce high-grade glycerine triacetate. The main use of this product is as a plasticizer in the production of cigarette papers and filters.

VII. MISCELLANEOUS

There are literally scores of uses for glycerine in miscellaneous applications. For example, its strong hydrogen-bonding properties gives it excellent solvent compatibilizing properties for adhesives and pastes. Glue or gelatin mixed with glycerine renders the mass plastic. Glycerine changes the glue composition to a stable product that remains flexible and strong.

One of the important natural resins is rosin. Rosin is combined with glycerine and used as ester gum. The ester gum is resistant to water, the rosin itself is not, and in addition has a high acid value. When you react rosin with 12–15% glycerine at about 250°C, it becomes an important additive to formulations based on cellulose nitrate and acetate, vinyl resins, and rubber, both natural and synthetic. The alkyds previously discussed can be blended with a great variety of polymeric resins. They are excellent bonding agents for cellulosic films, rubber, wood, textiles, and wood.

Glycerine is also used in air-conditioning and refrigeration systems. Its properties of smoothness and high viscosity make it a natural lubricant, and its hygroscopicity contributes to its use as a plasticizing agent. Other useful properties such as low temperature fluidity and high boiling point are also important. These attributes find use in the production of viscous dust-trapping fluids used in filters. The glycerine-treated filters can be water washed and when clean, retreated with glycerine solution.

In the automotive industry glycerine was once used as antifreeze. However, less expensive materials have taken its place. In other aspects of automotive manufacture and maintenance glycerine does find uses. Formulation for emulsions used for cleaners and polishes do contain glycerine. Glycerine cuts down streaking and rubbing marks. Glass cleaners used in automobiles frequently contain glycerine to enhance "antifogging."

Glycerine also finds use in electrical appliances, such as electrolytic condensers and neon lights. Glycerine is used frequently in processes for electrodeposition and treatment of metals. Resins that contain glycerine are also used as electrical insulating agents and binders. Glycerine is used in the manufacture of electrolytes for electrolytic condensers. These electrolytes are combinations of glycerine with boric acid or sodium borate and are used for making semidry electrolytic condensers. There are many modifications of these formulations.

Glycerine is used in many lubricant applications that range from fine instruments to powerful diesel engines. It has a wide range between freezing point and boiling point which makes it very effective as a lubricant under a wide range of conditions. It is particularly useful at low temperatures. It is also low in fire hazard and is nontoxic. Its GRAS status makes it very useful in food processing equipment and in lubricating diagnostic and surgical instruments.

Glycerine is very stable in contact with hydrocarbons and other organic solvents and vapors that may remove other types of lubricants. This makes it adaptable in pumps that process propane and butane. Sealing of the glands is accomplished by using a reservoir containing glycerine that surrounds the stuffing box and prevents the gas from reaching it. Glycerine may also be combined with potassium soaps for use in stuffing boxes. Glycerine is adaptable in the production of grease lubricants which are resistant to hydrocarbons and suitable for low temperatures. The antifreeze properties of glycerine and its low toxicity make it a valuable lubricant in food production machinery. In ice cream production the low temperature that glycerine can withstand makes it a good lubricant for the conveyors.

The use of glycerine in the rubber industry are many and varied ranging from use as lubricants to its miscibility with water and its stability over a wide temperature range. Its plasticizing and softening properties are important while some applications put emphasis on its value as a pressure-transmitting agent.

The low toxicity and nonvolatility of glycerine make it possible to use glycerine as a lubricant in hydraulic brake systems on vehicles. These properties also make it useful in rubber works in mixtures for surface treatment of unvolcanized but partly processed rubber compounds in keeping partly processed rubbers from sticking during storage. When combined with water and talc the formulation may be applied as a slurry before the mixture dries. The glycerine, being hygroscopic and nonvolatile makes the lubricating talc more adherent to the rubber surfaces. Glycerine can, and is used, in the production of puncture-sealing compounds for rubber.

Glycerine is used as a textile-conducting agent in sizing, lubrication, and softening of all kinds of yarns and fabrics. In fiber production, glycerine is used in spinning, twist setting, knitting, and weaving. The glycerine is oily and smooth enough to lubricate yarns without being sticky or tacky.

The humectant properties of glycerine make it suitable to be useful in various ceramic mixtures and formulations. Water is better formulated into the ceramic mixture and when the ceramic is kilned the glycerine burns out without leaving any carbonaceous residue.

In glass production glycerine has many varied uses. In etching glass surfaces glycerine provides good solvent systems for active fluorine compounds and provides the proper body or viscosity so the solution will remain in the proper position. Also, the hygroscopic glycerine formulation retains moisture. Pastes of glycerine, water, and abrasives will remove scratches on glass surfaces by rubbing and can leave the glass surface smooth and polished.

VIII. FOOD

The high boiling point and other properties of glycerine makes it useful to the food industry; it allows the elimination of ethanol which is commonly used in many food preparations. Glycerine is used in flavoring materials and curing salts and acts as a plasticizer in scores of casings and coatings which were developed for the meat industry. Both synthetic and natural casings often contain glycerine. The glycerine increases the flexibility of the casings, ease of handling and increases the shelf life.

Glycerine is also an ingredient in cork liners for bottle caps. The cork is treated with glycerine to stabilize the moisture which maintains softness and flexibility of the cork. This prevents shrinkage and maintains a tight seal. Whole corks are sometimes treated with glycerine by soaking before inserting the cork in wine bottles. It is firmly established that glycerine is nontoxic and can be used in the area of direct contact freezing; glycerine, glycerine salt and glycerine inverts/sugar solutions have been found very satisfactory. Aqueous glycerine solutions are used for this type of freezing because of their suitable viscosities and good heat transfer properties; they are also noncorrosive, resistant to fermentation, and will retain natural colors. They do not cause excessive rupture of cells at cut surfaces and produce natural-looking products. Glycerine is used to freeze fish before rigor mortis set in to reduce the amount of ice formation, resulting in less tissue protein denaturation. Addition of glycerine to peanut butter reduces oil separation and increases the stiff texture of the butter. Usually it is added after grinding of the peanuts and it does not alter the taste. When 4% glycerine is added to shredded coconut it acts as a softener and humectant and keeps the coconut from drying out when the package is opened. Cake recipes also use glycerine to preserve moisture and retard staling. The percent of usage will vary, but frequently can be as high as 10%.

Glycerine is useful in production of monoglycerides (as an intermediate). Monoglycerides are stablizers and emulsifiers for many food products. The fats and fatty acids used for the production of monoglycerides are many and varied and the raw material used will depend on the desired properties of the end product. The resulting product will be a mixture of mono-, di-, and triglyceride, but will be mainly of the mono variety. These monoglycerides are surface-active materials, impart both oil-soluble and water-dispersible properties. As good emulsion stabilizers, they are used extensively in margarine products to improve stability and reduce spattering on heating. In shortening they increase plasticity. In dough mixtures they increase shelf life and permit richer formulations. Salad dressings, candy, food coatings, and frozen desserts almost always contain monoglycerides in their formulation. In white bread, the monoglycerides act as a softening agent and an antistaling agent. The excellent solvent properties of glycerine make it useful in many food flavors and extracts; this solvent property allows the elimination of part or all of the alcohol used in these preparations. Vanilla flavorings and chocolate syrups contain glycerine to improve body and smoothness. Flavor pastes and powders may also contain glycerine. Many food colors contain U.S.P. glycerine. It is completely nontoxic and is accepted by the Food and Drug Administration as a food component, except where specific food standards fail to list it as an optional ingredient.

Because glycerine is a product of alcohol fermentation, it is present in beer (0.09–0.18%) and in wine (up to 10%). Glycerine often is added to distilled liquors to improve body and smoothness. A small amount of glycerine added to mixed drinks improves the flavor and produces a smoother blend. In soft candies, such as jelly beans, glycerine is frequently used to prevent drying and graining. It maintains the soft texture and fine grain in many candies, especially fudge. The use rate can be as high as 10%, based on the weight of the sugar. Usage can go as high as 15%, depending on the desired softness. Cake icings may also contain glycerine to prevent brittleness.

Glycerine is used in dried fruits, applied by dipping or spraying, to reduce stickiness and inhibit surface crystallization.

Index

Printed and bound by CPI Group (UK) Ltd, Croydon, CR0 4YY

23/10/2024

01778237-0006